The diminishing populations of African and Asian elephants call to mind the extinctions of other elephantlike species, such as mammoths and mastodonts, that occurred more than 10,000 years ago. The purpose of this book is to examine the ecology and behavior of modern elephants to create models for reconstructing the lives and deaths of extinct mammoths and mastodonts. The sources for these models are long-term continuing studies of elephants in Zimbabwe, Africa. These models are clearly described with respect to the anatomical, behavioral, and ecological similarities between past and present proboscideans. The implications of these similarities for the lives and deaths of mammoths and mastodonts are explored in detail.

The importance of this book is primarily its unifying perspective on living and extinct proboscideans: The fossil record is as carefully examined as is the natural history of surviving elephants. Dr. Haynes's studies of the situations in which African elephants die (sometimes in great numbers) are unique and can provide crucial insights into ancient proboscidean bone collections.

Mammoths, mastodonts, and elephants

Mammoths, mastodonts, and elephants

Biology, behavior, and the fossil record

GARY HAYNES

Department of Anthropology
University of Nevada

PUBLISHED BY THE PRESS SYNDICATE OF THE UNIVERSITY OF CAMBRIDGE
The Pitt Building, Trumpington Street, Cambridge, United Kingdom

CAMBRIDGE UNIVERSITY PRESS
The Edinburgh Building, Cambridge CB2 2RU, UK   http://www.cup.cam.ac.uk
40 West 20th Street, New York, NY 10011-4211, USA   http://www.cup.org
10 Stamford Road, Oakleigh, Melbourne 3166, Australia

© Cambridge University Press 1991

First published 1991
First paperback edition 1993
Reprinted 1999

Printed in the United States of America

A catalogue record for this book is available from the British Library

Library of Congress Cataloguing-in-Publication data is available

ISBN 0 521 45691 6 paperback

Contents

v

Preface

There is perhaps no inquiry in the whole range of
Natural History more fascinating and romantic than
that which deals with the Mammoth and its
surroundings.

—H. Howorth, *The Mammoth and the Flood* (1887)

Time on earth has nearly run out for African and Asian elephants. Only a few pockets of survivors remain in the wild. Some 10,000–11,000 years ago other related elephant-like species met their end under circumstances that have yet to be fully understood. Those extinct forms – the genera *Mammuthus* and *Mammut* – were so similar to the modern elephants of Africa and Asia in terms of anatomy, ecology, and behavior that it is sensible to discuss them together, and yet somehow there were sufficient differences among the proboscideans that while the majority were dying out, two species managed to survive the end of the Pleistocene. One purpose of this book is to suggest some ways that we can look for the possible parallels between that great dying-out and the twentieth-century decline of modern elephants.

The first part of this book emphasizes the many similarities between the life conditions of past and present proboscideans, one of the goals being to strengthen the case that actualistic studies of modern elephants will accurately guide us in finding new meaning in bone assemblages left by extinct proboscideans such as mammoths and mastodonts. After briefly discussing proboscidean taxonomy, I point out the anatomical similarities among mammoths, mastodonts, and elephants, as well as behavioral similarities suggested by body size and shape.

Chapter 3 points out important aspects of elephant social behavior and use of habitats, especially the manner in which such characteristics affect reproduction, mortality, and the nature of death assemblages. Chapter 3 also indicates how the demographic attributes of modern elephant bone assemblages reflect important aspects of the behavior and ecology of living elephant groups. These meaningful characteristics of bone assemblages can also be found in fossil bone deposits, where they may have the same implications concerning the lives and deaths of mammoths and mastodonts.

Chapters 4 and 5 provide a detailed look at actualistic studies of modern elephants in Africa. Data from these studies provide keys for unlocking more meaning from fossil bone assemblages, as I hope becomes apparent in later discussions. Chapter 6 reviews fossil sites in an attempt to define patterns in the record from Plio-Pleistocene to Recent times.

The final section of this book looks at the problem of extinction. In spite of the great similarities in body size and shape, life history, and social behavior among the proboscideans of interest here, some species became extinct, and some did not. This is a rather famous scientific puzzle. Scientists want to lay the blame on one set of controlling factors or another, but have yet to reach consensus. One book like this will not set the problem to rest once and for all, but it does offer suggestions for the future research needed to address the issue.

In the Appendix, I describe the dentition of extinct and living proboscideans and discuss new and previously published data reflecting the close correlation of an animal's life-age and the sequencing of tooth eruption and skeletal maturation. The major aim is to show the reader how the age of a proboscidean individual may be determined, regardless of taxon represented, and how the mammoths, mastodonts, and elephants lived extremely similar lives.

Note on the spelling of Russian words and names

I have spelled most Russian words and site names so that American English-speakers can pronounce them to sound more like the Russian. In many cases my changes from the conventional spelling (which uses arbitrary alphabetical rules, such as the Library of Congress transliteration system) involve replacing some letters with others – for example, "Kostenki" or "Kostienki" becomes "Kostyonki," and "Berelekh" becomes "Berelyokh"; in these cases the "yo" is pronounced with a long *o*, more in line with proper Russian pronunciation. I feel justified in doing this because the Russian and English languages use different alphabets, and it seems more rational to preserve sounds than to convert symbols.

Exceptions to the spelling changes are people's surnames, because many have already been catalogued in English libraries using spellings whose English pronunciations are unrelated to the correct Russian – for example, "Semenov" (whose name is pronounced Sem-YOH-nof) or "Herz" (whose name is pronounced more like Gertz).

I have not let stand the usual misleading English spelling of the name of the well-known Russian scientist and fantasy writer Yefremov (pronounced Yeh-FREH-muf), usually spelled "Efremov" for some reason.

Acknowledgments

The fieldwork and research reported here were carried out over more than a decade, and I owe huge debts of gratitude to many individuals. To the people of Zimbabwe, and especially to the director, assistant directors, wardens, rangers, ecologists, scouts, technicians, and laborers of the Zimbabwe Department of National Parks and Wild Life Management, I am grateful for assistance, encouragement, and companionship. I want to thank (in alphabetical order) Rob Clifford, Clem Coetsee, Drew Conybeare, Ian Coulson, Colin Craig, David Cumming, William Deviyas, Doug Evans, Steve Edwards, Debbie Gibson, Claudius Hove, Mike Jones, Stanwell Malinganiso, Rowan Martin, Morven Mdondo, Ajibu Mhlanga, Norman Monks, Willie Nduku, Peter Ngwenya, T. A. J. O'Hara, Adrian Read, Paul Read, Ian Sibanda, Million Sibanda, Russell Taylor, Maggie Taylor, and Glen Tatham. Some of these people are no longer with the Department of National Parks. I thank their families for memorable hospitality while I was visiting in the field. Gary MacDougal and family were also good friends in Zimbabwe.

In 1982 I happened to be shown a letter from Jehezkel Shoshani to Dennis Stanford, and because of it I made a start in studying elephants in Africa. I thank Hezy for the good luck that he brought.

For assistance during my studies of American bone collections I thank Larry Agenbroad (Northern Arizona University); Dale Guthrie, Carol Allison, and James Dixon (University of Alaska Museum); John Dallman (University of Wisconsin, Madison, Museum of Zoology); Kurt Hallin (Milwaukee Public Museum); Clark Howell (University of California, Berkeley); Michael Novacek (American Museum of Natural History); Dennis Stanford, Robert Purdy, and Clayton Ray (National Museum of Natural History); Frank Whitmore (U.S. Geological Survey); Jeff Saunders and Russ Graham (Illinois State Museum); Dave Gillette (Utah Division of State History, Antiquities Section); Klaus Westphal (University of Wisconsin, Madison, Geology Museum); and Calvin Smith (Strecker Museum).

I greatly appreciate the kindness of the following people who shared ideas and unpublished data with me: David Reese, Hezy Shoshani, Rick

Potts, Louise Roth, Jack Fisher, Richard Klein, Mary Stiner, Jeff Saunders, Kate Scott, Drew Conybeare, Mike Jones, and David Cumming.

I am extremely grateful for the warm and generous reception afforded me at the USSR Academy of Sciences' Zoological Institute in Leningrad and the Paleontological Institute in Moscow. I thank N. K. Vereshchagin, T. Vereshchagina and her family, V. Ye. Garutt, A. Tikhonov, I. E. Kuz'mina, N. Garutt of the Geology Museum (LGU), I. Dubrovo, and Yu. Reshetov for their interest and aid. I especially thank Aleksey Tikhonov for giving me his laborer's coat when I was suffering from the cold in Leningrad, in spite of the fact that no coat-check attendant ever wanted to hang it up properly.

In some ways this book follows the model established by Vadim Garutt in his 1964 book *Das Mammut*, and I hope he accepts my emulation as the high praise it is intended to be.

I thank Adrian Lister in Cambridge, England, for his interest and openness about his own research and his helpfulness with this book. I wish to express my gratitude to Kate Scott and her husband Richard Cropper for their willingness to entertain my wife and me in Oxford. Antony Sutcliffe of the British Natural History Museum [formerly the British Museum (Natural History)] was very helpful during my several trips to London.

Over the course of five years of association with the Smithsonian Institution I relied on Dennis Stanford for nearly every necessity, including financial support, an office in which to work (and sometimes sleep), and a part to play in Paleo-Indian studies. I am also obliged to Betty Meggers and Mildred and Waldo Wedel for advice and counsel. Jean Fitzgerald made life interesting in the place from time to time. Members of the Anthropology Department art staff contributed in a large way to the studies; Ellen Paige assisted in the African fieldwork for two seasons and drew several maps and figures; Robert Lewis and Marcia Bakry also contributed several figures.

In Reno, Tom Duggan drew most of the artwork, and I am grateful to him for appearing at just the right time. The Photographic Laboratory at the University of Nevada, Reno, printed many of the book's photographs in a timely and helpful manner.

For reading numerous revised drafts of chapters or the whole manuscript, Jeff Saunders, Paul Martin, Mike Jones, and Adrian Lister should be rewarded with more than just my thanks, which they have in great volume. Drew Conybeare also read, reread, and read again several chapters, and yet he still speaks pleasantly to me, for which I am beholden. C. V. Haynes read part of one chapter in an earlier form, as did John Speth. Diane Gifford-Gonzalez bravely endured parts of my first attempt at an important chapter. Adrian Lister and V. Louise Roth read the Appendix.

Financial support for the research came mainly from the National Geographic Society; I was awarded fieldwork grants each year from 1982 to 1988. To that organization I am obliged and grateful for sustaining my work in lean times. The Smithsonian Institution gave me a predoctoral fellowship in the Anthropology Department in 1979–80. I was also partly supported by the Smithsonian Institution Paleo-Indian Program (Department of Anthropology) while working as a contractor for Dennis Stanford from 1981 to 1985. The University of Nevada, Reno, gave me a grant from the Graduate School (Research Advisory Board) in 1986 to pay for costs of preparing an early version of the manuscript. The U.S. National Research Council, Office of International Affairs, supported my studies in the USSR in 1987.

I met some very good friends for the first time during the fieldwork in Zimbabwe. To them I express my deepest appreciation and love for years of help in every way, from the passing on of their knowledge to the sharing of foolishness on the Kalahari frontier. This book would not have been possible without the contributions of David and Meg Cumming, Mike and Gay Jones, and Drew and Barbie Conybeare. Perhaps the strangest times (and worst hangovers) were had with Barney O'Hara.

I thank Susan Abrams for her editorial suggestions and Peter-John Leone for his faith and helpfulness at Cambridge University Press.

Janis Klimowicz has nurtured this manuscript through all stages of its creation, and it is to her that I am the most indebted for support and concern.

I am responsible for errors and misinterpretations.

I

PROBOSCIDEAN FLESH
AND BONES

1

Taxonomy: classification of fossil and living forms

1.1. Distribution

Today there are only two genera in the order Proboscidea: *Loxodonta* (the African elephant) and *Elephas* (the Asian elephant). Two other genera, *Mammuthus* and *Mammut*, disappeared from North America about 10,000 years ago, and in South America at that time three genera of gomphotheres (*Cuvieronius*, *Haplomastodon*, and *Stegomastodon*, also considered mastodonts) died out. Those five extinct genera and the two surviving genera were the only proboscideans present in the late Pleistocene (50,000–10,000 years ago) anywhere in the world. Mammoths (*Mammuthus* spp.) were widespread in Europe, northern Asia, North America, and central Mexico, but they never penetrated into South America. Mastodonts (*Mammut*) were found in North America, but members of the separate taxonomic family of animals popularly called gomphotheres were the only South American proboscideans. The genus *Elephas* was present in Africa and Asia, but never in the New World. *Loxodonta* was always restricted to Africa. No proboscideans were ever native to Australia.

1.2. Origins and evolution

Proboscideans are a relatively diverse order, as evidenced by the numerous extinct genera and species in the fossil record. Figure 1.1 shows a possible set of phyletic relationships among the elephant families of the order Proboscidea (after Maglio 1973; following Lister 1989; J. Saunders 1989 personal communication). The earliest members of the order appeared in northern Africa over 50 million years ago, during the early part of the Eocene epoch. The first forms were rather small by modern standards, but were elephantlike in some ways (Mahboubi et al. 1984). Over the next several million years, evolutionary divergence resulted in the development of mastodonts and gomphotheres, distinguished primarily by the shapes of their check teeth. By early Miocene times (about 20 million years ago), gomphotheres and mastodontlike forms had expanded their ranges out of Africa and into Eurasia and eventually North America. By the middle

3

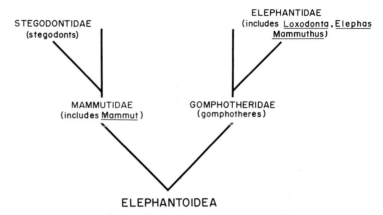

Figure 1.1. Phyletic diagram suggesting the relationships among different taxonomic families of the order Proboscidea.

Miocene (about 15 million years ago), the bones of numerous different subfamilies and genera had appeared in Africa, Eurasia, and North America, figuring prominently in local fossil records.

Throughout the Pliocene-Pleistocene fossil record the most common genus in southern and eastern Africa was *Elephas*, until about 400,000 years ago, when only *Loxodonta* continued to survive in Africa. The use of the generic term *Elephas* to refer to prehistoric European elephants is common in some literature, but it is probably an unjustified case of taxonomic lumping (A. Lister 1990 personal communication). Today *Elephas* is restricted to southern Asia, from India east to Sumatra and Borneo.

1.3. *Mammut*

Mastodonts are primitive proboscideans, insofar as types such as mammutids first appeared so early (by the early Miocene, 20 million years ago) and persisted in the fossil record until the end of the Pleistocene, 10,000 years ago, when they ceased to exist. However, there was considerable diversification within the mastodonts. Note that the term "mastodont," with the letter *t* at the end, refers to *Mammut* and also several other genera such as the gomphotheres. The term "mastodon" has been proposed at various times and by different authorities as a vernacular word referring specifically to *Mammut*, whereas "mastodont" would refer collectively to mastodons and gomphotheres. I use the more inclusive term (with the *t* at the end) even when discussing *Mammut* only, because proper English word formation from the Greek roots of the term requires the final *t* (J. Saunders 1989 personal communication). In this book, the word "mammutid" refers only to *Mammut*.

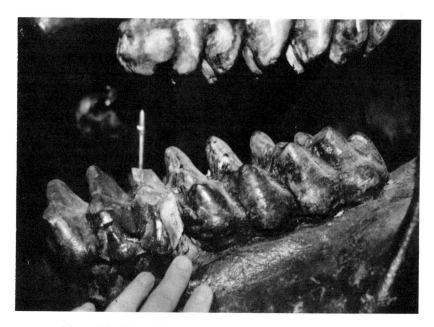

Figure 1.2. View of the left cheek teeth in the lower jaw of *Mammut americanum*, with the anterior side to the left.

The teeth of *Mammut* were characterized by rounded and pointed enamel-covered cones (Figure 1.2) that wore down with use and flattened out considerably. The teeth would have functioned effectively in clipping or crushing twigs, leaves, and stems, but would not have been suitably adapted for grinding hard or abrasive foods. The *Mammut* diet may have consisted of plants that grew around swamps and ponds in woodlands. Pollen and macrofossils of spruce and other conifers are associated with American mastodonts, but how prominently those trees were featured in the diet is difficult to judge (see Chapter 3).

1.4. *Mammuthus*

Animals classified in this genus appeared in sub-Saharan Africa during the middle Pliocene epoch (about 3–4 million years ago). By the end of the Pliocene and the beginning of the Pleistocene they were extinct in Africa, but widespread in Eurasia, where Soviet scholars classify the earliest members of the group as *Archidiskidon gromovi. A. gromovi* is regarded by some systematists as an early form of *A.* (or *Mammuthus*) *meridionalis* (A. Lister 1989 personal communication). Some species lived in relatively warm southern ranges, whereas others were associated with northern

Table 1.1.   *Enamel thickness in Mammuthus spp.*

Taxon[a]	Faunal age	Enamel thickness (last molar) (mm)
M. meridionalis	Irvingtonian	2.4–4.0
M. imperator	Middle Irvingtonian to middle Rancholabrean	2.5–3.2
M. columbi	Wisconsin glacial stage	2.0–2.3 (early) 1.5–2.0 (later)
M. primigenius	Rancholabrean	1.0–2.0

[a]These taxonomic classes are those that seem to be most widely applied by other workers, in spite of Kurtén and Anderson's correct insistence on using the earliest established designations. *M. imperator* in this table is called *M. columbi* by Kurtén and Anderson; *M. columbi* in this table refers to the species called *M. jeffersonii* by Kurtén and Anderson.

Source: Data from Kurtén and Anderson (1980).

latitudes. It has been suggested that the largest mammoths were the southern races, but clear evidence for this is elusive. Mammoths seem to have become progressively cold-adapted in Eurasia, successfully colonizing extreme northern Asia and Beringia, and eventually crossing eastward and southward into North America by at least 1.7 million years ago (Webb et al. 1989). Once into interior North America, a lineage of large mammoths, similar to *Mammuthus* (*Archidiskodon*) *meridionalis* of Eurasia, evolved into a species adapted to open habitats–the columbian mammoth, *M. columbi*. A second wave of migration from Asia into North America may have brought the relatively smaller woolly mammoth (*M. primigenius*). The distribution of this late Pleistocene species seems to have been limited to the northern regions of the continent, or near the glacial borders. However, there may be a gradient based on body size and tooth morphology that could indicate that *M. columbi* and *M. primigenius* were not descendants of two entirely separate dispersals, but were geoclinal or chronoclinal variants. No clear differences in postcranial morphology distinguish the two species; enamel thickness is considered partially diagnostic, but individual teeth and even parts of a given tooth have variable enamel thickness (Corner and Diffendal 1983; Kurtén and Anderson 1980) (Table 1.1).

Most researchers agree that mammoths were predominantly grazers. Increasing dental and cranial adaptations for the processing of grass set the mammoths apart from many other proboscideans, with the exception of *Elephas*. Their teeth are large, blocky masses made up of multiple enamel-covered and dentine-cored laminae (Figure 1.3); the laminae are held

Figure 1.3. View of the cheek teeth of *Mammuthus columbi*. Both left and right sides are shown in a broken lower jaw lacking an ascending ramus on the left. Anterior sides are to the left. The smooth-surfaced, peglike teeth on the left are both very worn and are being replaced from behind by the larger teeth.

together with cementum (see the Appendix for further descriptions of elephant teeth).

1.5. Pygmy or dwarf elephantids

Insular proboscidean populations in several widely different parts of the world during the Pleistocene evolved extremely small body sizes, as predicted by the "island rule" of theoretical biology (Lomolino 1985; Sondaar 1977). On the islands of Sicily and Malta, *Palaeoloxodon falconeri* (= *Elephas falconeri* in some literature) was about one-quarter the height and length of its probable mainland ancestor *P. namadicus* (= *E. namadicus*), whereas another dwarfed form (*P.* or *E. mnaidriensis*) was not quite so small (Caloi et al. 1989). In the northern Channel Islands of California, *Mammuthus exilis* was about half as big as its likely ancestor *M. imperator*. The dwarfed *P.* (or *E.*) *cypriotes* is known from over a dozen sites on the island of Cyprus (Bate 1903; Reese n.d.). There are other possible dwarfed forms, all of which show similar trends in character, such as relatively shortened distal limb elements, cheek teeth that are not as much reduced (compared with those of full-sized ancestors), and skeletal elements such as limb bones that are not as much reduced (Maglio 1973; Roth 1984, in press; Sondaar 1977). Dwarf stegodonts (which are not true elephants in the taxonomic sense), found on several islands in southeast Asia, may have been hunted by colonizing humans during the Pleistocene (Sondaar et al. 1989).

1.6. *Elephas*

This genus originated in sub-Saharan Africa during the Pliocene epoch, ranging throughout Africa and into southern Asia. A form that reached into Europe by the middle Pleistocene has been called *Elephas* in published literature, but the term *Palaeoloxodon* is perhaps more appropriate. As the teeth and skull of *Elephas* evolved throughout the long span of its existence, the overall trend was toward specialized features for processing an abrasive diet, specifically grass. As in mammoths, the skull of *Elephas* is high, with domed forehead, a development associated with a shift in the skull's center of gravity to facilitate grinding food in the mouth (Maglio 1973; Osborn 1942). Only one species survives today, with perhaps three subspecies (*E. maximus maximus* of Sri Lanka, *E. m. indicus* of India, Indochina, and Borneo, and *E. m. sumatranus* of Sumatra).

1.7. *Loxodonta*

Loxodonta is a genus that never left Africa. It first appeared about 4 million years ago, but over its evolutionary life the features of its teeth and skeleton changed very little. Members of this taxon are associated (perhaps only inferentially) with wooded habitats, where browse could have formed a major part of the diet, either seasonally or year-round. *Loxodonta* does not seem to have been a competitor with *Elephas* in Africa, where both genera originally appeared and coexisted through the Pliocene and most of the Pleistocene. *Loxodonta* has never been found in fossil assemblages that also contain *Elephas* in southern Africa, and only rarely has it been found with *Elephas* in sites elsewhere on the continent (Klein 1984). In fact, *Loxodonta* is believed to have been relatively rare in east Africa until after the late Pleistocene disappearance of *Elephas*.

There is only one surviving species, *L. africana,* and possibly two subspecies (*L. a. africana,* the typical savanna and woodland elephant, and *L. a. cyclotis,* the forest elephant of central and western Africa, a smaller animal with different tusk and body morphology). Soviet zoologists consider the two races to be separate species (V. Garutt 1987 personal communication).

2

Physical appearance: mammoths, mastodonts, and modern elephants

*Father elephant walks heavily..., without fear, sure of his
strength.... Elephant hunter... lift up your heart, leap and walk:
[there is] meat in front of you, the huge piece of meat, the meat that
walks like a hill.*

—*Gabon Pygmy song, translated by C. M. Bowra*
(Lomax and Abdul 1970:4)

2.1. Animals that walked like hills

Modern elephants are the heaviest land animals, and they are exceeded in
height by only one species: The giraffe. No other terrestrial animal alive
today weighs half as much. The two living species (*Loxodonta africana* in
Africa and *Elephas maximus* in Asia) are different in physical appearance
(Figure 2.1), but very similar in social behavior and biology. The extinct
taxa *Mammuthus* and *Mammut* may not have lived in the same habitats as
each other, but in spite of differences in behavior and appearance, their life
histories may not have been radically unlike those of the modern elephants.

The order Proboscidea includes the animals with trunks, but the Greek
roots of the word "proboscidean" (*pro-bosko*) literally mean "to feed."
Modern elephants, like the extinct taxa, possess an elongated trunk that
functions not only as a nose but also as a prehensile organ to grasp and
manipulate objects in the environment. At one time, some students of
proboscidean evolution considered the trunk a result of parallel evolution
in the mastodonts and elephants, but it is now recognized that a common
ancestor bequeathed the organ to all its descendant taxa. The trunks of the
various forms, though fundamentally similar, were somewhat different for
the various taxa. An Asian elephant's trunk has one fingerlike projection on
the front part of its tip, whereas an African elephant has a front "finger" and
a broader rear projection; in the mammoth, the trunk tip had two fingerlike
projections, one at the front and one at the back (Flerov 1931; Garutt 1964).
No mastodont trunks have been preserved, so the shape of the tip is
unknown. Trunk tips are so sensitive and mobile that elephants can use
them to sort and pluck single stems of grass from the ground, but they are
also tough enough to allow heavy lifting. Another use for the trunk is as an

10

Figure 2.1. Profiles of three different proboscidean species. From the left are *Elephas maximus* (275 cm tall at the shoulder), *Loxodonta africana* (360 cm tall), and a suggested reconstruction of *Mammuthus columbi* (340 cm tall).

aid in drinking: Water is drawn up into the trunk by inhaling through the trunk, and it is then transferred to the mouth for swallowing.

For mammoths, mastodonts, and modern elephants the skeleton is comparatively inflexible and is characterized by vertically oriented legs and a rigid, nearly horizontal spine offering support for a heavy body. The upper and lower bones of the leg align vertically with each other when the limb is extended, and thus the mass of the animal is carried on legs that are like columns or pillars. Elephants do not "run" in the usual sense of the word; at top speed there is very little oscillation of the leg about a pivotal point, and there is no free-flight phase in which all feet are off the ground at the same time (Gambaryan 1974). The maximal rearward and forward extensions of the moving legs are restricted, so the legs are nearly always under the body (Hildebrand and Hurley 1985).

The neck is short, but the added length of the trunk extends the animal's reach outward by 1–2 m and makes up for the difficulty of getting the mouth down to the ground for feeding or drinking. The head is not very mobile and cannot be turned far sideways. As a result, in order to see to the rear, an elephant must move its whole body, as noted by Benedict (1936:65). The captive elephants that Benedict studied were troubled by sounds and movements behind them that they could not easily investigate (Benedict 1936:75). Wild elephants also show an excitable alertness regarding unfamiliar sounds or smells behind them, and they shift position very quickly to inspect distractions.

Wild elephants sleep lying down, generally from one to three hours in the middle of the night (Moss 1988), as did the captive elephants studied by Benedict (1936:73). They stand to urinate or defecate. Elephants are light sleepers, quickly alerted by disturbances; for an example, see G. Haynes (1987a). They may also doze while standing, day or night. During hot, dry weather, when traveling long distances from forage to water, they may sleep lying down in daylight.

A male elephant has a large, erectile penis that weighs 50 kg or more (Hanks 1979:78) and has no penis bone. The testes are internal. A female elephant has an erectile clitoris and a vulva whose opening is directed nearly forward between the rear legs. In order to fit inside the external opening of the vulva, the male's penis is **S**-shaped when erect, and it is long enough to drag along the ground when the animal walks.

The Asian elephant possesses a domed forehead in profile, whereas the African elephant has an angled forehead (Figure 2.1). The Asian elephant usually carries its head so that the top is above the level of its shoulders, but for African elephants the shoulders and lumbar part of the back may be the highest points in the body profile. The profile of *M. primigenius*, the woolly mammoth, would have been somewhat similar to that of modern Asian elephants, except that the mammoth would have had a higher shoulder "hump" formed by long vertebral spines and a covering mass of long, thick hair (see Section 2.3).

2.2. Sexual dimorphism

Mammoths and mastodonts, like modern African and Asian elephants, showed a great deal of sexual dimorphism. Full-grown males of *Loxodonta* may have stood up to 1 m taller than females and weighed nearly twice as much (Eltringham 1982; Hanks 1979; G. Haynes unpublished data; Laws, Parker, and Johnstone 1975). In general, large mammals show increased size dimorphism as compared with smaller mammals (Clutton-Brock and Harvey 1978), but there is also an adaptive significance to great differences in body size. So much sexual dimorphism in proboscideans reflects important behavioral adaptations, specifically in social behavior, as discussed in Chapter 3.

One implication of great differences in body size between the sexes is that sexual maturity will be reached later in males, because they require a longer period of time to acquire competitive abilities that will make their breeding attempts worthwhile (Barash 1977; Clutton-Brock 1983; Daly and Wilson 1988:143). Indeed, it has been found that elephant males continue growing for many years past the age when female growth levels off (Eltringham 1982; Hanks 1979). As described in the Appendix, some long-bone epiphyses in male elephants remain unfused to the diaphyses (reflecting continued growth) (Roth 1984) for a decade or longer after the age when the same epiphyses fuse in females.

The degree of dimorphism seen in modern elephants is similar to that for mammoths and mammutids. Relatively lesser dimorphism in different taxa may reflect less fierce competition among males for access to females, but I

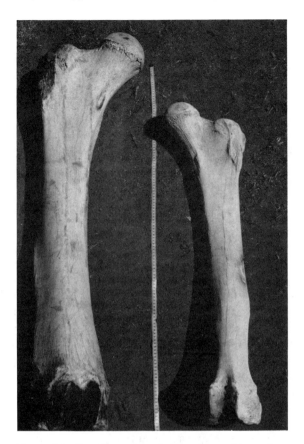

Figure 2.2. Size dimorphism in *Loxodonta africana* femora. To the left is a right femur of a male, and to the right a left femur of a female. Maximum length for the male femur is 1,300 mm.

think the differences in dimorphism among elephants, mammutids, and mammoths are not great enough to indicate significant differences in social behavior. Judging from bimodal bone measurements, the extinct males were taller and heavier than females, and their tusks were much larger. Dimorphism in columbian mammoths (*M. columbi*) was as great as in modern elephants.

It should be noted that not all proboscidean bone assemblages yield dimorphic bone measurements that resulted only from sexual differences, because adult males and females kept themselves segregated under normal conditions (see Chapter 3).

Modern adult male African elephants in Hwange National Park, Zimbabwe, stand about 20–40 percent taller than adult females; the femora and humeri in males (over fifteen years old) are about 17–25 percent longer than in females of similar age (G. Haynes unpublished data) (Figure 2.2). Laws et al. (1975) found that the mass of Ugandan elephant males was about 73 percent greater than the mass of like-age females, and male height was about 22 percent greater. A 307-cm-tall male weighed 4,743 kg, whereas a 252-cm-tall female weighed 2,737 kg (both aged about sixty years). The maximum possible weights and heights are greater than these figures, depending on the population and local conditions. For example, V. Wilson, as cited by Hanks (1979), dissected a male elephant in Zambia that weighed about 5,700 kg; its shoulder height was not reported, but it probably was taller than 307 cm, judging from its tusk weights.

Among the Berelyokh mammoth bones at the Zoological Institute of the USSR Academy of Sciences in Leningrad (Baryshnikov, Kuz'mina, and Khrabrii 1977; Vereshchagin 1977; see also Chapter 6 and the Appendix), length differences of 14–26 percent were found among adult femora and humeri, probably reflecting sexual dimorphism (G. Haynes unpublished data). Different *Mammut* skeletons from late Wisconsin contexts in the central United States also show such differences between (presumed) members of the two sexes. For example, the young adult "Denver" mastodont found in Indiana in 1903 (on exhibit at the Milwaukee Public Museum) would have stood about 17 percent taller than the similarly aged animal from the John Neath site in Wisconsin, curated at the University of Wisconsin, Madison, Museum of Zoology. Limb bones in the still-growing Denver male were 15–27 percent longer than those in the Neath animal, probably a female, which had nearly completed statural growth.

The *M. columbi* individuals at the large site in Hot Springs, South Dakota (Agenbroad 1984, 1990), are significantly larger than the *M. columbi* individuals at the geocontemporaneous Waco, Texas, site; a small sample of seven measurable humeri and five femora at Hot Springs (Agenbroad 1990) averages 20–30 percent longer than the very small number of the same elements at Waco (G. Haynes unpublished data). In this case, the size differences may be due as much to gender differences as to clinal variations. The Waco animals may be females, and the Hot Springs animals may be mainly males (Agenbroad 1984). Also (or alternatively), animals in both assemblages may be conforming to Bergmann's rule. Bergmann's rule is an empirical generalization to the effect that smaller body size is associated with warm and humid environmental conditions, whereas larger size is associated with cold and dry conditions (if the two environments are roughly equal in regard to forage or available nutrition).

Table 2.1 shows some approximate shoulder heights for male and female *Mammuthus primigenius, M. columbi, Loxodonta africana, Elephas maximus,* and *Mammut americanum,* as measured from mounted skeletons or carcass discoveries (Agenbroad 1990; Dutrow 1980; Harington, Tipper, and Mott

Table 2.1.   *Shoulder heights of fully grown Mammuthus primigenius,*
M. columbi, M. imperator, Mammut americanum, Loxodonta africana,
and Elephas maximus

Taxon	Shoulder height (cm)	
	Male	Female
M. primigenius	∼ 300–350[a]	∼ 230–250[a]
M. columbi	∼ 320–395[b]	?
M. imperator	360–390[c] (sex undetermined)	
Mammut americanum	∼ 275–300[d]	< 250
Warren mastodont	∼ 280[d]	
Rancho La Brea	∼ 190[c] (average, sex undetermined)	
Boney Spring	∼ 215–300[e] (both sexes?)	
Denver mastodont	∼ 270[a]	
Neath mastodont		∼ 230[a]
Loxodonta africana		
Hwange sample	> 310–350[a]	∼ 260–300
U.S. NMNH[f] (Fenykovi)	401.3[g]	
Jumbo	∼ 330–340[h]	
Elephas maximus	∼ 250–275 (record = 320)[i]	∼ 220–250[i]

[a] Original measurements; also see Garutt (1964) for *M. primigenius*.
[b] Harington et al. (1974), Osborn (1942), and Dutrow 1980; "*Parelephas jeffersonii*" and "*M. jeffersonii*" in Kurtén and Anderson (1980) are synonymous. Kurtén and Anderson consider Osborn's *P. columbi* synonymous with *M. imperator* and their *M. columbi*. Other sources, such as Graham (1986a), consider *M. columbi* distinct as the latest form of southern mammoth, taxonomically separate from both the larger *M. imperator* and *M. jeffersonii*.
[c] Harris and Jefferson (1985).
[d] Osborn (1936, 1942).
[e] Saunders (1977b).
[f] U.S. National Museum of Natural History. Fenykovi is the name of the hunter who shot this elephant, the largest mounted *Loxodonta* in the world.
[g] Best and Raw (1975). The maximum recorded *Loxodonta* shoulder heights are 442 cm, 417 cm, and 411 cm (S. Smith 1986), all from Namibia or Angola.
[h] Osborn (1942).
[i] S. Smith (1986).

1974; Osborn 1942). It is interesting to note that the shoulder height for modern Asian and African elephants (male or female) is very nearly equal to twice the circumference of one forefoot (Benedict 1936; Boyle 1929; Carrington 1958; J. Shoshani et al. 1982; Western, Moss, and Georgiadis 1983), which may also have been true for *Mammuthus* and *Mammut*.

2.3.   Body profile

The femur is the longest limb bone for mammoths, mastodonts, and modern elephants. In *Elephas, Loxodonta, Mammut*, and *Mammuthus* the

Table 2.2. *Limb articular lengths*

Specimen	Sex	Humerus + ulna (cm)	Femur + tibia (cm)	Tib:fem ratio
Mammuthus primigenius				
Adams (1799)[a]	M	100.0 + 77.0 = 177	120.0 + 67.5 = 187.5	0.56
Beryozovka (1901)	M	85.9 + 67.2 = 152.2	103.0 + 59.0 = 162.0	0.57
Taimyr (1948)	?	85.0 + 65.5 = 150.5	105.5 + 55.2 = 160.7	0.52
Sanga-Yuryakh (1903)	F?	72.5 + 56.3 = 138.8	91.5 + n.a.[b] = ?	—
Schmidt (1867)	M	98.0 + 70.0 = 168.0	n.a. + 67.0 = ?	—
Hatnga (1977)	M	∼ 99.0 + ∼ 70.5 = ∼ 169.5	112.0 + 70.0 = 182.0	0.62
Alaikha (1839)	M	101.5 + n.a. = ?	113.0 + 68.0 = 181.0	0.60
Mammuthus columbi (= *Parelephas jeffersonii* of Osborn 1942)				
Type specimen	M	114.0 + ∼ 75.0 = ∼ 189	123.0 + 69.8 = 192.8	0.57
Huntington Reservoir (Utah)	M	110.0 + ∼ 84.0 = ∼ 194	125.5 + 74.0 = 199.5	0.59
Mammut americanum				
Warren mastodont	M	95.0 + 65.8 = 160.8	106.0 + 71.0 = 177.0	0.67
Peale mastodont	M	∼ 86.0 + n.a. = ?	∼ 109.0 + ∼ 61.0 = ∼ 170.0	0.56
Denver mastodont	M	∼ 88.0 + 60.5 = ∼ 148.5	∼ 104.0 + 66.5 = ∼ 170.5	0.64
Loxodonta africana				
Makololo 2[c]	M	111.0 + 81.0 = 192.0	133.0 + 78.5 = 211.5	0.59
Guvalala	M	108.0 + 84.0 = 192.0	133.0 + 78.0 = 211.0	0.59
Shabi Shabi (a)	F	85.5 + ∼ 65.0 = ∼ 150.5	106.0 + 64.0 = 170.0	0.60
Shabi Shabi (b)	F	88.0 + 64.0 = 152.0	103.5 + 60.5 = 164.0	0.58
Shabi Shabi (c)	F	83.0 + n.a. = ?	101.5 + 63.0 = 164.5	0.62

[a]Name of specimen, with year of discovery in parentheses.
[b]Not available.
[c]Names of sites where found in Hwange National Park, Zimbabwe.
Sources: Garutt (1954, 1964), Osborn (1936, 1942), Peale (1803), Vereshchagin and Nikolayev (1982), Zalyenskii (1903), G. Haynes (unpublished data).

length of the humerus and ulna combined (foreleg length) is less than the length of the femur and tibia (rear-leg length) (Garutt 1964; G. Haynes unpublished data; Roth 1984; Saunders 1970). Of course, the addition of the scapular length makes forelimbs longer than rear limbs.

Some proboscideans had relatively higher forequarters than others. Different limb proportions in the various species of *Mammuthus* have not been well studied (Dutrow 1980; Gambaryan 1974: 174; Osborn 1942; Roth in press). The relative lengthening of the forelimbs in some taxa may have been an adaptation to feeding, locomotor efficiency, or some other selective influence, or it may simply have coevolved with large tusk size; the longer humeri functioned to keep the tusks from dragging on the ground. The larger tusks had evolved as a product of sexual selection (Cluttton-Brock 1988; Darwin 1871).

According to Kubiak (1982:282), the limbs of African elephants are longer than mammoth limbs because modern elephants are evolved browsers; the added height is explained as an aid to feeding from trees. However, measurements of proboscidean limb lengths do not show a simple lengthening of legs in *Loxodonta*, the taxon that relies more on browse. The rear limbs of *Loxodonta* are longer relative to the forelimbs than was the case for *M. primigenius* (Table 2.2). The forelimbs (not including scapulae) of woolly mammoths were 4–8 percent shorter than rear limbs, whereas in the African elephant the forelimbs are 12–14 percent shorter than rear limbs. In other words, the browsing *Loxodonta* has not raised its feeding end up (to get to the treetops?), relative to its rear end, as much as did *Mammuthus*. The greater length of *Loxodonta* legs must have more to do with mobility than with feeding specialization.

Relatively longer tibiae contribute to the lengthening of rear limbs in taller mammals. The tibia:femur ratios for all proboscidean taxa of interest here are variable but quite similar (Table 2.2). The tibia:femur ratio has been thought to reflect locomotion speed in phyletically related land vertebrates (Bakker 1980; Coombs 1978), although later studies have suggested that lengthening of distal leg bones may correlate with endurance rather than acceleration (Hildebrand and Hurley 1985). The relative lengthening of certain limb bones may be a necessary evolutionary development required by an increase in proboscidean body size, and not a "specialized" adaptation (McMahon 1973, 1975a, b).

The differences in limb proportions between mammoths and elephants may have to do with biomechanics, efficiency of locomotion, or speed of traveling, rather than with feeding. There is no clear consensus that browse is always preferred in the diets of African elephants, although they may eat proportionately more woody vegetation than woolly mammoths would have been able to, because of the relative scarcity of trees in local habitats during the Pleistocene (Guthrie 1982; Hanks 1979; Sikes 1971). Some male

Figure 2.3. Right side of a male *Elephas maximus*. Note the convex outline of the spine, the domed forehead, small ear, and sloping back. Photographed in the Taronga Zoo, Sydney, Australia.

woolly and columbian mammoths were taller than modern adult African and Asian elephants, so the differences in limb lengths are not direct reflections of statural differences relating exclusively to species-typical diets.

The body profiles of mammoths and modern elephants are unlike each other, partly as a result of different limb proportions, and partly as a result of differences in the lengths of vertebral spines (Frade 1955: Figure 748; Garutt 1964: Figure 57). In *Mammuthus* the longest spinous processes are at about the position of the front shoulder and are slanted backward, creating a humped outline sloping to the back end. This hump is not a stored fat deposit, but is formed by the vertebrae and a mass of hair covering the shoulders (Garutt and Dubinin 1951). In *Elephas* the spines at the front shoulder are also the longest, but they do not slant so much toward the rear or decrease in length going rearward as abruptly as in *Mammuthus*, so they create an outwardly rounded back outline (Figure 2.3). In *Loxodonta* the longest vertebral spines are located at the front shoulder and decrease in length to the lumbar region, where they lengthen again, creating a dished profile (Figure 2.4). The profile of *Mammut*, judging from reconstructed skeletons, appears to have its highest point at the front shoulder, created by vertebral spines with a mild rearward tilt. The back has a very slight dishing of the spine.

Figure 2.4. Right side of a male *Loxodonta africana* in Hwange National Park, Zimbabwe. Note the concave ("dished") outline formed by the spine, the smooth curve of the forehead, and the large ear.

The sloping back of *Mammuthus* often seems exaggerated in illustrations or fleshed-out reconstructions (e.g., Augusta and Burian 1964). The rear limbs of the skeletons that have been measured are too long to have created a body slope as great as is often reconstructed, based on prehistoric art. In an effort to conform to the Paleolithic images of mammoths in profile, skeletons mounted in museums have sometimes had their rear limbs bent at unnatural angles, resembling bears or cats in their posture, with too great a separation of fore and aft rear legs, and knees too far bent. The probable reason for exaggerations like these is that Upper Paleolithic carvings and paintings depicted *Mammuthus* with huge forequarters and very small rear limbs (Figure 2.5), an image frequently reproduced in modern artistic reconstructions. Vereshchagin and Tikhonov (1987 personal communication) suggest that the Pleistocene depiction of the mammoth's shape was an example of artistic license, emphasizing the animal's front end, which was the important and most memorable part seen by human hunters during head-on confrontations. Table 2.2 gives measurements of limb-bone lengths for several complete *M. primigenius* skeletons; note that the rear limbs are not as short as artistic depictions suggest.

Although it is possible every now and then to read reports of landscape features in the New World shaped "like a mastodont" (e.g., Zavala 1987),

Figure 2.5. Engraved outlines of Upper Pleistocene *Mammuthus primi-genius* from the caves Font-de-Gaumes (left and right figures) and Les Combarelles (top figure) in France. Redrawn from Plate 42 of *Prehistoric Man* by Augusta and Burian (1960), published by Paul Hamlyn (London).

supposedly sculpted by early people, no undisputed Paleolithic art survives to show us an American mastodont in the flesh. Archeologists tend to consider objects reportedly depicting mastodonts as either forgeries or artistic images of animals other than proboscideans. For example, an effigy mound (called the "Elephant Mound") located in the state of Wisconsin is supposed (by some) to have depicted a mastodont (or perhaps a mammoth), although the feature that gave it its elephantlike outline, namely the trunk, in fact may have been only drifted sand accumulated around one end of an otherwise unidentifiable quadrupedal figure (C. Thomas 1985:92). It appears that mastodonts (and in fact mammoths, too) did not make the same strong impression on prehistoric American artists as mammoths made on Eurasian artists.

The bone and body proportions of *Mammut* were different from those of *Mammuthus*, *Loxodonta*, and *Elephas*. Mastodont limb bones are notice-ably thicker and wider and proportionately shorter. Compared with the Asian elephant, *Mammut* was a heavyset animal, with a deep chest, relatively short legs, a long back and wide pelvis (Kurtén and Anderson 1980), and a head oriented much like that of *Loxodonta*, more nearly horizontal and without the doming seen in *Mammuthus* and *Elephas*. Its shoulder was the high point of the body profile. Its teeth were far different from the teeth of *Mammuthus*, *Elephas*, and *Loxodonta* (see Chapter 3 and the Appendix). Its feet seem to have been wide compared with modern elephant feet (Osborn

1936). Quite possibly the greater length and width of the body reflected an increased gut capacity, allowing for greater food intake as compensation for a low-quality diet, consisting partly of coniferous or other woody browse (see Chapter 3).

The long axis of the mammoth skull is oriented more nearly vertically than is the case for *Mammut* or *Loxodonta*, a result of evolutionary changes in the jaws and teeth (Maglio 1972, 1973; Osborn 1942). The more "upright" head of *Mammuthus* has a domed crown that in profile is much higher above the neck than is the top of the skull of *Loxodonta*, but similar to the shape of the head and neck in *Elephas* (Garutt 1981). This high dome also contributed to the great difference in overall body profile of *Mammuthus* (Garutt 1964, 1981).

2.4. Body size and mass

Stature need not directly reflect body mass, because the limb bones of adult mammoths and mammutids were quite robust and usually had greater diameters than those of modern elephants, yet their stature was not always as great. The midshaft diameters of mastodont and mammoth limb bones may be about 20 percent greater than the diameters of the same (and longer) elements in African elephants (G. Haynes unpublished data).

Figure 2.6 is a graph of femoral measurements from American mastodonts [sample from Boney Spring, Missouri (Saunders 1977b)], woolly mammoths [sample from Berelyokh, Siberia (original measurements, and Baryshnikov et al. 1977)], and African elephants [sample from Hwange National Park, Zimbabwe (original measurements)]. Notice the separation among the samples, all of which probably include males and female. A small sample of *M. columbi* femora has been measured, and when graphed the points fall between those for woolly mammoth and mastodont. A sampling of measurements published for other proboscidean taxa such as *Palaeomastodon* (Andrews 1906), *Zygolophodon* (Tassy and Pickford 1983), and *Eozygodon* (Tassy and Pickford 1983) indicates that the earlier mammutids had thick femora of lengths similar to those of *Mammut americanum*.

Among African elephants, shoulder height varies with the cube root of body mass (Eltringham 1982; Laws et al. 1975). The manner in which shoulder heights (or body lengths) vary with body mass can be compared with the patterns for other land animals, and from this it is possible to extrapolate from least-squares regression lines in order to predict body mass for other proboscideans if shoulder heights, body lengths, or limb-bone circumferences are known (or vice versa). Further discussions are available: Anderson, Hall-Martin, and Russell (1985), Economos (1981), who reports on an unpublished study, Peters (1983), Roth (1990), Schmidt-Nielsen (1984), F. Smith (1980), R. Smith (1984), and Western (1979). However,

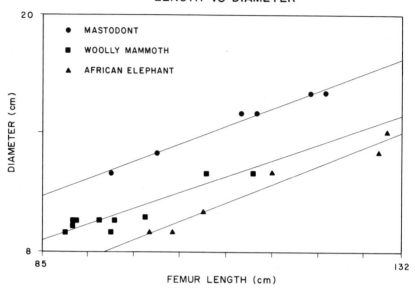

Figure 2.6. Graph of femoral lengths *versus* diameters for a sample of *Mammut americanum* (circles), *Mammuthus primigenius* (squares), and *Loxodonta africana* (triangles). The regression polynomial of line 1 (*Mammut*) is $(-1.730E + 00) + (1.475E - 01) * X$(variance $= 3.689E - 02$); the regression polynomial of line 2 (*Mammuthus*) is $(-2.745E + 00) + (1.336E - 01) * X$ (variance $= 1.922E - 01$); the regression polynomial of line 3 (*Loxodonta*) is $(-6.414E + 00) + (1.549E - 01) * X$ (variance $= 1.525E - 01$).

the geometric differences among the bodies of elephants, mammoths, and mastodonts may invalidate such estimates of body mass based on limb-bone sizes. Body shape (especially girth) does not vary in a simple way when body weight is decreased or increased. Larger animals usually are stockier. That generalization may hold true for single bones as well. As R. Smith (1984) emphasized, the use of allometric scaling as a predictive "technique" tends to reduce a general field of inquiry to the level of a cookbook; it should be a study of comparative biology of size (Schmidt-Nielsen 1984), not a method for predicting biological traits. It seems probable that a better approximation of body mass can be achieved using a method that Bakker (1986) devised for estimating dinosaur mass – building a scale model, immersing it in water, and weighing the displaced fluid. Most mammalian bodies weigh about 0.95 as much as an equal volume of water.

McMahon (1973, 1975a, b) described a theory of body-part scaling that pertains to animals of closely related taxa. McMahon argued that evolutionary enlargement of the body in related animals is accomplished with a regular "distortion" or change of bone or limb shape. When body size has increased, some anatomical parameters (such as limb-bone lengths) can be shown to have increased at different rates than other parameters (such as diameters of limb shafts; in other words, larger animals are not simply smaller animals that have been scaled upward, because body proportions must change, following predictable rules and patterns that reflect constraints on elastic stability and flexure of limbs. McMahon's explanation of body shapes is flawed (Alexander et al. 1979a; Garland 1983; Peters 1983:217; Schmidt-Nielsen 1984), because in some animal groupings bone mass and length do not conform to the predicted relationships, although the prediction is near to the reality.

The rules that McMahon described lead to minimum required measurements; sometimes animals do not obey the rules, and their limb bones are too thick for their predicted lengths. Bakker (1980:449–54; 1986) applied McMahon's theory of elastic scaling to limbs of mammals and dinosaurs and found that limb strength in large terrestrial quadrupeds was quite variable, nearly independent of body mass. He measured limb strength by graphing femur circumference against length. The variability is not directly due to differences in body size, because some large animals (such as *Loxodonta*) have "weak" limbs, but others of similar body mass have much stronger limbs.

Bakker suggested that limb strength correlates with gait. The greatest strength is seen in animals in which the gait has a suspended phase (when all four feet are off the ground at once). Modern elephants do not have a suspended phase, or period of free flight (Alexander et al. 1979b; Gambaryan 1974), and indeed according to Bakker's analysis (Bakker 1980:453) elephant limb strength is relatively weak. Modern elephants have two semi-suspended phases: when the two rear legs are off the ground together, and when the two forelegs are off the ground together (Gambaryan 1974:24, 166, 188).

On the other hand, the rhinoceros, which is large and heavy, has a suspended phase when it gallops, and its limb strength is scaled high (Bakker 1980, 1986). The top speed for a modern African elephant has been estimated at about 35 km/hour (about 26 km/hour for an Asian elephant), whereas top speed for a black rhino is about 56 km/hour (Bakker 1986; Coombs 1978; Garland 1983; A. Howell 1944). Limb-bone diameter relative to length is much greater when more thrust is applied to the limb; the thickening of bone increases the limb's resistance to bending or compression failure.

Late Pleistocene mastodont and mammoth limb bones were similar in length to bones of modern elephants, but their midshaft diameters were much greater (Figure 2.6). This disproportion in limb cross sections may also be noted in other late Pleistocene taxa, such as equids and camelids. Although modern and extinct forms are comparable in terms of stature, the leg bones of extinct forms were much more massive. According to allometric scaling formulas (Anderson et al. 1985; Roth in press; Western 1979), skeletal mass increases at a faster rate than body mass; in other words, skeletal mass accounts for a greater part of the total body mass in larger animals. If body mass increases by a factor of 2, skeletal mass increases by *more* than a factor of 2. Larger animals have comparatively heavier bones. However, Alexander et al. (1979a) have shown that the limb bones of *Loxodonta* are somewhat more slender than allometric scaling would predict. Yet the sizes of cranial and postcranial elements in the skeletons of mastodonts and mammoths indicate that in regard to stature, girth, and body length, they may not have been extraordinarily different from modern elephants. Why were their limb bones so much thicker?

Possibly the bones were thicker because mammoths and mastodonts ran either faster or more often than do modern elephants, although their pillarlike limbs were not suited to high-speed locomotion involving wide oscillations of the bones. The joints of elephant limbs do not flex greatly during the propulsive phase of locomotion, and the scapulae do not freely rotate (Alexander et al. 1979b; Gambaryan 1974:170, 187). In the extinct forms, as in the modern, the cnemial crest on the tibia was relatively small as crests go, indicating that the muscles attaching there (which flexed and extended the knee) were not as large relative to the leg's size as are comparable muscles in animals with suspended phases, such as horses or rhinos, which have large cnemial crests. Finally, whereas the distal limb elements of mastodonts and mammoths were less massive than the proximal elements, as compared with bones in modern elephants, the distal ends of the bones were nearly as massive as the proximal ends, which would not conserve energy well in the nonpropulsive "recovery" phase of a fast gait (in which the leg is pulled back and repositioned to push off again) (Coombs 1978). Thus, the demand for muscle power during a fast gait would have been outweighed by the problems of weight bearing. The few "improvements" for faster gait in mammoths and mastodonts (relative to modern elephant anatomy) could not have overcome these problems.

According to the formulas of Anderson et al. (1985) (which are not intended to predict unknown values, but rather to describe existing patterns), proboscideans with femoral circumferences on the order of those recorded for *Mammut* (see Table 2.3A and B for some published measurements) may have had body weights about twice those of modern *Loxodonta* that are about 10–20 percent taller at the shoulder. The Warren mastodont from New York (Warren 1855) may have weighed about 1.8 times as much as Jumbo, the famous African elephant exhibited by P. T. Barnum; the largest mastodont from Boney Spring, Missouri (see Chapter 6), may have

Table 2.3A. *Mastodont femur lengths and midshaft circumferences*

Taxon	Maximum articulation length (mm)	Midshaft circumference (mm)
Mammut americanum		
Late Pleistocene (North America), Boney Spring, Missouri[a]	Range 940–1,216	Range 379–507 (least circumference)
	940	379
	1105	483
	1216	507
Warren mastodont[b]	1060	432
Cambridge mastodont[c]	914	330
Denver mastodont (unpublished data)	1040	365
Paleomastodon beadnelli		
Fayum, Egypt[d]	875	Not reported
Zygolophodon atavus[e]	882	Not reported
Eozygodon morotoensis		
Early Miocene (East Africa)[e]	1060	Not reported

[a] Saunders (1977b).
[b] Osborn (1936).
[c] Warren (1855).
[d] Andrews (1906).
[e] Tassy and Pickford (1983).

weighed about 2.8 times as much as Jumbo. These predictions seem excessive and unlikely, and they are in fact unreliable, because the extinct proboscideans had body geometries different from those of living mammals.

Mammoths and mastodonts need not have had a suspended phase in one of their gaits, but perhaps did subject their bones to much greater stress. Quite possibly the added weight of a thick hair coat, relatively taller forequarters, and larger, spreading tusks necessitated stronger legs.

A less probable explanation for the greater thickness of mastodont limb bones is that it was an evolutionary response to spending much of the time in moderately deep water, suggested (but not proven) by the presence of fossil finds in lacustrine sediments and by the discovery of possible aquatic adaptation of mammutid hair (Hallin 1989; Hallin and Gabriel 1981). The increased density of the limb elements may have overcome body buoyancy and allowed them to submerge without struggling against a tendency to float. Undoubtedly mastodonts did not spend their lives submerged, but heavier limbs may have given them an occasional advantage for keeping contact with the bottom while wading through deep water. Their feet were

Table 2.3B. *Mammuthus and Loxodonta femur lengths and midshaft circumferences*

Diaphysis length (mm)	Total (maximum) length (mm)	Midshaft circumference (mm)
M. primigenius (USSR Academy of Sciences Zoological Institute, Leningrad)		
700	880	277
805	985	307.5
745	945	302.5
750	925	300
780	940	282.5
722.5	895	300
720	890	292.5
730	890	302.5
910	920	285
880	1,065	370
880	1,065	380
920	1,125	375
Loxodonta africana (original measurements, Hwange National Park, Zimbabwe)		
830 (estimated)	990	287.5
850	1,020	285
905	1,060	317.5
980	1,150 (estimated)	370
1,100	1,290	400
1,110	1,300	430

wide and squat, perhaps more so than modern elephant feet, a sensible adaptation to walking on muddy or boggy ground.

A possible ancestor of the early Eocene proboscideans was semiamphibious, similar in form to *Moeritherium* (Maglio 1973). However, mammoths undoubtedly were not aquatic animals, and yet their limb bones were also thicker in diameter than the bones of modern elephants (although less so than mammutid bones), so the massiveness need not have to do with aquatic habits.

2.5. Size reduction ("dwarfing")

Toward the end of the last glacial advance in the Pleistocene, mammoth body size and stature apparently decreased. Such a response to climate has been postulated for earlier phases of the Pleistocene as well. Garutt (1965) suggested that mammoth stature was adjusted in synchrony with environmental changes and that mammoths of interstadial age were relatively tall, possibly in response to improved foraging conditions, whereas mammoths living at the end of the Pleistocene were relatively short, possibly in response to climatic deterioration after 18,000 years ago (Saunders 1987). The radiocarbon dates that provide the basis for this proposition may be open to some question; see Garutt (1965:124) and Heintz and Garutt (1964), and note (1) the extremely large standard deviations of the dates, (2) the discordant multiple dates from single specimens, and (3) dates near the maximum age limit for the radiocarbon method.

The postglacial body-size decrease in taxa that survived the late Pleistocene extinctions is recognized as a general evolutionary trend (Kurtén 1968; Marshall and Corruccini 1978), and Guthrie (1984) proposed that the trend applies to extinct taxa as well. He speculated that some relatively small mammoths in European, Asian, and American Paleolithic sites are males, and not females as has usually been assumed. Guthrie thought that some male mammoths in the Berelyokh site in northeastern Asia (Vereshchagin 1977) (see Chapter 6) were so small and their tusks so "vestigial" that they were originally mistaken for females (Guthrie 1984:270). There is a bimodal distribution of limb-bone sizes in this collection (Baryshnikov et al. 1977); marked size differences also distinguish the lower jaws and teeth of males and females.

Whatever the sex ratio in the Berelyokh collection, many adult animals fit into the same size range as some earlier mammoths, although on the small end of the scale. For example, bones similar in size to many from Berelyokh came from a carcass found on the Sanga-Yuryakh River (Vollosovich 1909); this animal is quite small. The carcass was dated four times, with ages ranging from 29,000 to 44,000 years (Heintz and Garutt 1964), spanning the end of a cold period and a longer time of climatic amelioration, not an interval when exceptional dwarfing would be expected. If the animal is female, its size is much the same as that of other Siberian female mammoths. If it is male, it is remarkably small. Its tusk circumference at the alveolus is about 24 cm (see Table 2.6).

Guthrie (1984:270) mentioned a "dwarfed" mammoth from the Gönnersdorf site in Germany; among five mammoth bones from the site is a femur 94 cm long and broken tusk tip whose robustness may indicate that the animal was male (Poplin 1976:50). The femur is about the same length as femora of adult female mammoths in Siberian bone collections. Male femora from that time period (13,000–11,000 years ago) measure about 10–15 percent longer than female femora (Figure 2.7). Poplin (1976, 1978) and Schild (1984) interpreted the Gönnersdorf fauna as reflecting a relatively rich environment, not one that would have led to male mammoths dwarfed to the stature of females. The Gönnersdorf mammoth may be a normal female, not a dwarfed male.

However, my point here is not to dispute that some size reduction occurred at the end of the Pleistocene. I think it is possible to distinguish dwarfing from dimorphic size differences in bone assemblages, but such a

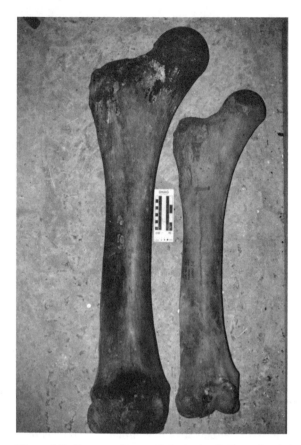

Figure 2.7. Femora from adult male (left) and adult female (right) *Mammuthus primigenius* from the Berelyokh site, USSR. Photographed in the Zoological Institute of the USSR Academy of Sciences in Leningrad. Note that shaft diameters are relatively greater than in *Loxodonta*, although the size dimorphism is similar.

distinction has not been reliably made. It should be possible to compile clear demonstrations of stress that led to dwarfed body size, such as finding growth-arrest lines in mammoth bones, age profiles diagnosing high subadult mortality, or broken tusks resulting from intraspecific aggression (Conybeare and Haynes 1984; G. Haynes 1987c; see also Chapter 4).

When body size reduction did occur in woolly mammoths, it must have resulted from a shortening of the growth season, or a decrease in habitat quality (Guthrie 1984; King and Saunders 1984; Marshall and Corruccini 1978). Several populations of North American caribou (*Rangifer tarandus*) are smaller than Bergmann's rule predicts, because during the lengthy and harsh northern winters their basal metabolic rate drops 25 percent or more, and they have a period of no body growth in order to conserve energy; see Calef (1981) and the references therein. If late Pleistocene winters became harsh, mammoths, too, may have stopped growing and become ontogenically dwarfed.

Dwarfing may also have resulted from hunting by humans (Edwards 1967; McDonald 1984). It has been proposed that in the face of pressures put on by hunters choosing large individuals to slay, smaller body size would have been selectively more fit than larger size. Also animals that did not grow as much in their maturational history could have reproduced sooner, thus maintaining population levels in spite of limitations imposed by being hunted (Kurtén 1964; Marshall 1984; McDonald 1984). However, an equally feasible alternative model of human predation predicts that smaller, more manageable individuals would have been targeted.

Because the small adult mammoths and mastodonts that have been discovered may be either normal females or dwarfed males, it is difficult to determine the distinctive characteristics of the sexes. In my opinion, the best way to distinguish males and females is by reference to two sets of data from bone collections: First, there should be clearly dimorphic size differences (which will be rare in some mass assemblages, because of sexual segregation during life); second, the epiphyseal fusion scheduling of one dimorphic group should be distinct from that of the other group (see the Appendix).

2.6. Male and female shape differences at a glance

Male African elephants possess a distinctive head shape compared with females; in males, the curve of the forehead is smoothly formed by the top and front of the head, whereas in females the curve is more abrupt and angular. The head in males is more massive and bulky compared with the rest of the body, and the tusks are much thicker.

In the Asian elephant, the head of a male is also more massive than that of a female, because of enlarged parietals and occipitals and a bulging trunk base; the female head is more square and slender. Females very rarely have tusks. The outline of the back in female Asian elephants is straighter than in males, where it is more convex (McKay 1973).

Male mammoths had more massive skulls and much thicker tusks and were considerably taller than females. Differences between male and female mastodonts are not clearly known, but were probably along the same lines as in other taxa – larger heads, thicker tusks, and taller bodies in males.

As discussed earlier, adult male mammoths were about 15–30 percent larger than adult females; this goes for their overall body stature, as well as for the size of each bone element. In dealing with bones, as opposed to live animals, clear dimorphism may not appear in many assemblages, because

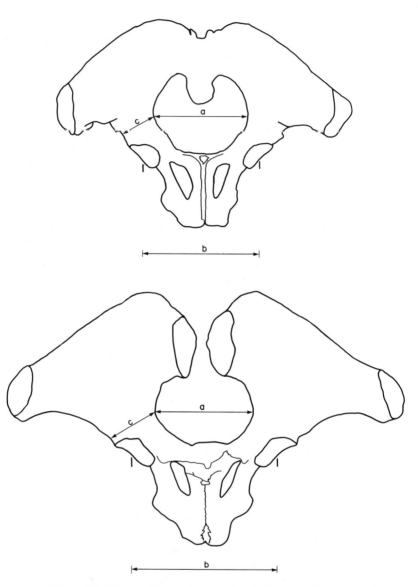

Figure 2.8. Pelves of female (top) and male (bottom) *Loxodonta africana* from Hwange National Park, Zimbabwe. Measurement "a" refers to the maximum width of the pelvic canal; "b" refers to the width between midpoints of the acetabula; "c" refers to the thickness of the ilium at the point where measurement "a" is taken, with the thickness being measured perpendicular to the long axis of the ilium.

males and females do not often associate in life. In a decade of searching for elephant bone accumulations in wild areas of Africa, I have never found a modern bone assemblage dominated by adult males, but I have found many that were dominated by females and young (see Chapter 4). Hence, I would guess that most of the larger mass bone deposits of elephantids (fossil or modern) were produced mainly by mixed herds containing few or no adult males, although individual males could have died in the same places. The unusual circumstances of entrapment may make Hot Springs, South Dakota, an exception (see Chapter 6).

There is little sexual distinction in terms of traumatic bone fracturing – females break limbs and ribs as frequently as do adult males (G. Haynes unpublished data). Among *Bison*, broken ribs resulting from fights are far more often found in adult males, but that trait does not seem to be as reliable for distinguishing elephant males and females.

Osteological distinctions between the sexes involve shape as well as size differences. None of the shape differences have been well studied, but the morphology of the innominate bones has been a focus of some recent interest (Coope and Lister 1987; Lister in press). In *Loxodonta*, adult males have a pelvic canal that is equal in size to the canal in adult females, but the maximum widths, heights, and thicknesses of the pelvic bones themselves are much greater. For example, the distance between midpoints of adult male acetabula may be about 50 cm, when the maximum width of the pelvic canal equals about 40 cm. In an adult female whose birth canal is 40 cm wide, the distance between midpoints of acetabula is only about 40 cm (see Figure 2.8 for an illustration of these differences). The thickness of the iliac "neck" or pedicle, measured at the point of the maximum width of the birth canal, never exceeds about 16 cm in adult females, whereas in males over the age of twleve years this measurement is rarely less than 16 cm. Similar differences are seen with *Mammuthus* innominates (Lister in press), as well as those of *Mammut* (G. Haynes unpublished data).

2.7. Skin and hair

There is great variability in the skin thicknesses of modern elephants and well-preserved mammoths. The skin of African and Asian elephants is thinnest on the trunk, breast, groin, ears, and legs (Sokolov 1982a); on one captive Asian female, the thinnest body skin measured 1.0 cm thick on the medial leg (J. Shoshani et al. 1982), whereas on wild African elephants the skin of the outside surface of the foreleg is thinnest (not considering ears and trunk) (Sikes 1971). The thickest skin on the captive Asian animal's carcass was 3.2 cm on the back, and the thickest skin on wild African elephants was about the same, but was located along the back and flanks. The cornified skin on the soles of the feet may be much thicker than the skin elsewhere on the body.

According to Pfizenmayer (1939), the thickest skin preserved on the frozen carcass of the Beryozovka mammoth (*M. primigenius*) was about 1.9 cm, relatively thin when compared with modern elephant skin. Although Garutt (1964) estimated that the average thickness of skin on woolly mammoths was about 3.0 cm (which is less than the average thickness of modern elephant skin), the thinnest skin on preserved mammoths has measured 2.0 cm thick, which is much thicker than the thinnest modern elephant skin. The preserved skin on the front of the head of the Adams mammoth (Adams 1807; Dubinin and Garutt 1954; Tilesius 1812), though missing epidermis in places, is 1.3–1.5 cm thick (G. Haynes unpublished data); Section 2.12 describes this specimen and several others. The skin of a mammoth calf's head and one foreleg, recovered from Fairbanks Creek, Alaska, in 1948 (Anthony 1949), is about 0.6 cm thick (Péwé 1975:98) and virtually hairless, perhaps because of the pressured-water jets that exposed it.

The rough, dark skin of modern African and Asian elephants appears to be almost bare of hair, except on the chin, the outside ear openings, the tail tip, the eyelids, the knees, the elbows, and the trunk; however, there are numerous black, bristly hairs scattered over all the body surface. Different individuals are hairier than others. Body hairs on the carcass of one captive Asian female ranged in length from 0.3 cm to over 26 cm (J. Shoshani et al. 1982). The longest hairs usually grow from the tail near its tip; these hairs may break off in old age. Deraniyagala (1955:75–7) described the locations of hair tracts and detailed changes in the hair coat over time in the Asian elephant.

The captive Asian elephant studied by Benedict (1936:270) was exposed to temperatures of 7–15°C for long periods of her life, but hair growth was not stimulated by the cool weather, although such growth is a common reaction in other mammals. Wild African elephants in Zimbabwe live through one to three months of cold nighttime temperatures that often dip below freezing, and yet as a rule they do not develop thicker hair coats than those living in warmer conditions in west and central Africa. Neither captive nor wild elephants show much discomfort in cold weather (Benedict 1936:91; G. Haynes unpublished data), indicating that they have a wide comfort zone for air temperatures, although the ears are sensitive to cold and can be damaged by frost (Benedict 1936:159).

The woolly mammoth was entirely covered with a thick coat of hair, right down to its toes, but not including the soles of its feet (Ryder 1974; Vereshchagin 1977; Vereshchagin and Baryshnikov 1982). The hair coat consisted of very dense woollen underfur 5–15 cm thick from which moderately long and very long straight hairs emerged (Sokolov and

Sumina 1982; Vereshchagin 1981b). The longest straight hairs were up to 100 cm long and grew on the tail and foreleg. Even the trunk was covered with thick hair; the hair on the trunk of a small calf (the Magadan mammoth, also called the Kirgilyakh mammoth, as described in Section 2.12) was 5 cm long (Stewart 1979). Lower-leg hair on the Sanga-Yuryakh carcass (Vollosovich 1909) was 15–18 cm long; above the knee it was up to 35 cm long; on the back the hair was 40–45 cm long (Pfizenmayer 1939; Zalyenskii 1909). Sokolov and Sumina (1982) measured hair lengths on the subadult Yuribei mammoth and found the longest hair on the tail, with decreasing lengths on the foreleg, thigh, and belly; the shortest hair grew on the groin and side of the body.

The mammoth's hair functioned to maintain body temperature, of course, which in modern Asian elephants is around 36–37°C (Benedict 1936:139). Perhaps one of the most important functions of the hair coat was to prevent heat loss after mammoths drank water, which may have been relatively chilled in the cooler conditions of the Pleistocene. Benedict (1936:90, 176) found that "Jap," a captive Asian elephant, tolerated cool temperatures ($\sim 12°C$) without discomfort, but after she drank water (even at a temperature of around 23°C) she shivered and was chilled. A considerable amount of body heat was used to warm the water in her body. Mammoths would have needed water often, as do all elephants, and if it was frozen or in the form of snow, warming it in the stomach would have put a great burden on the body, which a thick hair coat could have reduced.

Modern northern terrestrial herbivores such as caribou normally do not find the cold temperatures of their ranges uncomfortable, because their hair is so effective in conserving body heat. Under very cold conditions (when temperatures are around $-35°C$), caribou raise their activity levels to create more body heat. Mammoths could have done the same. Under harsher conditions, such as lower temperatures and high winds, caribou seek shelter behind trees or ridges. Their main insulation in middle and late winter is not subcutaneous fat (Calef 1981) but a good hair coat, the most efficient adaptation to very cold conditions. The greatest winter-related threat to caribou is ice crusting, the formation of ice over snow that has been softened by warm temperatures. When this happens, animals cannot break through the snow to feed, in which case the superiority of the hair coat is irrelevant.

The original color of mammoth hairs is open to question. A summary of different opinions has been published by Augusta and Burian (1964:32). Most of the preserved skin specimens have black, brown, reddish brown, or yellowish hair growing from them, but the colors were somewhat altered after death because of degradation and slight decay before freezing (Garutt 1964; Gillespie 1970; Möbius 1892; Ryder 1974; Sokolov and Sumina 1982; Vereshchagin 1977). Péwé (1975:100) reported that the keratin from one

Alaskan mammoth's hair was morphologically in good condition nearly 32,000 years after the animal's death. Sokolov and Sumina (1981, 1982) found excellent preservation of hairs from the Magadan mammoth calf (dated about 44,000 years old) and the Yuribei mammoth (dated about 9,600 years old, a figure in which I have little confidence) (Yevseyev et al. 1982), although some cuticle was lost.

Like young African and Asian elephant calves, mammoth calves were covered with a woollier, fluffier coat that probably was somewhat lighter in color than the adult coat. As the animal grew older, the coat darkened, and the hairs became coarser. The hair recovered with the small Magadan mammoth calf was chestnut-colored. The calf was about seven to eight months old when it died (Vereshchagin 1981b).

Specimens of what appears to be carbonized skin holding together bundled and fine hairs interspersed with hollow, coarser hairs may be the only preserved *Mammut* soft tissue currently known. These specimens were recovered by Kurt Hallin in association with cranial fragments found near Milwaukee, Wisconsin (K. Hallin 1989 personal communication). The preserved guard hairs are hollow, a common enough trait in many mammals, including woolly mammoths and African elephants. The underfur appears similar to that of semiaquatic animals such as otter and beaver (Hallin 1983, 1989; Hallin and Gabriel 1981), in that it is very fine and wavy, and grows in dense bundles.

Fortuitous discoveries of mastodonts were quite common in the nineteenth century; some of those finds were poorly reported, but in a few cases claims were made that in addition to bones, soft tissue was also preserved, although Eiseley (1945, 1946) cast doubt on those claims. Those early reports seem to have been designed (unconsciously or otherwise) to encourage exploration of America's unknown lands, by fostering the idea of recently surviving mastodonts and mammoths. Such "reports" tended to be mythic, folkloric, or just plain misunderstood and exaggerated. For example, according to Howorth (1887), Shawnee Indians near the Ohio River discovered the trunk of a mastodont still attached to the skull; in fact, the original report of the Indians' discovery described a skull with a long "nose," referring to the bone, not to soft tissue, as quoted by Simpson (1942). Howorth reported another preserved trunk in Illinois, although there was no documentary or referential support given for that example. Koch (1839) reported finding preserved skin with mastodont bones in Missouri, but Koch was a showman dedicated to making a living from his finds, and his scientific objectivity (as well as reliability) may be doubted in our time (McMillan 1977).

Nevertheless, it is impossible to prove that all claims of preserved hair were fraudulent or in error. The hair usually was described as dun, a color defined in *Webster's Third New International Dictionary* (Gove 1967:701) as a "nearly neutral slightly brownish dark gray and ranging from red to yellow in hue." The length of the hair from different sites varied from about 3.8 cm to nearly 18 cm (Howorth 1887:299; Warren 1855). Loren Eiseley (1946) argued that such early reports of coarse, dun-colored hair were in fact describing preserved filaments of a green algae.

2.8. Ears

Eltringham (1982) pointed out that the ears of Asian elephants are shaped like maps of India, and the ears of African elephants are shaped like maps of Africa. The ears of the woolly mammoth were surprisingly humanlike. Mastodont ears presumably were small, perhaps smaller than *Elephas* ears, because of the seasonally cold climate, although their shape is unknown.

The ears of *Mammuthus primigenius* were small compared with the ears of modern *Loxodonta* and *Elephas*. Complete ears have been preserved on at least two well-known frozen mammoths found in Siberia; one was a calf (the Magadan or Kirgilyakh mammoth), a nearly complete carcass when discovered in 1977 (Shilo et al. 1983; Vereshchagin 1981b; Vereshchagin and Dubrovo 1979), and the other was the so-called Adams or Lena mammoth, from which most of the soft tissue had been either destroyed by decay or eaten by northern scavengers. The Lena carcass was salvaged by an expedition from the Russian Imperial Academy of Sciences (Adams 1807; Dubinin and Garutt 1954; Garutt 1964; Tilesius 1812). Both ears are present on the small calf, a male, but only one ear is present on the Adams mammoth, a full-grown male (Figure 2.9). These ears probably had shrunk somewhat after death, but then they were excellently preserved in the cold, dry climate of Siberia. Vereshchagin (1981b) measured the ears of the small calf and found them to be about one-ninth to one-tenth the size of the ears of an African elephant calf, and one-fifth to one-sixth the size of an Asian elephant calf's ears. The mammoth calf's ears are relatively larger than the adult male's ears when compared with the other dimensions of the heads, but that difference may be due to age-related disproportions in the body parts of very young mammals.

The relatively small size of woolly mammoth ears was an evolved feature preventing heat loss in temperate latitudes. Northern mammals, such as the arctic fox, have ears smaller than those of their southern relatives. Benedict (1936:159) noted that the ears of Jap, the captive Asian elephant, had been damaged by frostbite. The larger ears of African elephants would be at great risk if the animals could not raise their activity levels and thus produce enough body heat to protect their ear tissue. Benedict (1936:158) measured skin temperatures on Jap and found that the ears were warmer after exercise or emotional upset.

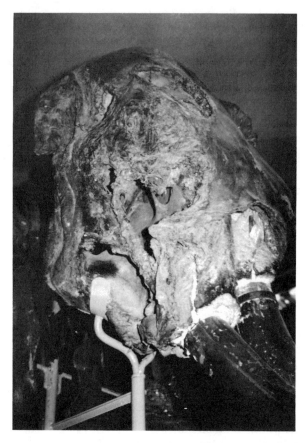

Figure 2.9. View of the right side and front of the head of the Adams (or
Lena) *Mammuthus primigenius* as mounted in the Zoological Museum
(USSR Academy of Sciences) in Leningrad. The tusks are not the originals.
Note the small ear to the upper left, and the skin preserved on the
specimen's head.

No columbian mammoth ears have been found, and so we probably shall
never know for certain how small the ears of other *Mammuthus* species
were. The earliest representatives of the genus *Mammuthus* originally
appeared in the tropics of Africa (Coppens et al. 1978; Maglio 1973; Osborn
1942), and because large species were present in southern latitudes during
the evolutionary history of the genus, it is possible that ears were larger in
the earlier forms, as an adaptation for cooling the body (fanning the ears
cools the blood in vessels near the skin).

It is well known that African elephants threaten or intimidate with head
shaking, nodding, trumpeting, and charges that stop short of making

contact (Douglas-Hamilton and Douglas-Hamilton 1975; Leuthold 1977; Moss 1976; Poole 1987); some charges do not stop short, and hard contact is made (Poole 1987). The ears figure prominently in these displays, being extended out from the body to give a larger appearance. The head is held up and often vigorously moved from side to side, which flaps the ears loudly, sounding like canvas snapping in a wind. The head/ear display may be used while standing still, or during brief forward rushes. Woolly mammoth ears were rather small to make much of a visual or auditory impact during display use, but mammoth threats may have had other aspects in common with those of modern elephants, perhaps involving tusk displays such as head shaking and nodding.

2.9. Osteology

Olsen (1972) is a good source of information about skeletal differences between mammutid mastodonts and mammoths. His monograph is amply illustrated with photographs on which diagnostic characteristics of the elements are indicated. Elephant osteology is discussed in several zoological publications, notably those by W. C. O. Hill (1953), Frade (1955), Grassé (1955), and Gambaryan (1974); see also Blair (1710a, b). Woolly mammoth osteology has been described in a few publications (e.g., Dubrovo 1982). Zalyenskii (1903) compared the Beryozovka mammoth and modern elephants in regard to osteology and dental morphology.

There is still much disagreement over certain aspects of proboscidean skeletal anatomy; for example, the numbers of ribs in late Pleistocene taxa have not been firmly established in the literature (Table 2.4). It may be that some variation was possible among different animals.

Klein and Cruz-Uribe (1984) specified nineteen pairs of ribs as typical of elephantids (encompassing modern genera and extinct *Mammuthus*); some mounted specimens of complete *M. primigenius* in the USSR Zoological Institute in Leningrad and the Paleontological Institute in Moscow contain nineteen pairs of ribs, but one specimen (the "Alaikha mammoth") contains twenty pairs. *Elephas* has nineteen pairs of ribs, as described by Blair (1710a, b) and counted in recent studies (J. Shoshani et al. 1982), yet mounted *Elephas* skeletons in Moscow and Leningrad contain twenty pairs. *Loxodonta* has twenty pairs of ribs, verified in recent studies (G. Haynes unpublished data; S. Shoshani, Shoshani, and Dahlinger 1986), although some individuals may have twenty-one pairs (Hill 1953). *Mammut* may also have had twenty pairs of ribs (Warren 1855), although Peale (1803) counted only nineteen pairs in his specimens, and the Neath mastodont from Wisconsin (UWZP 20500) had only eighteen pairs. Mounted museum specimens often have twenty pairs, some of which are reconstructed. Perhaps it is possible that one or two vertebrae that would ordinarily

Table 2.4. *Numbers of ribs and vertebrae*

Source	Ribs (pairs)	Cerv. vert.	Thor. vert.	Lumbar vert.	Sacral	Caudal[a]
Elephantidae						
Klein and Cruz-Uribe (1984)	19	7	19	4	4	—[b]
Elephas						
Blair (1710a, b)	19	7	19	3	5	29
Original measurements, Moscow State University Geology Museum	20	7	20	n.a.	n.a.	n.a.
J. Shoshani et al. (1982)	19	7	19	3	5	27
Original measurements, Zoological Institute, Leningrad	20	7	20	3	3	24+
Osborn (1942); S. Shoshani et al. (1986)	20	7	20	3	4	>12
Deraniyagala (1955)	19–20	7	19–20	3	4–5	29
Loxodonta						
Original measurements, Hwange National Park	20	7	20	3	4	>14
Mammut						
U.S. NMNH[c]	20	7	20	3	5	27
Peale (1803)	19	7	19	3	?	?
Warren (1855)	20	7	20	3	5	—

38

Original measurements, University of Wisconsin, Museum of Zoology	18	7	18	3	4	16+
Mammuthus						
Original measurements, Alaikha mammoth, Paleontological Institute, Moscow	20	7	20	n.a.	n.a.	n.a.
U.S. NMNH[d]	19	7	19	3	5–6	22
Original measurements, Beryozovka mammoth, Zoological Institute, Leningrad	19	7	19	4	4	>14
Original measurements, Adams mammoth, Zoological Institute, Leningrad	19	7	19	4	4	21
Osborn (1942) ("*Parelephas jeffersonii*")	19	7	19	4	5	12+
Osborn (1942) ("*M. primigenius*")	18–19	7	18–19	3–5	3–5	21

[a] Caudals are often manufactured in mounted specimens.
[b] No data available.
[c] U.S. National Museum of National History.
[d] This specimen was created out of several different finds.

develop as thoracics in some individuals instead become lumbar, thus lowering the count of ribs.

The numbers of caudal vertebrae likewise have not been confidently established in the literature (Table 2.4). Garutt (1964) and Osborn (1942) stated that in woolly mammoths there were no more than twenty-one caudal vertebrae (based on a single specimen found with preserved tail, the Beryozovka mammoth), whereas in modern elephants the tail contains twenty six to thirty three vertebrae (Garutt 1964; W. Hill 1953). Hence, modern elephant tails are much longer than mammoth tails. Caudal vertebrae are not robust bones, and they have been poorly preserved or recovered except under exceptional circumstances; also, many mounted specimens have lost tail vertebrae to visitors taking souvenirs. It is difficult to find samples of complete tails from mammoth, mastodont, or modern elephant.

A well-preserved mammoth foot from Siberia was X-rayed by Dubrovo (1981), and she pointed out that the metatarsals in *M. primigenius* were oriented much less vertically in life than they are usually reconstructed in museum-mounted skeletons. This orientation is similar to that of modern African and Asian elephants, in which the flexible soft-tissue pad cushioning the rear foot is smaller than the forefoot pad.

The fat pad in the foot of the modern elephant is a preferred food for some elephant-hunting people (Crader 1983; J. Wood 1868). If this pad is so much reduced in the rear foot, we might expect to find that rear-foot butchering and roasting in archeological assemblages was less frequent than front-foot utilization.

The scheduling of epiphyseal fusion is quite patterned in modern African and Asian elephants (Roth 1984) and provides a basis for determining individual animal ages. The samples of mammoth bones at the Zoological Institute in Leningrad and the Paleontological Institute in Moscow are rather loosely controlled (in terms of provenience and chronology), but fusion scheduling of their long-bone epiphyses seems to agree well with fusion sequences and scheduling in modern elephants (Lister in press). A sampling of Late Wisconsin mastodonts shows that their epiphyseal fusion followed the same schedule. This subject is discussed further in the Appendix.

2.10. Tusks

Modern elephants, like mammoths and mastodonts, grow a pair of deciduous tusks before growing the permanent set (Miller 1890a, b; Sikes 1971; Vereshchagin 1981b). These temporary tusks may reach a length of about 5 cm, and they remain in place about one year, after which time they are replaced by the permanent tusks growing behind them in the skull

(Eltringham 1982; Miller 1890a, b). At first, the tips of permanent tusks are covered with cementum over a layer of enamel, but as the cementum wears off, the enamel comes into wear and eventually disappears.

Both male and female mammoths, mastodonts, and African elephants have tusks. Most female Asian elephants do not, possibly a recently derived condition. Tusks are present in both sexes because they aid in feeding, and also because they aid in achieving dominance during feeding competition. A female with tusks can intimidate other animals and acquire a better spot for feeding or drinking. Male tusks are much larger than those of same-age females, but adult males usually segregate themselves from females, so there is little selective pressure for female tusks to increase to the size of those in males.

The microstructures and compositions of tusks from all late Quaternary taxa are similar (Brown and Moule 1977a, b; Miller 1890a, b), but there are some differences in the density of packing of dentinal tubules, the distinguishing structural elements of ivory (Espinoza et al. 1990; M. J. Mann 1990 personal communication). When samples of ivory from *Mammut, Mammuthus*, and modern elephants (*Elephas* and *Loxodonta*) are photomicrographically compared, it can be seen that the densest or closest packing occurs in *Mammuthus* and *Mammut*, possibly making their ivory more elastic (Saunders 1979), whereas modern elephant ivory is less densely packed, and hence possibly less elastic. The potential differences in tusk elasticity (not yet empirically proven) may have functional implications. Perhaps mammoths and mastodonts lifted heavy objects, fought more often, or pushed over more trees with their tusks than do modern elephants, or, as Saunders (1979) suggested, perhaps the greater density of dentinal tubules allowed the growth of larger, more curved tusks on the extinct species.

Mammoth and mastodont tusks grew spirally in tighter curves than do the tusks of modern elephants, although in fact many elephant tusks grow spirally. Table 2.5 gives a few measurements of *M. primigenius* tusks, with the first column showing "chord" lengths measured from the base to the tip on a straight line, and the second column showing approximate maximum length measurements taken along the greatest curve of the tusk. The great differences in the column measurements are due to the extreme curvature of mammoth tusks. The "index of curvature" is highest in specimens in which maximum length has increased greatly because of growth, while straight-line length (chord length) has not increased as quickly.

The later *M. primigenius* individuals were small elephants by modern African elephant standards, but their tusks were more impressive than those of many modern elephants because they were so large in comparison with the mammoth's body size. Maximum tusk girths are similar for *Loxodonta* and *Mammuthus* (both species), but the curvature of mammoth

Table 2.5. *Maximum lengths and chord lengths of some Mammuthus primigenius tusks*

Specimen	Straight-line (chord) length (base to tip) (cm)	Maximum length (base to tip, measured along curve) (cm)	Index of curvature[a]
Cracow (Bzianka find) (Kubiak 1980; Kunz 1916)	157.0	~200.0	0.27
Beryozovka (ZIN)[b] (original data)	92.0[c]	148.0[c]	0.61
Zoological Institute, Leningrad (Kunz 1916)	98.0	159.0	0.62
Stuttgart (Württemberg) (Kunz 1916)	~135.0 left	~270.0 left	1.00
	~130.0 right	~265.0 right	1.04

[a] Maximum length minus chord length, divided by chord length.
[b] Zoological Institute of the USSR Academy of Sciences, Leningrad.
[c] Measured from alveolus to tip.

tusks laterally and medially created a look of massiveness that the straighter tusks of modern *Loxodonta* do not achieve. The maximum possible circumferences of male *Mammut* tusks were also quite large, although somewhat smaller than *Mammuthus* measurements.

The tusks of columbian mammoths often were much larger than those of other mammoths and modern elephants. Table 2.6 lists some measurements made on tusks of several taxa. Figure 2.10 is a plot of modern African elephant tusk girths and age. Lengths are not given, because elephant tusks sometimes are broken during life (see Chapter 4), and even when not broken they are continually worn at the tip and along the shaft by use in digging, stripping tree bark, or fighting. Tusks of *Loxodonta* and *Elephas* may grow about 15 cm (or more) in length per year (Colyer and Miles 1957; Humphreys 1926; R. Martin 1983); growth rates for *M. primigenius* tusks apparently varied throughout life (Vereshchagin and Tikhonov 1986).

Elder (1970) studied African elephant tusks and found great differences between the parameters for male and female growth. Female tusks grow lengthwise throughout life, but decrease their rate of circumferential growth around the age of sexual maturity, whereas male tusks grow lengthwise and in circumference throughout life. Table 2.7 gives measurements of tusk girths with increasing age for a sample of *Loxodonta* from Hwange National Park in Zimbabwe. After puberty, the pulp cavity in the female tusk begins to "fill in," decreasing in depth over time; in males, pulp cavities continue to enlarge with age. Such differences also appear to be manifest in the tusks of male and female woolly mammoths from northern Siberia (Vereshchagin and Tikhonov 1986; G. Haynes unpublished data).

Table 2.6.   *Tusk girths in a sample of Mammuthus, Mammut, Loxodonta africana, and Elephas maximus*

Specimen	Approximate age (years)[a]	Sex	Girth (measured at lip) (cm)
Mammuthus primigenius			
Berelyokh	47	F	33
Beryozovka	?17[b]	M	44
Hatnga	38	M	48
Alaikha	Late 40s	M	41
Adams	48	M	45 (restored)
Taimyr	52	?	~40
Sanga-Yuryakh	~54	F	24
Mammuthus columbi			
Hot Springs	~47	M?	~60
Hot Springs	~23	?	42
Hot Springs	~41	?	~60
Hot Springs	~23	?	~60
Huntington Reservoir Dam, Utah	~58	M	~62
Mammuthus trogontherii			
Paleontological Institute, Moscow	Adult	?	~70
Mammut americanum			
Neath mastodont	15–25	M	~50
Denver mastodont	15–25	F	~30
Loxodonta africana			
Hwange (cull sample)	11	M	25
Hwange (cull sample)	17	M	32
Hwange (cull sample)	20	M	33
Hwange (cull sample)	21	M	37
Hwange (cull sample)	28	M	40
Hwange (cull sample)	38	M	41
Hwange (cull sample)	41	M	44
Hwange (cull sample)	43	M	43
Kenya[c]	?(adult)	M	62
	?(adult)	M	60
Kenya[d]	~55	M	~47
Kenya[e]	?(adult)	M	~62
Uganda[f]	?(adult)	M	~63
Elephas maximus			
Zoological Institute, Leningrad	? ~17/35[b]	M	~34

[a] Author's data, unless noted.
[b] See the Appendix.
[c] Maximum dimensions recorded by Sikes (1971).
[d] Ahmed, a small male with large tusks, who lived in Marsabit National Reserve (J. Shoshani et al. 1987).
[e] The heaviest tusks ever recorded (J. Shoshani et al. 1987).
[f] Third heaviest and thickest tusks on record (J. Shoshani et al. 1987).

Table 2.7. *Tusk girths, measured at lip line, for male Loxodonta from Hwange National Park, Zimbabwe, compared with male Mammuthus columbi and M. primigenius*

Age (years)	Loxodonta girth (cm)	M. columbi girth (cm)	M. primigenius girth (cm)
4.5	19	—[a]	—
5.5	16	—	—
8.4	17	—	—
11–12	25	—	—
	26	—	—
16	29	—	—
17	32	—	44
	33	—	—
18	34	—	—
20	32	—	—
	33	—	—
	39	—	—
	41	—	—
21	37	—	—
	38	—	—
22	—	∼60	—
26	35	—	—
28	40	—	—
31	39	—	—
33	42	—	—
37	36	—	—
38	41	—	∼48
41	44	—	41
43	43	—	—
47	—	∼60	—
48	—	—	45 (estimated)
58	—	∼62	—

Notes: Length is not presented here as a useful measurement, because although it increases as age increases, elephants recurrently break their tusks in fights or while feeding, or wear them down.
[a]No data available.

Mammoth tusks as long as 3.5 m or more (measured along the curve) have been found in Siberia; some African elephant tusks described in the literature have been reported to be nearly 3.5 m in length, but 2.5 m is exceptional (R. Cumming 1857; Kunz 1916:410–11, 428–31; Sikes 1966; S. Smith 1986). The longest Asian elephant tusks recorded were over 2.7 m, although 1.5 m is considered exceptional (Kunz 1916:427–8; S. Smith 1986). Mastodont tusks may reach over 2.5 m in length, but usually are much shorter. Single tusks from woolly mammoths have been found weighing

TUSK GIRTH MEASURED AT LIPLINE (MALES)

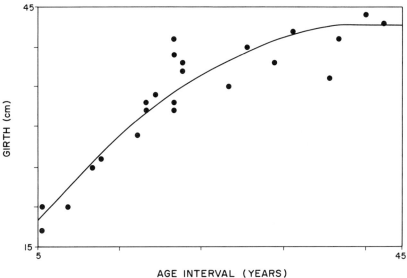

Figure 2.10. Tusk girth (the circumference measured at the lip line or the alveolar insertion) graphed against life-age for *Loxodonta africana* from Zimbabwe.

90–100 kg or more (Digby 1926; Garutt 1964; Kunz 1916). The largest tusk reported for an African elephant weighed about 107 kg (Eltringham 1982; Kunz 1916; S. Smith 1986); see J. Shoshani, Hillman, and Walcek (1987: Table 4) and the references therein for measurements of very large tusks.

Wear patterns on the ends of mammoth tusks have been described by Flerov (1931), Garutt (1964), and Vereshchagin (1977), among others. Garutt thought that tusks were used to scrape snow off the ground, in order to expose grass; wear facets on mammoth tusks often are deeply scratched, but also smoothed, a type of wear that could have resulted from scraping against gravel, grass, ice, and snow. Vereshchagin and Tikhonov (1986) suggested that tusks were used to chip away at icicles and snow on crevasse walls or rocks, in order to get water for drinking. Because mammoth tusks grew with a strong spiral twist, older wear facets continually moved around the tusk tips during normal growth, terminating on a different side of the tusk than where originally produced. Meanwhile, as the facets rotated, the tips were being worn down, thus smoothing over traces of wear that had not yet grown around to a different side. It is sometimes difficult to determine which wear facet on a mammoth tusk was the one in use at the time of death,

Figure 2.11. The tip of a broken tusk of a female *Loxodonta africana* in Zimbabwe, showing flake scars and rounding of old fracture edges.

and in fact it is probable that tusks had more than a single mode of use, and thus they had more than a single wear facet at any given time.

Modern African and Asian elephants also wear down their tusks to create facets that often are scratched. Such a facet may be on the lateral or medial side of a tusk. Flat facets may result when flakes break off the tusk tips, and then the scar is worn smooth (Figure 2.11). Elephants rub the ground and dig with their tusks to loosen mineral-rich soil; they use their tusks to break woody plants and strip bark, and they fight with their tusks. Some wear may result from abrasion against the trunk or items held in the trunk. Modern elephants do not scrape ice or snow, and yet their tusks may be as heavily scratched as mammoth tusks.

As far as I know, no one has published results of detailed wear studies on *Mammut* tusks. I have examined a few dozen tusks and have found some wear facets on the lateral side and some on the medial side. Some tusks have wear facets on both sides, or on dorsal (upward) surfaces at the tip. Perhaps mastodonts were "shovel tusking" for water or food, rubbing trees to strip bark, or using the tusks in conjunction with the trunk to manipulate vegetation, preparing it for placement in the mouth. Mandibular tusks rarely remain in the jaws of mature adults, but when they are found they usually show use wear, perhaps resulting from bark stripping for feeding.

2.11.　Conclusion

Extinct and living proboscideans clearly share important physical characteristics, many of them having interesting behavioral implications. Therefore, we can derive exceptional clues about the behavior of *Mammuthus* and *Mammut* from studies that compare them to living proboscideans in terms of anatomy, stature, body size, sexual dimorphism, and growth and maturational scheduling. In the future, perhaps more detailed research into these and other features will continue to provide a better understanding of the life of mammoths and mastodonts.

Additional knowledge of a different sort concerning extinct proboscideans can be inferred from observational studies of the social structure of living elephants. Chapter 3 examines the behavioral organization of elephant populations in the wild and attempts to model the social behavior of mammoths and mastodonts based on recent studies.

2.12.　Supplement: frozen mammoth carcasses

Published scientific descriptions of frozen mammoths date to the eighteenth century. Ides (1706), who traveled overland from Moscow to China, noted that during ice breakup on northern Siberian rivers the swollen waters often carried off segments of banks and bordering land surfaces, occasionally exposing nearly whole frozen mammoths. During the reign of Peter the Great, the scientist Tatishchev made the world's first studies of mammoths (Ivanov 1977; Tatishchev 1732). An account by the botanist M. Adams (1807) told the story of his official expedition to recover an exposed mammoth carcass that a local native had found slipping down an icy bank in 1796. That was the first find to be preserved in a museum (Tilesius 1812).

The recovery of mammoth bones and ivory became an enormous commercial enterprise in the eighteenth and nineteenth centuries. In the nineteenth century, tens of thousands of kilograms of mammoth tusks were sold each year worldwide out of Siberia. The ivory from at least 46,000 individual mammoths had been sold in Russian Siberia by about 1913 (Tolmachoff 1929). Nearly perfect tusks and well-preserved bits of mammoths are still found each year (N. Vereshchagin 1987 personal communication), sometimes nearly intact carcasses (I. Dubrovo 1987 personal communication) that are lost to dogs, erosion, or scavenging wild animals before they can be salvaged.

The Russian Academy of Sciences published announcements in 1860, 1880, 1924, 1938, and 1973 stating that if skeletons, bones, or articulated remains of mammoths or rhinoceroses were found in the north, the Academy would pay rewards of several hundred rubles and useful goods for the information (Gekker 1977). Rhinoceros carcasses were rarer than

Table 2.8.   *Radiocarbon dates for well-preserved Mammuthus primigenius specimens found in the USSR*

Specimen	Published ages
Lena mammoth	34,450 ± 2,500
	35,800 ± 1,200
Beryozovka mammoth	29,500 ± 3,000
	32,650 ± 2,500
	31,500 ± 2,000
	44,000 ± 3,500
Taimyr mammoth	11,450 ± 3,500

mammoth, but were known from several locales (Tolmachoff 1929:21–2; von Schrenck 1880). Fossil bones or carcasses of horse, bison, musk-ox, and moose probably were also widespread in northern river valleys, but were largely ignored by local inhabitants because of lack of commercial value. Many reports of mammoth finds are on record (de Baer 1866; Digby 1926; Garutt 1965; Herz 1902, 1904; Nordenskiold 1882; Shilo et al. 1983; Sokolov 1982b; Tilesius 1812; Vereshchagin 1981b; Vollosovich 1909).

Such carcasses were from animals that died under a variety of circumstances, but their deaths and the preservation of their bones share some taphonomic characteristics (Farrand 1961; Guthrie 1984; H. Lang 1925; Vereshchagin 1974). Ferrand (1961) pointed out that most frozen carcasses were not complete, but had been fed upon by scavengers before freezing and had suffered some slow decay of soft tissues. Such findings indicate that the animals did not die by freezing, but perhaps died suddenly by drowning or asphyxiation following burial in mud flows, caved-in river banks, or collapsed gully walls. Vereshchagin (1977, 1981b) believed that many fell through melting ice in the spring. A sample of radiocarbon dates for carcass tissues is presented in Table 2.8.

Some preserved carcasses have been studied and curated in museums, especially the Zoological Museum in Leningrad (which also houses the Zoological Institute of the USSR Academy of Sciences). I describe here four of the better-known finds, all of which I have examined in Leningrad.

The Adams mammoth

This specimen is also called the Lena mammoth, because it was found in the delta of the Lena River in Siberia (Dubinin and Garutt 1954). Much of the skeleton was preserved, although there has been considerable skillful reconstruction of missing and damaged bones. The carcass was first discovered in 1799 (Adams 1807; Tilesius 1812), and over the next few years it was progressively exposed by the thawing of the cliff edge in which it lay buried. Five years after discovery, it slipped down to the river's shore; its tusks were chopped out and sold for fifty rubles, but a merchant made a sketch map of the beast that was eventually seen by a botanist and member of the Russian Academy of Sciences, M. F. Adams. In 1806, Adams found that most of the soft tissue had been eaten by dogs and wild carnivores, although some hide and hair remained. The left forefoot, the tusks, and the tail were missing. Adams boiled the remaining bones clean. The mounted skeleton today has tusks that were purchased to fit into the restored skull (Pfizenmayer and Brandt, cited by Tolmachoff 1929); the left forefoot is a copy (apparently plaster) of the preserved right forefoot. The left innominate has also been reconstructed, as has the shaft of the left femur and the proximal end of the left tibia. Soft tissue is preserved on the right feet, including the toenails and a sole. The sole is smooth, not etched and deeply fissured as on elephant feet, because the thick horny layer has been lost.

The skin of the head is partly preserved, covering the front, top, and parts of the sides of the skull. The right ear is present, and the opening around the left eye is well defined. In December 1987, I climbed a ladder to examine this mammoth's skull, and when I put my face close to peer at it, I recognized the strong smell of elephant, lingering tens of thousands of years after death.

Skin and subcutaneous fatty tissue from the Adams mammoth were ^{14}C-dated at $34,450 \pm 2,500$ and $35,800 \pm 1,200$ years B.P. (Heintz and Garutt 1964).

This specimen is the largest woolly mammoth on display in the Zoological Museum in Leningrad. Its skeleton stands about 300 cm tall at the top of the scapula, 20% taller than the other mounted mammoth skeletons on display (the Beryozovka and Taimyr mammoths). It is an adult male whose age is about fifty years (based on criteria described in the Appendix). Epiphyses are fused on the vertebral borders of the scapulae; plates have fused to vertebral centra; but the sacrum does not appear to be fused to the innominates. All visible long-bone epiphyses appear to have been fused before death, supporting the age estimate.

The Beryozovka mammoth

This specimen is probably the best-known frozen mammoth, because photographs of the reconstructed ("stuffed") carcass in the pose in which it was found (Figure 2.12) have been widely published (Herz 1902, 1904). The carcass was discovered in 1900 by a local man in northeastern Siberia, on the banks of the Beryozovka River (usually spelled Beresovka in English). The Russian Academy of Sciences learned of the discovery in 1901; a zoologist (Herz, more correctly spelled Gertz in English), a taxidermist

Figure 2.12. View of the excavated carcass of the Beryozovka *Mammuthus primigenius* photographed in 1901 by a member of the Russian Imperial Academy of Sciences (Herz 1904).

(Pfizenmayer), and a young geologist (who was the only one of the three never to publish a book or articles about the find) led an expedition to the carcass. The unfortunate geologist could not endure the hardships of the trip by horse and reindeer sledge and never made it to the site (Pfizenmayer 1939; Tolmachoff 1929). In order to thaw body parts for separation, a log building was erected near the carcass, but photographs show that the carcass itself was not enclosed in this building during its excavation. The mammoth's remains were dug out in the open air. Within a month the carcass had been fully excavated by Herz, Pfizenmayer, and their laborers (Herz 1902; Pfizenmayer 1939). Some parts of the body were shipped frozen to St. Petersburg (now Leningrad), where they could be studied. Dogs ate the meat, and perhaps Herz and Pfizenmayer tasted it, but there was never a mammoth-meat feast from this carcass.

About one-third of the "stuffed" animal in the Zoological Museum is made up of the preserved skin from the find, and about two-thirds is restored (V. Garutt 1987 personal communication) (Figure 2.13). The tail and some soft tissue of the back and legs are well preserved. The skeleton is mounted nearby (Figure 2.14). Although according to different accounts the pelvis, the right scapula, one foreleg, and several ribs had been found

Figure 2.13. A restoration of the Beryozovka mammoth in death, mounted at the Zoological Museum in Leningrad. About one-third of the mount is covered by actual preserved skin and hair.

broken at the site (leading to speculation that the animal had fallen to its death into a crevasse or down a cliff), the mounted skeleton has a restored fractured femur, pelvis, and ribs; the long bones of the forelegs do not appear to have been broken, although fractured scapulae may have been skillfully restored. The pelvis clearly has been restored from fractured bones. Herz (1902) mentioned saving some pathological growths from the right scapula. but it is no longer clear what those were or what happened to them.

This animal, ^{14}C-dated at 29,500 ± 3,000, 32,650 ± 2,500, 31,500 ± 2,000, and 44,000 ± 3,500 years B.P. (Heintz and Garutt 1964), is a male, whose long-bone epiphyses were mostly unfused. The distal ends of the humeri and proximal ends of ulnae had fused shortly before death. The animal was about sixteen to eighteen years old at death, according to the age-determination criteria described in the Appendix; note, however, that Dr. Garutt (1987 personal communication) thinks the animal was closer to thirty-five years of age, based on his appraisal of tooth eruption. Professor Vereshchagin (1987 personal communication) speculated that the animal could be up to about fifty years old, based on counts of possible growth lines in the tusk (Vereshchagin and Tikhonov 1986).

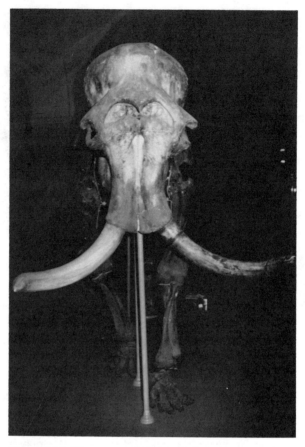

Figure 2.14. The mounted skeleton of the Beryozovka mammoth at the Zoological Museum in Leningrad. The right tusk is a cast, and the skull has been partially restored.

One original tusk is about 44 cm in circumference at the lip line, relatively large when compared with other tusks in the collections of the Zoological Institute. The entire penis was preserved extended, about 86 cm long although flattened from the weight of the carcass lying on it. The Beryozovka animal apparently was vigorously growing and was just past the age of sexual maturity at the time of its death (although there is a question about its life-age); its shoulder height was 255–260 cm, but it is difficult to say whether or not it would have reached the height of the Adams mammoth over the next twenty or twenty-five years of its growth. This animal's height is pictured in a diagram in Heintz and Garutt (1964) correlating Siberian mammoth stature with climatic episodes; I propose that the Beryozovka mammoth may not have been fully grown, and had it lived to maturity its stature might have been significantly greater than that attained at the time of its death.

This specimen's hair, skin, blood, stomach and mouth contents, and skeleton, as well as the surroundings in which it was found, have been well described in numerous publications (e.g., Herz 1902, 1904; Pfizenmayer 1939; Zalyenskii 1903).

The Taimyr mammoth

This is another nearly complete mammoth skeleton mounted in the Zoological Museum. The carcass was found in 1948 and recovered in 1949. The recovery expedition was described by Garutt (1964:42–4 and Figures 27–32). Dr. Garutt was a young zoologist at the Institute at the time of the discovery, and this animal was part of his practical introduction to the study of mammoths.

The animal's mounted shoulder height is similar to the height of the Beryozovka mammoth, but the Taimyr mammoth seems to have been an older individual. The progression and wear of its teeth correspond to those seen on the teeth of African and Asian elephants aged about fifty to fifty-five years (see the Appendix). Also, most (or all) long-bone epiphyses were apparently fused before death – it is difficult to be precise about this, because it was Institute policy for chipped or broken bones to be restored before mounting, and thus open sutures or epiphyses would have been carefully filled with plaster or other material colored to match the bone surfaces. The circumference at the lip line is 39–40 cm, relatively large for a female, and yet its stature seems relatively small for a male. The ^{14}C date on skin and tissue from this specimen is $11,450 \pm 3,500$ years B.P. (Heintz and Garutt 1964). If that date is correct, perhaps the animal is a small male whose stature reflects the overall trend toward body size reduction in late Pleistocene *Mammuthus* (Garutt 1965; Heintz and Garutt 1964). Alternatively, perhaps the animal was a large female, perhaps a sterile cow, similar to a few individuals found during elephant culling in Hwange National Park in Africa. Such sterile females grow much larger than other cows, have larger tusks, and do not ever conceive or bear young, because of hormonal imbalances such as excessive androgens.

The Magadan (or Kirgilyakh) mammoth (nicknamed "Dima")

This specimen is a small male calf, between six and twelve months old, discovered in 1979 by a bulldozer operator scraping ground for gold mining in northeastern Siberia (Shilo et al. 1983; Vereshchagin 1981b). This is one of the most nearly intact carcasses found so far [a later find of a smaller calf is perhaps more nearly complete (Tikhonov and Khrabrii 1989)], although one side of the body was sliced open by the bulldozer blade. A traumatic puncture on the lower part of one foreleg was interpreted as a wound made by a hunter's spear (Vereshchagin 1981b), but it could have resulted from the bulldozer's disturbance (Shilo et al. 1983). The calf's internal organs had been deformed or somewhat decayed; the penis was preserved. Most of the calf's thick coat of reddish fur has fallen out, and the skin has been turned a glossy black by the method of preservation (M. A. Zaslavsky 1987 personal communication). At the time of discovery, the skin was paler reddish brown, but unusually warm weather after the uncovering of the carcass thawed the remains, causing problems in the salvaging, which were overcome by refreezing the carcass, soaking it for a year in spirits, and then treating it with balsam preservative after dissecting out study specimens (M. A. Zaslavsky 1987 personal communication).

The calf was named Dima after a small stream near where it was found. It was called the Magadan mammoth by Vereshchagin (1981b) after the nearest significant city (N. Vereshchagin 1987 personal communication), and the Kirgilyakh mammoth by Shilo et al. (1983) after the nearest larger stream. The geology, geomorphology, and paleobiology of the site where it was found have been well described, as have its anatomy and morphology. A sample of dried muscle and blood was examined by Goodman, Shoshani, and Barnhart (1979:3), who found "almost perfectly shaped red and white blood cells," and extremely well preserved collagen fibril morphology, although the collagen proteins apparently had undergone oxidation or other processes of decay. Shilo et al. (1983) and Vereshchagin (1981b) described other laboratory analyses of tissue.

Little Dima has been hung up behind glass so that its feet are at eye level to school children who tour the Zoological Museum in Leningrad. The humanlike ear and shape of the head usually draw the attention of the crowds that come to stare at the carcass. Thick hair remains only at the bottom of the legs, but a cast of the body lying on its side has been covered with preserved hair, to restore the appearance it had before thawing.

These four examples of frozen carcasses found and recovered represent only a small selection of the known finds. Other discoveries include another very small calf found in late 1988 (Tikhonov and Khrabrii 1989), the Yuribei mammoth (Sokolov 1982b), the Hatnga mammoth (usually spelled Khatanga in English) (Vereshchagin and Nikolayev 1982), and the Sanga-Yuryakh mammoth (also spelled Sanga-Yurakh in the English literature) (Vollosovich 1909), as well as the nineteenth-century mammoths recovered by Bunge and Schmidt on behalf of the Russian Academy of Sciences (Digby 1926; Garutt 1964; Gekker 1977; Tolmachoff 1929). Yet the dozen or so salvaged remains probably are poor representations of the actual numbers of preserved carcasses that continue to be lost as they erode away, rot, or are devoured by animals and mutilated by ivory hunters (A. Tikhonov 1987 personal communication; N. Vereshchagin 1987 personal communication).

Sutcliffe (1985) pointed out that the more nearly complete mammoth carcasses date to two different time intervals: 30,000–40,000 years ago, and 13,000–10,000 years ago. Skeletons lacking preserved soft tissue are common from the intervening times. Sutcliffe suggested that soft tissue was better preserved in those two periods because there was more water available then to supply mud flows to cover carcasses. Poorer preservation was the rule in the intervening millennia because of drier climatic conditions.

Discoveries of intact carcasses are rare, because the land where such remains are found is so vast and lightly populated. It is frozen and snow-covered most of the time. Furthermore, local people who once avoided notifying the authorities of discoveries because of dread or superstitious fear of the mysterious animals (Czaplicka 1914:259) now neglect to notify officials because of a desire to avoid outside interference in local affairs, such as their right to hunt and trap, or to sell the ivory from mammoth tusks. Beginning in 1891, the paleontologist Chersky, and later the taxidermist-preparator Pfizenmayer, suggested that a full-time professional mammoth hunter be employed in Siberia to track down frozen carcasses so that they could be salvaged with minimal delay. A. Tikhonov (1987 personal communication) reports that a similar desire to have a professional carcass hunter working in northern Siberia endures among scientists of the Zoological Institute in Leningrad. Vereshchagin (1974) estimated that 90–100 tusks per year and hundreds of bones erode into the ocean at one collecting locality alone.

3

A referential model for understanding
mammoths and mastodonts: social
structure and habitat use by
modern elephants

...the mountainous mammoth, hairy, abhorrent, alone...

—Rudyard Kipling, "The Story of Ung" (1921)

3.1. Introduction

The elephant is a sociable animal. Wherever free-roaming animals are found, the two living genera, *Loxodonta* and *Elephas*, appear to have similar social structures. Evolutionary processes shaped the social behavior of elephants in the same way that such processes led to the development of other characteristics shared by all late Pleistocene proboscideans – for example, the presence of a trunk, a postgastric digestive system, specialized ivory tusks, and large size.[1] Great differences in the social behaviors of the extinct species seem unlikely; the two surviving taxa diverged from a common ancestry millions of years ago, and yet today they display behaviors that are strikingly similar. Their behaviors might therefore be similar to those of directly ancestral or closely related extinct forms. The two modern species share many characteristics of developmental biology (Roth 1984), diet (Olivier 1982), habitat preferences (Eltringham 1982), and so on, characteristics that could have been inherited from the common ancestral taxon that gave rise to *Mammuthus, Loxodonta*, and *Elephas* (Coppens et al. 1978; Maglio 1973). Therefore, a close look at the social behavior of extant elephants should provide a suitable guide for a first attempt at reconstructing the social behavior of extinct *Mammuthus* and possibly *Mammut*, although the very early phyletic appearance of mastodonts may mean that fewer behavioral traits were shared with elephants and mammoths.

The abundance and distribution of resources in the natural environment put limits on the social and reproductive behaviors of large mammals; for examples of studies confirming this simple but controversial statement, see

---

[1] Not all elephants were large, but the smaller forms (such as *M. exilis* and *M. falconeri*) developed their miniaturized size in response to isolated island life.

Ghiglieri (1987) on apes and early hominids or Clutton-Brock, Guinness, and Albon (1982) on red deer. Similarities between proboscideans thus depend no less on environmental conditions than on evolutionary "inheritances" from a common ancestor.

In this chapter I describe the social behavior of elephants and the demographic characteristics of well-studied elephant populations. I also discuss the known facts about proboscidean habitats and the ways in which those habitats are used by animals seeking resources and mates. I then speculate about mammoths and mastodonts, referring to the model established for modern elephants.

3.2. Communication

Elephants teach each other by behavioral example and direct communicated messages. They can convey alarm, anger, and other states through visual display and also through vocalized sound. Many day-to-day social processes in elephant populations may result from vocalized communication.

Elephants create a wide variety of vocalizations (Berg 1983) to communicate mood, intent, and desire. They make sounds that a human ear hears as screams, trumpets, growls, and rumbles. These calls have many different meanings. For example, when herd members become widely separated from each other during feeding or traveling, they rumble and call (sounds mistakenly called belly-rumbling – the noises in fact come from the throat and head) and thus maintain their contact when out of sight of each other.

Recently, Payne, Langbauer, and Thomas (1986) discovered that Asian elephants communicate by infrasonic noises (sounds whose frequencies are below the range of human hearing). Based on studies of sound attenuation in different environments, they postulated that elephants in grassy savannas would be able to perceive low-frequency sounds better than high-frequency sounds if the distance to the source were greater than 100–300 m. Elephants using low-frequency sounds are able to communicate over long distances, and groups of individuals can coordinate their foraging movements even when widely separated.

Poole and associates (Poole 1987) discovered that calls inaudible to human ears are produced by African elephants as well. Some infrasounds are created at very high pressure levels, over 100 dB, quite loud to ears that can perceive them. Of course, elephants can also perceive sounds at much lower levels (Heffner and Heffner 1980). The mystifying phenomenon of widely separated herds moving in the same direction at the same time has often been noted in different elephant ranges (R. Martin 1986 personal communication; Payne et al. 1986), and it can be explained by long-distance infrasound communication. Because it appears that all living proboscideans have the capacity to communicate with infrasound, it is likely that mammoths and mastodonts had the same ability.

3.3. Modern habitats and diets

The habitats of the two modern genera have not overlapped geographically in recent times. *Loxodonta* once was found over much of Africa, in an unexpectedly wide range of habitats: In southern Africa, elephants lived where there were trees and shrubs, because during dry seasons woody plants form the bulk of their diet. Elephants lived on the edges of hyperarid deserts, such as the Sahara and Namib, but they also lived in thick rain forests. The African elephant today is found in numerous national parks and game reserves, which rarely comprise more than 10 percent of any African nation's land area. The proportion of land where protected Asian elephants are found is even smaller.

Elephas once inhabited much of southern Asia, from India east to extreme southeastern Asia. Although elephants are still found in this region, their range has been severely restricted by local extirpation and habitat destruction. The habitats they can tolerate are diverse, but the isolated pockets of surviving populations are found mainly in forests and woodlands.

Surviving populations of African elephants still have an extremely broad habitat tolerance, from near deserts to savannas to woodlands to tropical forests. *Elephas* and *Loxodonta* survive well on a wide variety of food, which is eaten in huge quantities, because their digestive tracts are relatively inefficient at converting plants into energy (Benedict 1936; Mitchell 1916). Up to 175–200 kg of plant material may be ingested daily by an adult male during twelve to fourteen hours of feeding (Guy 1974). Because they need to eat so much, elephants usually are on the move much of every day. Guy (1974, 1976) has found, however, that there are great differences in the amounts of food ingested throughout the year.

The digestive system is based on postgastric (hindgut) fermentation (Van Hoven and Boomker 1985; van Hoven, Prins, and Lankhorst 1981). The stomach is large, but serves mainly to store ingested food. Enzymes within the stomach partly break down vegetation, but most nutrients are extracted in the huge cecum and large intestine, where microbes ferment the food remaining after gastric processing (Van Hoven et al. 1981). These microbes are introduced into the gut of the young elephant by ingestion of adult dung (Guy 1977) or regurgitated stomach contents of older animals (Eltringham 1982).

Hindgut fermentation increases the nutritional value of a diet that may be seasonally poor, such as during dry seasons, when high-fiber, low-nutrient browse is eaten. Elephants do not ruminate; instead, their feeding strategy seems to be to eat great quantities of vegetation, ferment it and extract nutrients quickly, then excrete the undigestible fraction (which is relatively large), while continuing to ingest more food. Processing time is brief, about twenty-four to fifty-four hours, depending on the quality of the food eaten (Bax and Sheldrick 1963; Benedict 1936). Mature vegetation and wood are only about 10–40 percent assimilable, but young vegetation may be 40–70 percent assimilable (Peters 1983:151). The captive Asian elephant studied by Benedict digested less than 50 percent of the hay she was fed.

Although not as efficient, in many ways, as the ruminant system, hindgut fermentation probably is better for extracting adequate energy and nutrition from poor-quality grass and browse, because food moves through the system more quickly (Eltringham 1982). Van Hoven et al. (1981) discovered that although fermentation rates of volatile fatty acids in elephants were only about half the rates in ruminants, the amounts produced were sufficient to cover all the energy requirements for basal metabolism, feeding, and standing still.

Indirect evidence that mammoths had identical postgastric digestion exists in the form of preserved dung and stomach contents, as discussed later; much more direct evidence comes from the preserved viscera of frozen Siberian carcasses. There is no good reason to think that mastodonts had a different system of food processing, although their diets are not clearly known.

Other aspects of anatomy contribute to the elephant's wide habitat tolerances; for example, they are not limited to particular habitat types by the height of vegetation, because they can feed from ground cover, shrubs, and trees up to 5 m above the ground. Elephants use their trunks and tusks to aid in feeding, which further increases the types of food they can eat. For example, they break or push over trees in order to feed on otherwise inaccessible vegetation at the top, and they can also shake loose seedpods or fruits that are out of reach. In addition, bark can be stripped and roots dug up for eating. Their diet is one of the most inclusive possible in any habitat.

In the rather vast literature about feeding habits there is disagreement over the preferred diet in different elephant ranges. Wheelock (1980) and Bell (1985) discussed some of the contradictions. Bax and Sheldrick (1963) stated that woody vegetation was sought by African elephants only in drought years, whereas Laws et al. (1975) reported that woody material from regenerating trees or shrubs was preferred food. Buss (1961) observed African elephants selecting grass over other foods, leading him to conclude that open grasslands must be the natural habitat of the African elephant; yet according to Glover (1963), because only *green* grass is nutritionally adequate, elephants more often seek other food sources. Guy's studies (1974,

1976) of elephants in Zimbabwe indicated an overwhelming preponderance of browse in the diet during the dry season, but more time was spent grazing than browsing in the wet season. Olivier (1982) interpreted the studies in east Africa as indicating that African elephants predominantly feed on grass, whereas browse is important only in the dry season. Yet Sikes (1971) was certain that the diet of wild elephants in east Africa had a high proportion of grass only when the animals were displaced or compressed into new ranges; otherwise they were naturally browsers. Field (1971) observed that African elephants eat plants in roughly the same proportions as are available locally; where woody plants are abundant, they form a greater part of the diet, especially in the dry seasons when grass is less nutritious, although browse is never dominant in the diet. Bell (1985) referred to a study of carbon-isotope ratios in elephant bones in which it was found that the degree of browse versus graze in the diet was a function of woodland density; elephants in closed forests had a diet of 100 percent browse, whereas elephants in open areas had a diet at 50 percent browse (van der Merwe, Thorp, and Bell 1988). Du Toit (1986), in response to a debate over the need for browse in *Loxodonta*'s diet, pointed out that several field studies clearly showed that elephants depend on browse for essential dietary lipids.

According to Olivier (1978, 1982), grass is the preferred food of Asian elephants living in forested habitats, and woody plants are eaten only to supplement the diet. However, browse makes up a larger part of the diet than does graze. McKay (1973) found that Sri Lankan elephants ate mainly grass, but they also fed on a wide range of woody plants.

It is no longer possible to describe the optimal or ideal preferred habitat of wild elephants, because virtually all surviving populations inhabit special protected lands such as national parks, which are limited ecosystems. These lands generally lack open access in and out for the resident animals. Such areas usually are managed as tourist attractions, meaning that large numbers of tourists are encouraged to visit them throughout the year. To accommodate tourists, and to continue drawing more and more visitors every year, roads, lodges, and viewing areas have been developed in many parks and reserves. Also, the numbers of animals often have been allowed to increase unchecked, to give a good show to the visitors (although some countries implement more realistic management plans, involving population reductions when necessary, to avoid habitat degradation).

Osborn (1925:4) wrote of the "insatiable *Wanderlust*" of elephants, an urge to find better forage combined with a desire to gratify their curiosity about new environments. According to Sikes (1971:241), elephants have "intrinsic migratory urges" that used to keep them wandering widely, urges that are no longer satisfied because of the compression of elephant populations into circumscribed areas. In the view of Sikes (1971), a range that includes gallery forests, grassy floodplains, and mixed swamps is the ideal type of habitat for African elephants. In such a habitat, many kinds of foods would be available at different seasons.

In Sikes's reconstruction of the recent past, before the days of European control and the human population eruption in Africa, east African elephants seasonally migrated from gallery forests to savanna woodland, or from high to low elevations and back again. Mating and calving took place where shade, water, and cover were most abundant.

3.4. Migrants or nomads?

I use the word "migration" here to refer to seasonal, long-distance travel that involves mass movements of populations from one habitat or zonal community to another, and back again in another season. In this usage, "migration" refers to periodic, directed movements; migrations are not exploratory, but are goal-directed. "Nomadism" refers to movements that may also be relatively long-distance and seasonal, but involve separate groupings of animals within local populations traveling to preferred points that may be geographically different for different groups. True migrations would therefore involve movements much greater than the "foraging radius" of adult animals (Pennycuick 1979); also, migratory animals (in this usage of the term) have no clearly identifiable "home range" that stays consistent from year to year. Migration involves linear movements from one distinct habitat to another, and back again; nomadism involves regular use and reuse of several different points within one or more habitats, a journeying without defined beginning or end (the Greek word *nomos* means pasture; nomadism is the use of different pastures). In migrations, whole populations move to different feeding areas; in nomadism, different groupings aggregate at different fixed points, such as feeding areas, but only temporarily. Groupings usually are segregated; resources draw different groupings together, not the movements to such resources. The literature on animal movements is large, and my definitions would not satisfy many authorities; the issues involved are complex and worth studying. Especially recommended are R. Baker (1978) and Heape (1931).

In many ranges, elephant movements are restricted by the geophysical orientations of land surfaces (A. Conybeare 1985 personal communication; R. Martin 1983). Elephant populations do not habitually cross mountain ranges, deserts, or other large natural features.

Elephants occupy rather large home ranges, in which certain areas or resource points may be favored at certain seasons. Lifetime movements of individuals or herds may define enormous home ranges, although in the short run the area visited may be quite small (Moss 1988). In small national parks, such as Lake Manyara in Tanzania, family-unit home ranges vary in size from $14\,km^2$ to $52\,km^2$ (Douglas-Hamilton 1973). In larger reserves, such as Tsavo National Park in Kenya, home ranges vary in size from $294\,km^2$ to $3,120\,km^2$. Elephants living in well-watered habitats that provide good feed throughout the seasons move far less than do elephants in poorer habitats (Eltringham 1982; Leuthold 1977; Leuthold and Sale 1973; R. Martin 1978; Olivier 1978).

The migratory proclivities of elephant demes are arguable; compare Leuthold (1977) and Sikes (1971). Some scholars have accepted the theory that recent elephant populations once regularly migrated long distances, but evidence in support of this presumption is mainly inferential or based on hearsay; see, for example, Roberts (1951) and Sikes (1971). Long-distance migrations may have been unnecessary and hardly adaptive, at least in the absence of human disruption. Migration requires the "total physiological attention" of a mammal (Meier and Fivizzani 1980:260). Migrating animals must be able to store great amounts of energy, and the metabolic costs of locomotion (as well as pregnancy, lactation, and calving on the move) must be outweighed by the benefits of the trip. Indeed, very few terrestrial mammals migrate long distances. Some caribou (*Rangifer tarandus*) populations in North America travel up to $600\,km$ between seasonal ranges (Kelsall 1968), but that is an exceptional case. Famous African migrators such as wildebeeste (*Connochaetes taurinus*) migrate shorter distances than do American caribou, probably no more than $250\,km$ in the east African Serengeti Plain (measured between end points, rather than following day-to-day travels on the ground) (Sinclair and Norton-Griffiths 1979).

Radiotelemetric studies carried out in Hwange National Park, Zimbabwe, recorded movements of no more than about $100\,km$ by male or female elephants (A. Conybeare 1988 personal communication), some moving back and forth over that distance seasonally. In Namibia, elephants apparently move regularly up to $80\,km$ between water and feeding areas. One of the few elephant ranges left with sufficient space to allow long-range movements is in northern Botswana, where a major telemetric study of movements is being conducted to help answer questions about migrations. Some elephants in northern Botswana (where population levels are extremely high) move hundreds of kilometers between seasonal ranges, probably because of the high population density.

Allometric scaling (i.e., regression formulas derived from comparing life-history parameters with body size in different taxa) predict a foraging radius of about $48\,km$ for an animal weighing $2,600\,kg$, approximately the mass of a full-grown female African elephant (females may be the ecologically optimal size, as opposed to the sexually selected larger males). "Foraging radius" refers to the distance an animal can move to find food without bioenergetically burdening itself (Pennycuick 1979).

Males weighing 5,000–6,000 kg scale to a foraging radius of over 60 km (Western 1979). These theoretical expectations compare well to the actually observed movements of elephants in unrestricted habitats.

Many authors believe that elephants were compelled to migrate long distances during certain seasons because of the presumed destructiveness of their feeding habits. Modern African elephants literally push down the forest (Buechner and Dawkins 1961; Caughley 1976; D. Cumming 1982; Guy 1976; Lamprey et al. 1967; Laws et al. 1975; Owen-Smith 1987; Thomson 1975). Elephants can flatten many trees each day, in efforts to strip bark or feed off the top leaves. Elephants also kill grass by their method of feeding; grass clumps sometimes are pulled out of the ground, struck against one foot to shake off the dirt, and then eaten, thus completely killing the plant. By contrast, trees can recover from being broken or heavily browsed, and even trees that have been pushed over may put out new growth from the roots. Grass is the preferred food only for a limited part of the year; once it has dried, elephants rarely select it as food, because its nutritional quality has decreased sharply (Rattray, Cormack, and Staples 1953). Elephants eat shrubby and arboreal forage in all seasons of the year. It is clear that elephants eat the best available food and that diets vary from range to range and season to season, and hence habitats are protected from overuse.

Another possible reason that many authors have believed that elephants migrated long distances every year could be the reports alleging scarcity of elephants in ranges where formerly they had been seen in some numbers. Nineteenth-century ivory hunters such as F. C. Selous (1881, 1893) remarked on the disappearance or scarcity of elephants from one year to the next. Such narratives may be the records of elephant depletion resulting from heavy ivory-hunting pressure (D. Cumming 1981:96), not objective observations of normal elephant movements.

Thousands of elephants were shot every year in the late nineteenth and early twentieth centuries; millions of pounds of ivory were imported into Great Britain, the United States, and other western countries each year to feed the cutlery trade and to manufacture buttons, billiard balls, piano keys, and other household objects (Kunz 1916).

It is also probable that late-nineteenth-century travelers in Africa were witnessing increased restrictions on elephant movements resulting from harsh weather, such as localized droughts in the 1880s and 1890s, decades that were also remarkable for severe weather (of differing effects) in Europe (Cornwell 1982; Mingay 1977) and North America (MacInnes 1930). If there were droughts during those decades, elephants would have been "tethered" to diminishing numbers of water sources, thus completely altering normal patterns of nomadic movements and perhaps creating abnormally dense crowding in certain regions.

Much of the literature presuming to document long-distance elephant migrations provides only post hoc accommodative arguments to explain local extirpation. Elephant demes undoubtedly were disrupted by hunters in the difficult last decades of the nineteenth century, and occasional reports of large elephant groupings on the move in the days before game reserves in Africa most likely resulted not from population assembly preceding long-distance migrations but from the bunching-up behavior of smaller herds in reaction to hunting pressure, especially in reaction to the killing of herd matriarchs (Hanks 1979:131; Laws et al. 1975).

No scientific studies have verified that elephants seasonally moved long distances, from one separate region to another, as part of their normal migratory patterns (Leuthold 1977:201–2).

3.5.   Social behavior and home ranges of modern elephants

Elephants do not defend territories; the home ranges of different social groups usually overlap. Home ranges may be large or small, depending on resource accessibility. Movements within the home ranges are determined according to the availability of water or vegetation. An elephant may live up to sixty years or longer, and during its lifetime it is always learning where to find food, water, or other elephants. It lives among relatives that pass down knowledge gained over generations of earlier learning about the environment.

Modern elephants are largely gregarious animals and associate with each other in herds that may contain up to thirty or more members. Single bulls and small bands of adult males usually keep to themselves, except when individual animals are in season, a state called "musth" (a term originally applied in India) (Poole and Moss 1981), and seek females for mating. Females in estrus may also leave their herds to seek bulls.

The social behavior of elephants is complex, and because elephants are so long-lived, recent studies and those currently under way are only beginning to unravel the complexities of their lives.

The research of the Douglas-Hamiltons among the 400–500 elephants of Lake Manyara National Park in Tanzania (Douglas-Hamilton and Douglas-Hamilton 1975) confirmed the impressions of earlier fieldworkers (e.g., Buss 1961; Laws and Parker 1969) that elephants organize themselves into familial and kin groups, made up mainly of adult females and their offspring. Sexually mature males rarely associate with these groups for long periods, although at any given time small numbers of adult males usually can be found associating with some family or kin groups. Later research at Lake Manyara National Park (Weyerhaeuser 1980) has indicated that elephant social groupings may split up or unite, but that the solidarity of most family-unit members remains unchanged over long periods of time.

Unfortunately, more recent study of the Lake Manyara elephants indicates very heavy losses to ivory poachers – fewer than 200 animals were alive in 1987, and most of the older adults were gone (Douglas-Hamilton 1988). Moss (1988) studied the 650 elephants in another small African park, Amboseli, and found the same basic social structure as at Lake Manyara (before the kill-off).

The smallest (and only truly stable) unit in the family group is the adult cow and her sexually immature offspring. This is termed the "family unit." Births usually are spaced about every three to nine years, according to unpublished records on file at Hwange National Park (also see Douglas-Hamilton 1973; Hanks 1979; Laws et al. 1975; Malpas 1978; Moss 1988; Williamson 1976). Sexual maturity is reached by females at different ages in different habitats, but generally at about twelve to fifteen years (Eltringham 1982:Table 4.1). Laws (1966) proposed that maturity is delayed when population density is high; members of a very dense elephant population in Uganda matured at around eighteen to twenty-two years of age. A recent study of three different South African elephant populations suggests that reproduction rates may be significantly depressed if the major food plants available to elephants provide an inadequate intake of phosphorus or if calcium intake is excessive as compared with phosphorus (Koen, Hall-Martin, and Erasmus 1988). A calcium–phosphorus imbalance in association with phosphorus deficiency may be a population-limiting factor that can lead to longer interbirth intervals, delayed maturity, and increasing age before first conception.

During an expected life span of about sixty years a normal elephant cow can bear between five and fifteen offspring, and the female offspring will stay in company with the mother after reaching sexual maturity. Males depart from the group and live independently or in loose, temporary associations with other mature bulls. The Douglas-Hamiltons (1975) witnessed maturing males being poorly treated by domineering females in mixed herds, a factor that undoubtedly encourages males to leave the herds. Eltringham (1982) and Moss (1988) suggested that bulls also tend to leave of their own volition on reaching maturity.

The female offspring themselves bear young, and those animals bear young when mature, and so on, producing a social unit that can be considered matrilocal and matrilineal, to borrow terms from the study of human society. Two or more matrilines may associate for varying times, although group size and membership is quite flexible daily. Such associations are true kinship groups, whose elder cows are sisters, cousins, or half-relatives (Douglas-Hamilton 1972). Moss (1988) termed these kin-based aggregations "bond" groups. Douglas-Hamilton and Douglas-Hamilton (1975) illustrated a family tree for one kinship group at Lake Manyara, and Laws et al. (1975) speculated on the relatedness of members of several family groups shot in Kabalega Falls National Park, Uganda. The study by Moss (1988) in Amboseli National Park, Kenya, which has lasted over a decade and a half, provides detailed information on social and kin-based groupings. Ongoing research in the small national parks of east Africa, such as Amboseli, will provide us with records of bond-group membership through several generations, if the elephants survive heavy poaching and habitat changes.

Larger kinship groups develop, but may not be stable over time. "Clans" are family units that are related, but rarely meet in groups that include the entire membership. Separate matrilines come and go in everyday group- ings, attaching themselves to other groups when convenient or preferred. Moss (1988) and R. Martin (1978) noted that cow-calf groups were often fusing and fissioning, although regularly associated animals usually were found near each other.

Bond groups may feed, migrate, and drink as coordinated units, led by an adult female, who is the "matriarch," or ancestor of many members. She leads by example, initiating travel and other daily activities during the yearly round of life. A cow becomes a leader by surviving all the educational experiences of her own life, during which time she will have explored a large area in company with her family. She will have learned where to find good feed in different seasons of the year, where to seek water in dry times, when to begin traveling, and so on. Not all adult females are group leaders or matriarchs, but cows may be both followers and leaders, because their own offspring follow them while they follow their mothers or a matriarch.

Modern African elephant females do not always associate in matriarch- led mixed herds. Smaller separate family groups may be common (R. Martin 1978; Moss 1988). Some mixed herds may have cooperative leadership, suggested by age similarities among elder females of the groups, as discussed later.

3.6. Patterns in data from a large population

The culling of elephants in Hwange National Park, Zimbabwe, has recently provided the largest body of data available anywhere on the demography of a healthy elephant population. Much of the analysis is still under way, but here I report results from preliminary studies.

The Hwange elephant population had been growing steadily over several decades, and in 1983 an aerial survey established that there were more than 20,000 animals in the park (D. Cumming 1983 personal communication). Field studies had already indicated that high densities of elephants were accelerating changes in the park's flora and fauna (A. Conybeare 1983 personal communication; D. Cumming 1983 personal communication), and to prevent further habitat destruction the Zimbabwe Department of

Table 3.1. *Numbers and percentages of animals shot in the three years of the Hwange cull*

Age class (years)	1983			1984			1985		
	Male	Female	Total	Male	Female	Total	Male	Female	Total
0–12	717	704	1,421 (70%)	1,319	1,565	2,884 (70%)	781	842	1,623 (67%)
13–24	81	293	374 (18%)	151	581	732 (18%)	139	343	482 (20%)
25–36	10	113	123 (6%)	19	232	251 (6%)	7	158	165 (7%)
37–60	2	117	119 (6%)	13	212	225 (5%)	0	146	146 (6%)
Total	810	1,227	2,037 (100%)	1,502	2,590	4,092 (100%)	927	1,489	2,413 (100%)

National Parks and Wild Life Management initiated large-scale culling in 1983. Between 1983 and 1987, over 9,000 elephants in more than 250 herds were destroyed, mainly in the park's eastern and north central regions, according to the unpublished records on file at Hwange National Park. None were disturbed in the seeps areas or wilderness blocks of the west (see Chapter 4). The numbers used in this analysis do not include data from the culls of 1986 and later. The sample analyzed here comprises 8,468 animals shot in 1983, 1984, and 1985; the total number of groups shot in this sample was 234. Table 3.1 shows numbers and proportions of animals of different age and sex classes shot in the culling.

The death samples from the Hwange cull are "catastrophic," in that their age profiles are the same as or representative of the live populations contributing to them. During the culling, all grouped elephants were considered part of one herd, and every member was shot, although at times a few calves were captured alive. The age of each animal in all culled herds was determined by examination of molar progression and wear (see the Appendix), which were checked against shoulder heights and growth curves.

3.6.1.   Herd size and structure

Modern elephant herds may contain from two to fifty or more animals (Douglas-Hamilton and Douglas-Hamilton 1975; Laws et al. 1975; McKay 1973; R. Martin 1978; Moss 1977, 1988; Olivier 1978), according to unpublished data on file at Hwange National Park. For convenience here, I use the word "herds" to refer to elephant groupings that contain adult females, and the word "bands" to refer to temporary groupings of adult males. A hypothetical average-sized herd contains six to fifteen members or more, based on observations made in different habitats. Herd membership is fluid daily and seasonally, but core groups are cohesive units that begin splintering when they grow too large for the available resources or when social relations become strained because of conflict between individual personalities. In general, progressively larger herds are led by increasingly older matriarchs (Table 3.2 and Figure 3.1).

The ages of all adult females in each culled herd were determined and compared to see if age differences between the matriarch (the oldest female) and the other females had any patterning correlated with herd size. Table 3.3(A–C) shows the number of herds in which the difference in age between the oldest female and the next oldest was greater or smaller than one standard deviation from the mean. Table 3.4 shows the means of age differences between matriarchs and next oldest females in cases where the ages differed by more than one standard deviation from the mean age of matriarchs in the herds of different size classes. Table 3.5 shows the wide

Table 3.2.   *Mean age of oldest female in shot herds*

Cull year	Herd size						
	1–9	10–19	20–29	30–39	40–49	50–59	60 or more
1983	25.7	42.8	45.3	50.7	51.4	48.0	50.6
	(3)[a]	(6)	(11)	(15)	(6)	(11)	(4)
1984	47.5	43.7	45.7	46.4	50.4	47.6	50.7
	(2)	(12)	(18)	(26)	(16)	(11)	(17)
1985	38.0	43.1	48.8	51.3	50.2	49.3	53.6
	(1)	(18)	(25)	(11)	(7)	(7)	(7)

[a] Number of herds shot.

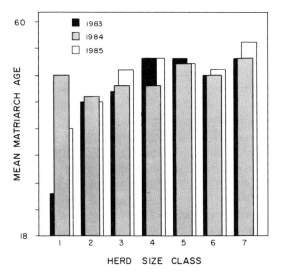

Figure 3.1. Graph of herd size class *versus* age of the oldest female present (matriarch age) for three samples taken during culling in Hwange National Park, Zimbabwe. The data from the smaller 1984 herds are least clear in showing the probable trend of increasing age with larger herd size. The herd size classes are 1 = 9 or fewer members; 2 = 10 − 19 members; 3 = 20–29 members; 4 = 30–39 members; 5 = 40–49 members; 6 = 50–59 members; 7 = 60 or more members.

range of differences possible between a matriarch's age and that of the next oldest female in the herd.

Groups that contained nine to twenty-nine animals were considered to be of "medium" size, although fifteen is probably an optimal number for a "medium" group. Groups with fewer than nine members were considered

Table 3.3A. *Numbers of herds led by old matriarchs (a matriarch whose age was more than one standard deviation above the mean age of matriarchs in the sample)*

Cull year	Herd size			
	1–9	10–29	30–59	60 or more
1983	$0(n = 2)$	$3(n = 17)$	$1(n = 2)$	$0(n = 4)$
1984	$0(n = 2)$	$2(n = 30)$	$6(n = 54)$	$3(n = 17)$
1985	$0(n = 1)$	$8(n = 43)$	$3(n = 25)$	$0(n = 7)$

Table 3.3B. *Numbers of herds led by young matriarchs (a matriarch whose age was more than one standard deviation below the mean age of matriarchs in the sample)*

Cull year	Herd size			
	1–9	10–29	30–59	60 or more
1983	$0(n = 2)$	$4(n = 17)$	$3(n = 32)$	$1(n = 4)$
1984	$0(n = 2)$	$4(n = 30)$	$6(n = 54)$	$1(n = 17)$
1985	$0(n = 1)$	$5(n = 43)$	$1(n = 25)$	$2(n = 7)$

Table 3.3C. *Numbers of large and small herds led by similarly aged oldest females (females whose age differences were within one standard deviation of the mean age of matriarchs in the sample)*

Cull year	Large herds[a]	Small herds[b]
1983	$26(n = 36)$	$12(n = 19)$
1984	$27(n = 32)$	$55(n = 71)$
1985	$25(n = 33)$	$31(n = 44)$

[a] More than 29 members.
[b] 1–29 members.

"small." Groups containing more than twenty-nine members were considered unusually "large."

Stochastic events may create an illusion of patterns in demographic observations of elephant herds, and readers should remain cautious when examining these regularities. Here I describe some patterns and offer my own impressions of their possible meanings.

Table 3.4. *Means of age differences (years) between oldest females and adult females that were more than one standard deviation older or younger than the mean matriarch age*

Cull year	Smaller groups (1–29)		Larger groups (30–60)	
	Old matriarch[a]	Young matriarch[b]	Old matriarch	Young matriarch
1983	12.8	8.3	3.3	2.5
1984	22.0	8.4	15.3	4.8
1985	19.9	8.1	16.1	2.7

[a]Old matriarch: The oldest female of the herd was more than 1 standard deviation older than the mean age for matriarchs.
[b]Young matriarch: The oldest female of the herd was more than 1 standard deviation younger than the mean age for matriarchs.

Some smaller and medium-sized groups (two to twenty-nine members) in the Hwange cull had unusually old matriarchs, but in each of these cases the matriarch was much older than the next oldest female in the group (Table 3.4, first column). The observed range of age differences in the smaller groups led by very old matriarchs was 12.8–22.0 years. An unusually old matriarch is here defined as one that is more than one standard deviation older than the mean age of matriarchs in the sample. Given that the age at sexual maturity is about twelve years, and the interbirth interval is about three to nine years, these old matriarchs appeared to be leading their daughters or granddaughters rather than their sisters or cousins of the same age set.

Some of the smaller and medium-sized groups were led by unexpectedly young matriarchs. The age differences between these young matriarchs and the next oldest females ranged from 8.1 to 8.3 years (Table 3.4, second column). In these cases, the main relationship may not have been between mother and firstborn daughter. Apparently these "young" matriarchs were in company with sisters or cousins of similar age sets.

In the larger groups (those with more than twenty-nine members), matriarchs generally were more than forty-five years old. Some groups were led by unexpectedly younger matriarchs that were in company with other females of very similar ages; the main differences ranged from 2.5 to 4.8 years (Table 3.4, fourth column), indicating that the young matriarchs may have been leading their groups through cooperative association with like-aged cousins. Very old matriarchs also were found leading large groups, and these animals were in company with either sisters or daughters, or possibly cousins. The differences in ages between oldest females and next oldest ranged from 3.3 to 16.1 years (Table 3.4, third column).

Table 3.5. *Age difference between oldest female and next oldest female in the herd*

1983		1984		1985	
Herd size 1–29	Herd size >29	Herd size 1–29	Herd size >29	Herd size 1–29	Herd size >29
> 10.7	22.7	7.9	> 10.7	> 18.0	> 18.0
> 0.7	2.4	24.9	7.1	8.9	2.4
10.7	10.7	> 18.0	13.6	30.8	15.4
5.1	14.9	> 27.5	13.1	10.4	7.9
12.8	1.5	1.9	14.1	> 15.8	26.2
5.5	9.1	15.4	6.7	0	20.2
17.5	7.9	> 8.6	3.2	6.0	4.0
25.3	6.4	21.5	31.4	24.0	5.9
> 15.8	24.8	32.7	9.5	17.3	20.9
17.6	6.7	13.3	13.9	> 15.8	10.7
14.9	6.6	16.8	3.8	16.9	0
6.4	5.5	5.0	15.4	> 13.6	13.3
7.9	10.9	30.8	9.5	4.4	13.8
22.3	6.4	25.3	9.5	5.0	4.7
5.9	0.8	9.5	15.7	24.4	2.6
2.9	8.3	17.5	3.2	12.5	6.6
> 15.8	16.4	5.6	14.4	18.0	3.4
6.9	3.1	16.0	15.9	8.4	6.7
7.1	2.6	9.5	7.5	24.1	2.4
	6.3	> 25.1	4.8	16.9	0.8
	9.3	2.6	3.4	5.6	0
	5.9	25.3	9.3	13.9	3.1
	0	1.6	9.1	20.2	18.0
	8.5	6.9	0	> 15.8	0
	0	11.7	6.6	2.2	8.6
	0	22.0	25.3	7.1	7.9
	3.2	2.6	7.9	18.1	6.7
	1.9	7.9	0.8	18.9	11.5
	3.1	3.1	3.2	12.9	8.5
	9.3	7.9	7.1	20.9	7.1
	6.4	4.8	11.7	4.4	5.6
	6.4		2.6	28.3	5.9
	3.8		7.3	18.1	
	0		7.1	11.7	
	5.9		9.3	0.8	
	2.6		5.9	28.3	
			12.3	15.4	
			8.3	2.6	
			10.9	8.9	
			4.8	13.8	
			8.5	2.6	
			9.5	15.4	

Table 3.5. (*cont.*)

1983		1984		1985	
Herd size 1–29	Herd size > 29	Herd size 1–29	Herd size > 29	Herd size 1–29	Herd size > 29
		·	8.5	5.1	
			2.4	7.1	
			6.3		
			0		
			11.5		
			8.9		
			0		
			5.9		
			2.2		
			10.7		
			8.3		
			2.4		
			8.5		
			4.0		
			1.6		
			9.1		
			6.4		
			18.9		
			12.3		
			0		
			0		
			9.1		
			19.4		
			8.5		
			9.5		
			6.6		
			2.4		
			0		

Herd fissioning may result from the death of a matriarch or other old female, or from social disagreements and feeding competition. In the face of tensions or environmental pressures on herds, the more distantly related offspring of a dead matriarch probably would be more inclined to split from larger groups than would the matriarch's daughters (A. Conybeare 1988 personal communication). As well, the death of an old female that was not the matriarch could motivate a break of her offspring from the rest of the group. Patterned age differences may result from such splitting of groups. In Hwange, it appears that the breakup of large herds into new, autonomous units often resulted in the creation of three different types of groupings: (1) small splinter groups (led by several young females that possibly were the original matriarch's daughters or granddaughters), (2) the original matriarch's much-reduced group (led by the original old female and one daughter or granddaughter, but including few or no equal-age cousins and younger sisters), and (3) larger splinter groups led by several older females of the same general age group (probably cousins, or daughters of the original matriarch's sisters).

In general, very large groups are led either by one very old female in company with much younger relatives or by several relatively younger females. A relatively young matriarch in a large group is in company with at least one other female of similar age, and the group appears to be led by a "team" of matriarchs, perhaps the daughters of a deceased old cow or the daughters of sister cows.

Medium-sized groups led by relatively young matriarchs contain teams as well, probably made up of cooperating sisters or cousins. When a very old matriarch leads a medium or small group, the next oldest female is most often much younger, and leadership seems to be by clear "seniority."

The difference in age between the matriarch (the oldest female) and the next oldest female (Table 3.5) varied widely in the Hwange cull samples, indicating that many different kinship relationships are possible between matriarchs and followers.

3.6.2.   *Differences in behavior of males and females*

In mammals, sexual selection (Darwin 1871) leads to evolutionary development of special adaptive attributes used by males in courtship or competition for females. Examples of such attributes include the bright colors of the peacock's tail, adapted to charm the peahens (Clutton-Brock 1983; Daly and Wilson 1988; Darwin 1871), and the exaggerated tusks on male elephants, adapted to outfight other males during mating season. Elephants usually fight by pushing each other head-on, or by running together to knock trunk bases and tusks (Leuthold 1977; Poole 1987); see Chapter 4 for a description of another kind of fighting. Figure 3.2 shows the interlocked tusks of two mammoths that died as a result of head-on fighting. They probably starved to death while locked together.

In many species, males must compete for access to females, and the competition becomes directly confrontational when access is limited by lengthy female "absence" from males, most often resulting from the necessary investment females make in raising their young. Female elephants may be inaccessible to males for a large proportion of their lives: Their gestation period is nearly two years, followed by about two more years of lactation; the interbirth spacing may be four years or longer, during which time a female is generally unavailable for mating. Males attempt to increase their Darwinian fitness when they compete for females (Bateman 1948;

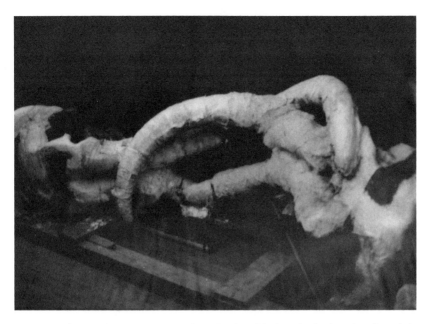

Figure 3.2. Two *Mammuthus* skulls with interlocked tusks, found in Nebraska. Specimens in the collections of the University of Nebraska Museum, Lincoln.

Clutton-Brock 1983), even when the competition is violent; see Poole (1987) for photographs of violent elephant bulls. Natural selection has therefore favored traits such as larger size and highly developed armament in males. Also, some female African elephants actually prefer to mate with the largest males (Moss 1988).

Additionally, smaller female size may be a sexually selected trait; smaller females should mature sooner than larger ones and therefore should reproduce sooner. However, that selective pressure may be balanced against an opposite influence: Large females give birth to large males, whose size favors them as breeders. Thus, their reproductive success in part depends on their mothers' large size, which is a selective advantage (Clutton-Brock et al. 1982).

Males are larger than females in those species that feature intense sexual selection. In species with great size differences between the sexes there is also an increased sex difference in mortality (Clutton-Brock and Harvey 1978; Clutton-Brock 1983). Males may die in combat with other males, and they also suffer from greater conspicuousness to predators (such as ivory hunters), as well as physiological difficulties (e.g., anemia or suppression of the immune system) brought on by elevated adrenal and gonadal functions during breeding (and competing) episodes. The males in competitive species are expected to die younger than females.

Sexual dimorphism in body size and tusk dimensions is therefore an indicator of social behavior in elephants. Animals with the greatest degree of sexual dimorphism may also show the greatest degree of polygyny in the breeding system (Barash 1977:160; Clutton-Brock 1983; Clutton-Brock and Harvey 1978). In a polygynous system, a male's reproductive success can be measured by the number of females with which he mates and the degree to which other males do not mate with them (Clutton-Brock 1988). So-called harem systems are the best examples. Modern elephant males do not keep harems, but they do compete for dominance and the right to mate with receptive females. Males do not defend territories in which females are exclusive "property," nor do they defend the females. Instead, they tolerate the presence of other males, but maintain their priority of access to receptive females. The fact that females in herds do not have synchronized estrus allows males to monopolize a relatively small number of females out of a potentially unmanageably large number (Clutton-Brock and Harvey 1978).

Elephant bulls live apart from mixed herds and from the majority of other males, at least for that part of the year when they are in season (musth) and are inclined to fight (Jainundeen, McKay, and Eisenberg 1972; Poole 1987; Poole and Moss 1981; Rasmussen et al. 1984). Not all polygynous species show the male–female segregation pattern, but it is clearly evident in modern elephants. Geist (1974) suggested that male and female ungulates sharing the same home ranges develop different feeding habits to avoid food competition, a pattern that would be especially important for species whose females have long gestation periods, times when high-energy food is necessary for fetal growth. It is adaptive for male elephants to live apart from females, preventing direct competition. Sexual dimorphism may thus reflect the different diets of males and females. Larger size may be an adaptive advantage for males, which can travel farther than females to find forage (Pennycuick 1979). The females are limited in how far they can travel by their attachment to the young (hence smaller) animals in the herd.

Recent studies by Moss and Poole (Moss 1988; Poole 1982, 1987; Poole and Moss 1981) in east Africa have shown that healthy males over twenty-five years old have periods of sexual and aggressive activity, and once a year they experience periods of heightened sexual activity for times ranging from one week to several months. The heightened state is called musth, and it sometimes leads to temporary reshuffling of immediate dominance hierarchies among males (Poole 1982). During musth, androgen levels increase in the blood plasma and urine (Poole et al. 1984; Rasmussen et al. 1984).

Different bulls come into musth at different times of the year. Older males (those in their late forties) may be in musth for three to four months; males in their early thirties are in musth for only a week (Poole 1982, 1987). Bulls in musth can dominate even larger nonmusth bulls and are more successful in mating. Musth was first documented in Asian elephants (e.g., Jainundeen et al. 1972), where the temporal gland concentrates androgens (testosterone and dihydrotestosterone), as in African elephants. The temporal gland is active in both male and female African elephants and probably functions to communicate an elephant's presence by olfactory means; in Asian males, the gland communicates the musth state (Moss 1988).

Barnes (1982) and others have speculated that under certain conditions where elephant density is low and their distribution is clumped in response to clumped natural resources, with wide separation between clumps, there may be an adaptive advantage for mixed herds to have "sire" bulls, or bulls that are most active sexually and are seeking mates more often than others. The idea of a single sire bull became established in the literature orginating from early accounts of ivory hunters or travelers in elephant country, but it has not found support in recent scientific studies (e.g., Douglas-Hamilton 1973; Laws et al. 1975; Leuthold 1977; R. Martin 1978, 1983; Moss 1977, 1982, 1988). However, it is possible that under unusual circumstances, such as the first exploratory "migrations" (or dispersal) of mammoths out of Asia on their way to populate North America, basic social groupings (mixed herds separated from bulls) may have included males permanently "attached" to mixed herds, as a result of achieving dominance, staying on the move, and seeking mates (Barnes 1982). In mosaic environments (as opposed to zonal environments), stable male-female groups do not occur except as fortuitous encounters.

At any time in modern elephant ranges, which are mosaics, there do seem to be *temporarily* attached adult males moving with many mixed herds (Laws et al. 1975; R. Martin 1978; Moss 1988). These may be males in season, or merely animals feeding in the same locations. Adult females are never very far from potential mates, thus increasing the chances that estrous individuals will be able to breed, in spite of general segregation of the sexes. R. Martin (1978) reported that some mixed cow-calf herds visited bulls in the bulls' habitual home ranges, probably seeking mates when in estrus.

Elephant populations in every range have seasonal peaks in conceptions and births, although individual females may conceive and give birth at any time of the year if environmental conditions are favorable (Eltringham 1982; Laws et al. 1975). Estrus cycles may be synchronized among individual females in each family (Moss 1983, 1988), but not necessarily throughout bond groups or matrilines. It is therefore advantageous for males to be near females in mixed herds at all times of the year, because one or more may be in season. The presence of adult males cannot be better explained than as a reproductive strategy, because males do not assist in raising the young – they do not provision the offspring, and they are not

Table 3.6.   *Mean percentages of adult males in culled mixed herds*

Cull year	Herd size						
	1–9	10–19	20–29	30–39	40–49	50–69	60 or more
1983(56)[a]	18	14	4	5	4	4	9
1984(102)	7	4	5	5	5	7	5
1985(76)	11	8	8	6	11	10	7

[a] Number of herds shot.

necessary to protect the young against predators. Females pay a price in terms of competition for food while the males are in company with the herd, but apparently the price is not too high when weighed against the benefits of having sperm available when estrus cycles begin. This explanation is not contradicted when adult females drive adolescent males out of their natal herds, because in fact these young males are clearly inferior and subdominant breeders that have not yet established themselves as prime mates (in terms of advertised vigor, such as tusk and body size or aggressive behavior). The young males are prevented from mating with closely related females, and thus the potential detrimental effects of interbreeding are avoided.

The proportion of adult males associating with matriarch-led herds does not vary directly with herd size. There are always low percentages of adult males in mixed herds; the recorded range in the Hwange cull was 0–18 percent of the total (Table 3.6). The means recorded in the different cull years varied from 5.2 to 9 percent. The culling began at different times of the three years, perhaps accounting for the wide variation in the proportion of adult males shot in the herds. Wet-season group movements are greater than dry-season movements, and more contact between male bands and mixed herds is to be expected early in the year, during the wet season.

Data from other elephant culls in Africa (Laws et al. 1975) indicate that the average age of adult bulls associating with mixed herds is younger than the average age in all-male herds (Eltringham 1982). The average age for adult bulls in the Hwange cull also appears to be relatively young, although the data have not been fully analyzed. Those bulls may have been subdominant hangers-on who sneaked into company with mixed herds whenever older and dominant bulls were not in musth or had left the herd's company.

3.6.3.   *Proboscidean demography*

Reporting age structures by twelve-year classes provides adequate information by comparative purposes. The two-year age classes reported in earlier studies (Conybeare and Haynes 1984) are difficult to apply to some elephant skeletons. Differences between methods of age determination also make age assignment to the two-year classes very problematical. The twelve-year age classes separate important intervals in the life history of the elephant, as observed by numerous biologists (e.g., Moss 1988) and determined by studies of skeletal maturation. Animals younger than thirteen years old usually are not sexually mature (Eltringham 1982), but are growing quickly. Animals in the next age class (thirteen to twenty-four years old) continue to grow, but already have reached sexual maturity; some long-bone epiphyses begin fusing during this interval, indicating a slowing of growth (G. Haynes unpublished data; Roth 1984). During the interval from twenty-five to thirty-six years old, many more long-bone epiphyses fuse, although male statural growth continues into the next age class; both males and females apparently continue body-length growth, as indicated by unfused vertebral plates, sacral joints, and epiphyses; see Moss (1988) for descriptions of "long" adults. Female shoulder heights nearly level off in this interval. This age interval is the time when some females establish themselves as group leaders or matriarchs, and some males establish themselves as dominant bulls. During the next interval, thirty-seven to forty-eight years, male statural growth levels off, and males are most active reproductively and agonistically (Poole 1987); females continue their reproductive activity, but by the end of the interval are passing beyond prime condition. During the last interval, forty-nine to sixty years, animals of both sexes pass into old age and sometimes senescence, although females may continue to reproduce; arthritic and degenerative bone conditions are common in skeletons of animals from this interval, and females in this age class are relatively more vulnerable to drought stresses than are younger females.

The Hwange National Park cull data indicate the kinds of demographic characteristics to be expected in large groups or local populations of African elephants. The age structure of the cull is presented in Table 3.1; all-bull herds were avoided in the culling, because they are almost impossible to destroy as units, unlike mixed herds. Bulls, when confronted by hunters, attempt to flee in all directions; mixed herds cluster around the older females. Therefore, the cull age profile in Table 3.1 underrepresents adult age classes, because mixed herds contain very low numbers of adult males, which leave mixed herds when they reach sexual maturity. Table 3.7 and Figure 3.3 present a "corrected" age structure that evens out the male–female ratios in adult age classes and simulates the age distribution of a sample in which sex ratios remain 1:1 throughout all age classes. This simulation may not necessarily present a truer picture of the Hwange population, because adult males probably have higher mortality rates than adult females, leading to underrepresentation in the whole population.

Table 3.7. *Proportions of animals of each age class that may have been present in the Hwange population, based on the cull sample (which has been "corrected" for avoidance of adult males)*

Age class (years)	1983	1984	1984
0–12	57%	60%	57%
13–24	24%	22%	23%
25–36	9%	9%	11%
37–60	10%	8%	10%

Table 3.8. *Proportions of different age classes in six herds shot on different days during the Hwange cull*

Year	Herd size	Approximate percentage in each age class				
		0–12 yr	13–24 yr	25–36 yr	37–48 yr	49–60 yr
1983	14	64	14	7	14	0
1983	74	58	28	9	3	1
1984	8	75	13	0	0	13
1984	16	75	6	13	0	6
1984	16	81	0	6	6	6
1984	21	67	10	19	5	0

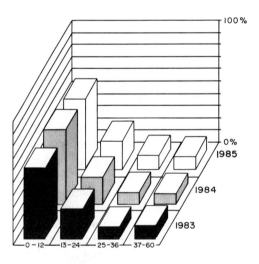

AGE INTERVAL (YEARS)

Figure 3.3. Simulated age profiles for elephant populations in Hwange National Park, Zimbabwe, based on data from culled samples.

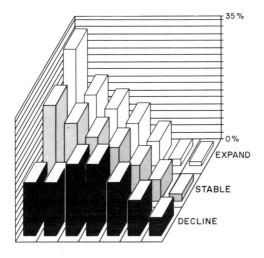

AGE INTERVAL (YEARS)

Figure 3.4. Predicted age profiles for populations in three different states – expansion, stability, and decline. Younger age classes are to the left, older to the right.

Figure 3.4 presents a diagram of the expected age distributions for three simulated populations – one that is in decline, one that is stable, and one that is expanding (Kurtén 1964; Voorhies 1969). Note that the Hwange-population age profile looks like the profile for an expanding population.

Age profiles for six herds that were shot as units on six different days during the culling season (May–September) are presented in Table 3.8 to illustrate the variability in age distribution.

3.6.4. *Unusually large groups*

Very large mixed herds, as well as all-bull groups, often were seen in east Africa during the nineteenth and twentieth centuries (S. Baker 1866; Barker 1953; Laws et al. 1975; Rushby 1953, 1965; Sikes 1971). Their existence is now generally recognized to have been a consequence of disturbances to local elephant populations (Hanks 1979; Laws et al. 1975). The disturbances came in the form of heavy hunting, habitat degradation due to fires, drought, and overuse by the elephants themselves, and compression of elephants into parks and game reserves. Even before major European incursions into large parts of the African interior, indigenous horticulturalists and pastoral nomads put pressure on local elephant populations, although Moss (1988) did not think that elephant spearing by Masai had significant effects on elephant behavior, a conclusion open to question.

Large associations of elephant family groups still occur in all habitats and are not always a result of outside disruption. These occasional aggregations are characteristic of kinship-based organizations called "clans" or demes (Douglas-Hamilton and Douglas-Hamilton 1975; Eltringham 1982; R. Martin 1978; Moss 1988). Members of such elephant clans or demes may come and go and join groups at will, if there are no physical barriers to movements; new members may also join from outside. Moss (1988) thought that elephants in fact prefer to associate with large numbers of fellow clan members, but can do so only when resources are abundant.

Day-to-day groupings within the demes vary widely. Free intermingling of several families may reshuffle the membership daily. Different matriarch-led nuclear groups may or may not undergo such flexible changes in group composition daily. R. Martin (1978), Leuthold and Sale (1973), Moss (1988) and others have shown that the intermingling of different groups occurs in response to changes in the availability of food or water. Wet-season movements generally are much greater than dry-season movements (e.g., Taylor 1987; Jarman 1972), as elephants range over wider areas and come into more frequent contact with other members of the regional population. Groupings of over 100 elephants that stay together for more than a short time are considered almost pathological (Eltringham 1982; Hanks 1979; Laws et al. 1975), but temporary groupings of 30–100 animals may be seen at watering points, in choice feeding areas, or along favored trails in nearly any African elephant range with a sizable population.

3.7. Possible measures of demographic "health"

The "recruitment rate" of a population is the proportion of animals not yet sexually mature (Eberhardt 1971; Sinclair and Grimsdell 1982). In this analysis I use the term to refer to the cohort of nursing young, or the animals under two years of age.

Recruitment varies among elephants from one population to another, depending on many external and internal factors. Culled samples from Kabalega Falls National Park and the Budongo Forest, both in Uganda, had very low recruitment rates (about 2 percent) because of range contraction and habitat degradation (Laws et al. 1975). On the other hand, recruitment in the growing population of the Luangwa Valley in Zambia (Hanks 1979) was about 9 percent, several times greater than those in the Ugandan samples. Recruitment in the growing Hwange population was around 10–13 percent.

With these numbers in mind, I estimate that at least 5 percent of a healthy proboscidean population should be unweaned calves. Catastrophic samples of healthy populations (Klein and Cruz-Uribe 1984; Kurtén 1964; Voorhies 1969) should contain the same proportion.

A related demographic parameter is the number of adult females that have succeeded in raising offspring that are themselves just reaching the age of breeding. This proportion has also been designated as recruitment (Sinclair and Grimsdell 1982), or the proportion of the herd joining the ranks of reproductive activity. To avoid confusion, the proportion of immature Hwange elephants compared with the proportion already mature is here designated the "replacement rate." This proportion should be greater than the recruitment or nursing-calf proportion. As in other mammalian taxa, one sometimes sees female elephants living under normal conditions with no nursing young at foot, and yet they are physiologically capable of producing calves. Such absence of subadult animals is therefore an indicator of environmental stress, heavy predation, or inaccessibility of fertile bulls.

The replacement rate may be a less ambiguous indicator of herd health than the recruitment rate, because adult bulls intermittently join and leave mixed herds, and their presence or absence may have a significant effect on the relatively small numbers that recruitment calculations give. A variation of more than a few percentage points significantly alters the meaning of the recruitment calculation. On the other hand, the subadult–adult ratio can vary much more widely because of the comings and goings of adult males and still not qualitatively change its meaning in terms of the population's demographic condition.

The replacement rate in the Hwange sample is nearly three subadults per adult, which is probably high for elephant populations in general. Stable populations will have lower proportions, and declining populations even lower.

Samples that show low proportions of subadults are destined to become "old" and are reproductively doomed as a population, for it would have been the missing younger age classes that would have reproduced in the future and kept the population alive. If they are missing, the population simply grows older and older and eventually dies out.

Healthy, growing populations of elephants contain 30–50 percent subadult animals, or more (Table 3.7). Proportions higher than that are possible, but probably are very rare for whole populations. Death samples taken from elephant populations may be biased by certain types of death processes, such as severe drought, resulting in exaggerated proportions of subadults in the bone assemblages (see Chapter 4) (Conybeare and Haynes 1984; G. Haynes 1985a,b, 1987a,b).

Analysts should suspect that fossil or modern death samples with subadult proportions significantly greater or smaller than the numbers cited here resulted from mortality events or processes that selectively killed particular age classes far out of proportion to their representation in the population. For example, selective predation on adults by ivory-hunting humans can create death samples containing few subadults.

3.8. Differential male mortality

The potential male population present in Hwange National Park but not shot in the cull can be simulated to suggest the demographic characteristics of this part of a proboscidean population. First, for these calculations I assume that there is a 1:1 ratio of males to females in all age classes of the population. This assumption is not supported by observed facts about elephants – male elephants have higher mortality rates than females – but no figures are available on age-specific mortality.

During all culls, there was nearly a 1:1 ratio in the subadult age category (zero to twelve years) (Table 3.1). To derive figures maintaining that ratio throughout the older age categories, I simply added whatever was necessary to bring the totals of shot males up to equality with the total of females shot in each age class. I then converted the numbers added in each category into proportions of the total and graphed them (Figure 3.5). What this graph represents is males that were assumed present in the Hwange population but were not shot because they were not associated with mixed herds. As can be seen, the graph is shaped rather like a catastrophic age profile for an expanding or stable population, with the exception of the zero value in the subadult age category (zero to twelve years). The zero value is there because there was a 1:1 sex ratio for the animals shot in the cull, and nothing had to be added to bring the ratio to equality. If the proportion of males in the subadult category were added to the age profile, it would be an exemplary catastrophic graph.

In the absence of a cull, it is possible to infer the kind of age profile that normal male mortality would create. It would be weighted toward younger age classes, because the most dangerous time in a male elephant's life seems to be the period immediately after sexual maturity is reached (twelve years old or older). During this time, males are driven out of mixed herds or leave on their own, and they live alone or in small temporary bachelor bands. They have little experience in dealing with drought or other unusual environmental stresses because they simply have not lived long enough to learn much about their home range. They must find food and water by exploration or memory, or else attach themselves to older bulls who will educate them by example in survival strategies.

Older males in season are fiercely competitive, and for the duration of musth they make poor teachers for younger males. And if younger males come into season themselves, they expend much of their energy fighting, bluffing, or otherwise behaving aggressively for periods that may be weeks

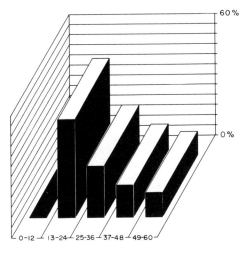

AGE INTERVAL (YEARS)

Figure 3.5. Simulated age profile for male elephants that were not shot in the Hwange National Park culling (1983–5).

long. Males also channel a lot of energy into continued growth of body and tusks, expenditures that probably are not as heavy as the huge expenses females make in pregnancy and lactation, but that are much riskier in terms of potential fitness payoffs. Females make lengthy parental investments in their offspring (twenty-two months of gestation, over two years of nursing, plus a potential lifetime of teaching the female offspring and at least twelve years of teaching males within matrilineal herds). Males invest nothing in raising offspring, but they invest everything in gaining the right to mate with as many females as are receptive. They may lose condition during their musth season, and perhaps sicken and die if the strain is too great. They are prone to infection (as a result of tusk wounds), anemia, suppression of the immune system, and other problems resulting directly from seasonally raging hormonal activities, perhaps every year of their adult lives.

Figure 3.6 is a simulated age profile for an all-male death assemblage created by noncultural mortality processes such as stress, predation, or intraspecific aggression. This graph is based on the Hwange numbers and simulations, but it has been brought a bit more into line with a stable (unexpanding) population's expectable proportions. It includes a proportion of animals in the youngest age class, because some maturing animals will leave mixed herds at tender ages, only to fall prey to large carnivores, such as lions, or perish in accidents, such as getting stuck in mud at water holes. Once past the first difficult years of living away from the natal herd, males are relatively safer from predators, and the intensity of their musth

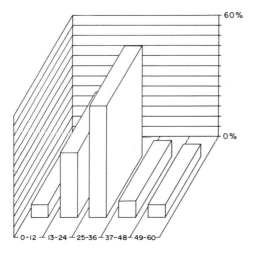

AGE INTERVAL (YEARS)

Figure 3.6. Predicted age profile for an all-male death assemblage averaging accumulated mortality over an extended time interval.

reaction is moderate for a couple more decades. Thus, mortality may not be especially great in the interval from twelve to twenty-four years of age.

During the next period of life, male musth becomes lengthier and much costlier in metabolic terms (Poole 1987). The figures here are impressionistically derived, so it is pointless to continue manipulating them. There is enough plausibility in the simulations that the distinctive shape of the age-profile graph should be kept in mind when real fossil assemblages are analyzed.

3.9.   Reconstructing mammoths and mastodonts based on modern elephants

In order to "reconstruct" the behavior and ecology of extinct mammoths and mastodonts, I have selected specific biological parameters to explore theoretically and by reference to possible living analogues. The biological variables of interest are (1) body size and its relationships with life-history parameters (gestation period, lactation duration, age at first reproduction, time spent feeding, life span), (2) digestive anatomy and its effects on feeding strategies and movements, (3) diversity and distribution of foods that can be utilized, (4) tolerance of water shortage, and (5) liability to predation (Clutton-Brock and Harvey 1978).

Allometric scaling (Peters 1983; Western 1979) predicts that mammoths,

Table 3.9. *Estimated biological parameters scaled to different possible body masses*

Parameter	Female body mass (kg)		
	2,600	5,000	7,500
Birth rate (per annum)	9%[a]	6.5%	~5%
Gestation time (days)	660	850	1,100
Age at first reproduction (years)	~10	~14	~18
Life span (years)	~60	~85	~110

Note: These figures are only theoretical, and in fact may be impossible; these are very rough estimates resulting from the fitting of imaginary points onto regression lines published in several sources cited in the text.
[a] 7% actual.

mastodonts, and elephants had extremely similar life-history parameters;[2] if the extinct taxa were heavier than modern elephants, as their more massive bones suggest (see Chapter 2), their reproductive rate would have scaled lower than that for elephants, their feed intake would have been higher, and their gestation period would have lengthened. Their foraging radius would not have been increased, and their locomotion speed would have been the same as or less than that of modern elephants, because their limbs were usually shorter – hence they would have needed to eat more, and yet would have been capable of less extensive movements in search of food than are modern elephants. Their populations would have grown or recovered from decline more slowly than do those of modern elephants. Table 3.9 shows how different body masses scale with biological parameters, following formulas published by Peters (1983) and Western (1979).

3.9.1. Mammoth and mastodont feeding strategies

By being large, proboscideans have been able to adapt to foods that are relatively difficult to digest, but are abundant. Large mammals have lower energy requirements per unit of body weight than do small mammals, and they are able to process their food more slowly, although they require a greater total intake.

Paleontologists have long recognized that the woolly mammoth's molars were adapted to foods different from those selected by modern elephants,

---

[2] Note that these scaling studies rely on mathematical formulas developed to *describe* observed relationships between variables in living animals. All formulas make allowances for variation, with the result that whenever relationships are expressed with exponents, there is a range of exponential values. Hence, upper and lower limits of predicted variables may be widely distant.

implying that there may have been less browse in the mammoth's diet. A fictitious account of one mammoth carcass identified pine and fir shoots and chewed fir cones in the stomach; unfortunately, that account was not widely perceived as fiction, and the coniferous browse diet was accepted by many nineteenth-century scholars (Carrington 1957; Howorth 1887). Felix (1912) examined a wad of masticated food from the mouth of one well-preserved Siberian mammoth, as well as about 12 kg of swallowed food in the gut or stomach, and he found chiefly grass in the identifiable sample.

Augusta and Burian (1964:35–6) and Sutcliffe (1985:114–15) summarized the results of several studies of food fragments and pollen found with well-preserved Siberian mammoths, such as the Shandrin carcass (Vereshchagin 1975; Yudichev and Averikhin 1975) and the Beryozovka carcass (Herz 1902, 1904). Gorlova (1982) identified traces of larch, birch, willow, sedge, grasses, and moss in the stomach contents of the Yuribei mammoth (Sokolov 1982a). The preferred diet consisted of grasses and herbs in all samples, but woody plants and mosses such as *Sphagnum* featured prominently. A preference for grass would be in agreement with the interpreted specializations of teeth and skeleton (Maglio 1973; Osborn 1942).

However, regarding the correlation between tooth form and inferred diet, at least one recent study (Akersten, Foppe, and Jefferson 1988) of the diets of Pleistocene *Camelops* and *Bison* from Rancho La Brea pointed out that hypsodont cheek teeth (i.e., very tall, enamel-covered molariform teeth, somewhat analogous to proboscidean teeth) may *permit* a herbivore to eat an abrasive diet such as grasses, but clearly do not *require* such a diet. Both *Camelops* and *Bison* are conventionally assumed to have had grass diets, yet preserved food samples recovered from their teeth were mainly nonmonocotyledons – although their identity as woody or herbaceous dicots was not made known (Akersten et al. 1988). We may infer that *Mammuthus*, too, did not obligatorily feed on grasses. Koch (1988) found isotopic evidence in the dentinal hydroxyapatite from tusks of Great Lakes (U.S.A.) mammoths (*Mammuthus columbi*) that seems to indicate a much greater proportion of browse in their diet than had previously been thought; in fact, Koch's evidence suggests that mammoths browsed more than mastodonts.

Dung of *M. columbi* has been found in several dry caves in Utah, including Cowboy Cave (R. Hansen 1980) and Bechan Cave (Agenbroad and Mead 1987, 1989; O. Davis et al. 1985; Mead and Agenbroad 1990; Mead et al. 1986). About 300 m^3 of trampled mammoth dung was preserved in Bechan Cave; the deposit also contained seventeen complete or nearly complete dung boluses. An adult African elephant drops, on average, 11 kg of dung every two hours, or 100–150 kg every day (Guy 1975), so it would not take long for a small group of elephants (or mammoths) to leave behind a deposit as large as the Bechan deposit. In summer, a modern elephant's food intake is much greater than in winter or during dry seasons (Guy 1975), and so a small number of animals can produce much more dung in summer than at other times. The thickness of the deposits of mammoth dung and their trampled nature may reflect repeated seasonal use of the sites by mammoths over several years or, alternatively, may reflect short-term use by larger groups. If site use was constant over an extended period of time (or was seasonally repeated over a long time span), the diet reflected in the dung sample should be an indication of preferred feed for at least part of the year; if the dung layer was deposited in a brief period and resulted from a single visit or limited number of mammoth visits, then the dung reflects little more than a last supper. Mead et al. (1986) interpreted the radiocarbon dates for dung in Bechan (spanning nearly 2,000 years, from about 13,500 to 11,700 B.P.) as reflecting short-term use of the cave by mammoths periodically returning to shelter there.

The sampled dung from Bechan Cave consisted (by weight) of over 95 percent grasses, sedges, and rushes. Also in the sample were birch, rose, saltbush, sagebrush, blue spruce, wolfberry, and red osier dogwood (O. Davis et al. 1985). Nearly a third of the plant macrofossils were sedge (*Carex*), but Gramineae (grasses) made up the major portion of the diet (Agenbroad and Mead 1989). Hence, it appears that columbian mammoths in that locale were grazers (in certain seasons, anyway), but also selected a wide variety of forbs, shrubs, and woody plants, as do modern elephants. The sampled dung at another site, Grobot Grotto, contained mainly common reed (*Phragmites*) (Mead and Agenbroad 1990). There may have been fewer large trees growing in Pleistocene habitats supporting woolly and columbian mammoths, as compared with modern elephant habitats, but browse was essential to the diet, supplying nutrients that grass alone did not provide (Olivier 1982); see the references cited by du Toit (1986). As mentioned earlier, Koch (1988) reported that the carbon-isotope composition of Great Lakes mammoth tusks indicated that their diet changed seasonally, but was unexpectedly high in browse, as compared with the diet of mastodonts.

J. McDonald (1988 personal communication) found mixed woody twigs and wadded herbaceous masses at Saltville, Virginia, in the lower levels of a late Pleistocene lake clay (the larger proportion of the material was made up of nonwoody plants); the identification of these patches of preserved vegetation is not firm – they may or may not be mammoth dung.

The diets of North American *Mammut* can be only indirectly reconstructed. The Cambridge mastodont discovered in New Jersey (one skeleton and four skulls in association) (Warren 1855:108) and the so-called Warren mastodont in New York State had abundant twigs and other vegetation preserved with their bones (Warren 1855). Lying among the ribs of the Cambridge mastodont were quantities of a material resembling "coarse, chopped straw, mixed with fragments of sticks" (Warren 1855:108, quoting the farmer who excavated the site). Within the bone bed of the Warren skeleton were five to six bushels of crushed (not ground) twigs and other bits of plants. A large part of the plant matter was shaped as a convoluted column about 10 cm in diameter, directed through the pelvic orifice; it was considered to be fecal matter preserving the original shape of the intestine (Warren 1855:199). This skeleton was purchased by Dr. Warren, a Boston surgeon, after it had been mounted and exhibited by another physician (Drumm 1963:20).

Some twigs recovered with this mastodont were fresh-looking, whereas others were black and thoroughly carbonized, according to Warren (1855), quoting a letter from botanist Asa Gray. Other preserved masses of twigs and leaves were known from finds in New Jersey and elsewhere, and all were thought to be stomach contents. Drumm (1963:22) mentioned one find in New York that had larch twigs sticking to the teeth. Clipped twigs were found in Saltville, Virginia, in this century, in association with bones of mastodonts and mammoths; spruce and fir were identified in that sample (J. McDonald 1988 personal communication). Some of these and earlier finds may be the remains of last meals and gut contents, reflecting browse and possibly some graze in the diet, but some may be simply peat and vegetation incorporated into the bone beds.

Feeding strategies are adaptations associated with particular habitats, of course. Pollen and associated fauna indicate that mammoths lived in open habitats, perhaps meadows and steppes dotted with patches of trees and shrubs. Unfortunately, the fossil record is biased toward water holes, where plant communities are not necessarily indicative of the surrounding landscape. Pleistocene vegetation was mosaic steppe land in Beringia (Guthrie 1982, 1984; Hopkins 1982) and probably was mosaic in the interior of North America as well, as opposed to the zonal communities of today. Faunal associates often are "disharmonious" in terms of temperature tolerance (Graham and Lundelius 1984), but usually include forms inferred to be grazers, such as *Equus*, and mixed grazing/browsing forms, such as *Camelops*.

Saunders (1977b, 1984), King and Saunders (1984), Drumm (1963), and Dreimanis (1967) have examined the evidence from mastodont bone sites in North America, and all are in general agreement about the habitat of *Mammut americanum* in its final millennia before extinction: Where mastodonts were most numerous, the vegetation consisted mainly of coniferous forests containing bogs, ponds, and marshes. Mastodonts have also been found in sites where the environmental evidence indicates (1) open, mixed coniferous and deciduous woodland, (2) deciduous forest, and

(3) open woodland (Graham et al. 1981; Taggart and Cross 1983). Such sites are most numerous in the eastern United States, especially in the Great Lakes area, which could have been a late Glacial (Pleistocene) refuge.

Feeding strategies also correlate strongly with group size and social structure. There is little reason to doubt that herd size for mammoths was similar to what is seen with modern elephants. The modal size is estimated at two to nine animals, but groupings of thirty or more animals probably were also common. Some assemblages of mammoth and mastodont bones contain numerous animals of different ages, and they appear to be from single populations. It is difficult to prove exact contemporaneity of fossil individuals, but in group assemblages the age and sex proportions may be so similar to what is seen with modern elephants that it is logical to designate the remains as the bones of face-to-face live populations.

The energy requirements of large groupings would have been a burden on plant communities; their foraging tactics would have been destructive, especially if grass was eaten, because proboscideans pull up grass plants with the trunk, killing the plant. Most likely, small groups fed on the move every day, as do modern elephants.

Most late Pleistocene mastodont finds in North America involve single skeletons or small numbers of individuals. A few mass occurrences are known from artesian springs in North America (Saunders 1977b, 1984). In South America, a type of mastodont has been found in at least one mass site (Simpson and de Paula Couto 1957). Numerous individuals of a phylogenetically related taxon (*Anancus*) have been found in a Pliocene-age bone-bed deposit in Bulgaria (Tassy 1982), but for the most part such mass occurrences are uncommon anywhere in the world, and when found usually consist of little more than teeth from time-averaged accumulations. Therefore, a current view of late Pleistocene mastodonts in North America is that they were nonherding loners, behaving much like modern moose (*Alces alces*), a browsing ruminant of northern forests (Saunders 1977b). However, demographic analyses of the American mass occurrences (Saunders 1977b; King and Saunders 1984) suggest that in fact there were mastodont social groupings involving adult females, their young, and occasional adult males, similar to the groupings formed by modern elephants.

Also closely tied to feeding strategies are certain other parameters of life history, such as maturational age. The main clue to this attribute in extinct proboscideans is the scheduling of tooth eruption. Tooth eruption and wear are highly patterned processes that occur throughout the long life of a proboscidean. They are extremely similar in mammoths and elephants (see the Appendix). As well, body maturation, especially the sequencing of epiphyseal fusion, is a complex series of scheduled events, and the series are virtually identical in mammoths and elephants (Roth 1984) (see the Appendix).

Teeth in *Mammut* are morphologically distinct from those in mammoths and elephants, but the scheduling of eruption and the patterning of wear are quite similar to what is seen with other late Pleistocene proboscidean taxa. The sequencing of epiphyseal fusion has not been well studied in *Mammut*, but it appears closely similar to skeletal maturation in *Mammuthus* and modern elephants, strongly supporting the idea that life-history parameters were nearly identical.

One of the more important behavioral traits linked with feeding strategies is the extent of population mobility within fixed ranges. Pleistocene proboscideans were nomadic, moving widely within large home ranges throughout the annual round of life. Within these home ranges, subpopulations and local herd groupings used some areas more than others, including "fixed points" (Ewer 1968), where specific activities were carried out, and a network of trails led to and from the points. Fixed points were feeding areas, drinking sites, rubbing trees, wallows, and so forth. Some trails would have been heavily used, whereas others would have been much less often traveled (G. Haynes 1987a; Laws et al. 1975). The sizes of elephant home ranges were discussed earlier, and important types of fixed points are described in Chapter 4. Mammoths and mastodonts, like elephants, were long-lived animals, and they would have spent many years learning all about their home range and its resources.

Ongoing studies by Holman and associates (Holman, Abraczinskas, and Westjohn 1988) have been examining the possibility that mammoths and mastodonts made special treks to salt seeps or shallow sources of saline water, located in present-day southern Michigan, in order to eat minerals. Holman and associates suggested that if proboscideans migrated from relatively saltless areas in Ohio, Indiana, and Illinois to eat sodium-rich earth in Michigan salt licks, then trace elements present in fossil proboscidean bones and teeth from different states might be identifiable as derived from Michigan salt deposits, clearly indicating migration routes and distances traveled. Proboscidean fossils dated to the last millennia of the Pleistocene outnumber fossils from all other taxa in Michigan, although the vegetation of that time interval consisted mainly of coniferous forest, perhaps reflecting periglacial conditions that should not have been optimal for mammoths and mastodonts. Holman and associates suggested that proboscideans are found in such abundance at Michigan salt licks because they traveled north into the periglacial forests seeking minerals that were not available elsewhere. Other geologically similar areas rich in salts also contain large accumulations of proboscidean bones – an example is Big Bone Lick, in Kentucky.

Holman et al. (1988) proposed that a large influx of sedimentation

Table 3.10. *Relationships among feeding ecology, body morphology, and spatial organization in African bovids*

Parameter	Type A	Type B	Type C	Type D	Type E
Body size	Small	Small to medium	Varies; medium	Medium to large	Large
Main habitat	Closed (e.g., forest)	Fairly closed	Varies; open	Open	Mostly open
Feeding style	Very selective browsing	Fairly selective	Varies seasonally	Selective grazing	Unselective
Social grouping	Solitary or paired	Small groups; solitary adult males	Varies	Large (varies)	Permanent herds; also solitary males
Group stability	(No groups)	High	Varies	Varies	High
Home range	Small, stable	Small, relatively stable	Medium to large; seasonal movements	Large; large movements	High mobility, large range
Territory	Individual or pair	Male only, may be entire home range	Male only, part of home range	Male only, part of home range	None, but male hierarchy
Antipredator behavior	Concealment, freezing	Concealment, freezing, flight	Flight; defense of young only	Flight; occasional group defense	Flight; individual or group defense
Representative species	Duikers, dik-dik	Reedbuck, bushbuck	Impala, waterbuck, kudu	Wildebeeste	Buffalo, eland

Source: Adapted from Jarman (1974), augmented with data from Leuthold (1977) and Peters (1983).

resulting from glacial melt during the Late Glacial interval significantly altered the availability of salts, leading to the dying-out of proboscideans in the Great Lakes area. This is an intriguing hypothesis that finds some support in the biological literature on modern elephants – elephants apparently need dietary minerals such as sodium in relatively large amounts (Weir 1972), and they must balance their intake of some minerals with other minerals. For example, a high-calcium diet must be balanced by an increased intake of phosphorus, or else reproductive rates will be depressed (Koen et al. 1988).

Jarman (1974) found strong correlations among habitat, group size, feeding style, and social behavior for African antelopes. The distinctions he found may be useful for reconstructing the social behavior of mammoths and mastodonts. Table 3.10 summarizes these descriptions, as augmented by Leuthold (1977) and Peters (1983).

Modern elephants certainly are not the same as African antelopes in many aspects of biology and behavior, but particular attributes of their feeding style, habitat and home-range use, social behavior, and social groupings are identical with those of the largest antelopes, classified as type E. This does not automatically mean that mammoths and mastodonts were also type-E animals. Jarman's approach was not to reduce all variability in mammal behavior to five categories, but the categories do let us see the expected interplay of social and ecological features in the lives of large herbivores, and therefore they provide us with a place to begin interpreting extinct taxa. With Jarman's classification as a springboard, it is possible to make educated guesses about extinct proboscideans, based on what we already know about their diets, sexual dimorphism, differences in age at maturity (related to reproductive behavior), and habitats.

3.9.2. Migrants or nomads?

There is a high degree of sexual dimorphism in type-E animals, but it is surprisingly low in type-D taxa, which migrate, such as wildebeeste, hartebeeste, and blesbok (Jarman 1974:260). As with any classification system, the categories do not necessarily cover all possible animal taxa. Nonantelope species such as caribou (*Rangifer tarandus*) would fit most appropriately into a category between types D and E. Caribou migrate to find preferred feed and do not exhibit as much sexual dimorphism as many other cervids; for example, female caribou also have antlers, which no other cervid females have. But caribou diets are rather diverse. Whereas they preferentially feed on lichens in all seasons (Kelsall 1968), depending on its abundance, browse and other plants such as sedges and mushrooms are selected when lichens are unavailable. Their diverse diet and the fact that males do not defend territories qualify them for inclusion into type E; on the other hand, their migratory behavior and reduced male–female dimorphism suggest type D.

The social behavior of modern elephants is similar to that of caribou in some respects. Males do not defend territories, and the basic social grouping is a matriarchal cow-with-juvenile herd, sometimes encountered or joined by breeding males, which otherwise live in all-bull bands. Yet caribou migrate long distances, whereas elephants do not, probably because the seasonal differences in availability of forage are not so dramatic in the regions inhabited by modern elephants.

If snow cover were to be added to their ranges and there were increased seasonal extremes in forage availability, it is possible that elephants might be compelled to migrate. Therefore, perhaps mammoths and mastodonts migrated. The bioenergetics of long-distance migration are expensive for large animals such as elephants, especially subadults. Ontogenetic development is a lengthy process, and subadults do not achieve adult body size and leg length for many years. Small animals would be much slower travelers than adults. Fetuses also would be at tremendous risk, because mammoth and mastodont gestation was nearly two years, as it is in the living elephants; therefore, females would have had to migrate at least twice while pregnant.

Maddock (1979) pointed out that herbivore migrations fulfill more functions for the animals than merely getting them to food. Migrant zebra and wildebeeste in the Serengeti in east Africa move to the plains in the wet season, thereby gaining access to grass when its protein content is highest, but that also allows them to avoid muddy ground in the woodlands during the wet season. As well, they improve their defenses against predators during the peak birth season by moving away from tree cover where predators could hide.

Proboscideans also have peak birth seasons, but their relative invulnerability to predators would not necessarily be altered significantly by selecting more open habitats, unless human predation was systematically organized and posed a seasonal threat, which apparently was never the case in prehistoric Africa, judging from the striking scarcity of cultural bonesites. The broad feet and thick legs of elephants are less liable to become deeply mired in poorly drained woodlands than are the thin, hoofed appendages of wildebeeste or zebra. Thus, it would seem that in the absence of regular hunting by humans, food quality and dietary needs are the only strong motivations for proboscidean "migration." Such migrations, or special-purpose treks, might be made not only to find better-quality food but also to seek essential minerals for the diet. As mentioned earlier, laboratory studies are under way to test this possibility for Michigan mammoths and mastodonts by seeking trace-element similarities linking the salt sources and the chemical compositions of bones and teeth recovered in the region.

If "migrations" were undertaken by Pleistocene proboscideans (and other herbivores) in search of salt licks, it is likely that they were related to seasonal dietary needs, such as springtime mineral imbalances. Jones and Hanson (1985), in their study of geophagy (earth eating) by North American herbivores, found that minerals other than sodium chloride were important dietary constituents sought by animals visiting so-called salt licks during different seasons of the year. For example, magnesium is a mineral necessary in many biochemical reactions, such as the activation of numerous enzymes. Magnesium ions also affect neuromuscular excitability. Lactating animals respond quickly to magnesium deficiency, often suffering muscle tetany, going into convulsions, or developing cardiovascular lesions. Most of a herbivore's total body magnesium is tied up in the mineral phase of the skeleton, and so it is not rapidly available when dietary magnesium is in short supply. Magnesium deficiency can result not only from a lack in the diet but also from excessive dietary intake of potassium. One common cause of increased potassium intake is feeding from spring grasses undergoing early "flush" growth. Such cool-season growth is highly palatable to herbivores, but the relatively high levels of potassium in such plants (as compared with low levels of calcium and magnesium) are easily absorbed by the feeding animals, resulting in increased aldosterone secretion by the adrenal glands. Aldosterone increases kidney excretion of potassium, but also results in increased excretion of magnesium (Jones and Hanson 1985:207–8). Thus, animals that feed on spring grasses may find it necessary to seek out extra sources of magnesium, such as by trekking to salt licks to eat mineral-rich earth.

Almost no information is available on the minimum levels of different nutritional substances required for maintenance and production in wild elephants. Studies of other wild ungulates have shown that animals may migrate away from areas of relatively higher food productivity to reach mineral licks, especially during calving season (D. Williamson, Williamson, and Ngwamotsoko 1988). Koen et al. (1988) suggested that low levels of some minerals and trace elements may affect elephants' reproductive rates, such as by suppressing the signs of estrus that attract male mates. There is good evidence that proboscideans benefit from seeking out mineralized soils to ingest, even at the expense of optimized food intake. Moss (1988) has suggested that African elephants living outside Amboseli National Park, Kenya, traveled into the park to eat the salt found there, further supporting the idea that proboscideans will make special journeys to find minerals needed in the diet.

Knight, Knight-Eloff, and Bornman (1988) studied ungulates that visited lick sites in the Kalahari and found (1) that browsers more often than grazers visited licks, (2) that animals living in highly seasonal (wet/dry) habitats countered springtime acidosis by ingesting mineral soils during the time of abrupt dietary shift from dry forage to green sprouts, (3) that animals also countered the toxic compounds in plants, such as tannins, by eating clay particles, and (4) that a high sodium intake among herbivores in dry habitats was a regular strategy that enabled them to survive dehydration by protecting cell walls from damage during osmosis.

One currently popular view of late Pleistocene climates is that seasonal variations in temperature, precipitation, and stormy periods in northern latitudes were reduced, as compared with modern conditions. This reduced seasonality, a state called "equability" (Hibbard 1960), would have had the effect of lengthening the growing season for some vegetation and creating mosaics of vegetation types where today there are stands or zones of less diverse plant communities (Graham 1979; Graham and Lundelius 1984; Guthrie 1982, 1984; Hopkins 1982). Thus, the biotic conditions of the late Pleistocene may have been optimal for nonmigratory animals. A diversity of plants would have supported diversity of diet (Guthrie 1984), and mammoths would have had little need to travel long distances seasonally in order to find adequate forage.

I do not think that mammoths migrated hundreds of kilometers between winter and summer ranges. Other researchers have presented the case favoring migration of mammoths, both woolly and columbian. For example, Soffer (1985) based her migration model on expectations about the floral resources and the severity of Ukrainian environmental conditions during the last glacial maximum (24,000–16,000 years ago); mammoths simply had to migrate, following river systems to find suitable forage. There are meaningful biotic differences between the north and south extremes of the modeled migration route (Soffer 1985:170, 174, 178–81); steppe or meadowlike floodplains with riverine forests apparently were more extensive in the south. These were optimal habitats for mammoths. The more boreal forests of the northern part of the Russian plain had such vegetational patches within them during glacial periods. Yet Soffer postulated that mammoths migrated northward in summer to feed on deciduous forbs and shrubs, before the grasses in the south greened. This scheduling does not take full advantage of the greater summer grass cover in southern ranges. However, Soffer suggested that two months were spent in the cold north and eight months in the south (with two months spent traveling); hence the time spent in forested habitats was less than half the year.

If the food resources of the north and south were greening at significantly different times of the year, the southern ranges would have had a clear year-round advantage over northern ones, because they had more meadowland, an earlier greening season, and deciduous browse that had a longer growing season. Perhaps the presence of mineral soils in the northern part of the range could have motivated mammoths to leave the more productive south.

Churcher (1980) suggested that evidence to support the hypothesis of mammoth migration is available in the form of different age profiles in different North American sites, and also in the form of mixing of different mammoth species in single assemblages. Churcher reasoned that sites with large proportions of young animals (such as Friesenhahn Cave, in Texas) were in southern ranges because springtime calving took place there, whereas those with mixed associated taxa were in midlatitude ranges because overlap occurred between columbian mammoths migrating north and woolly mammoths migrating south.

Soffer and Churcher depended on few sources for the idea that elephants are migratory. For example, Sikes (1971) proposed that African elephants migrated rather long distances before they were restricted by human influence, but her sources were secondhand and vague, and in fact seemed to be reporting conditions of great human influence at times of disruption to elephant demes. Leuthold (1977) was of the same opinion regarding the discussion by Sikes.

The preponderance of young mammoths in the southerly sites to which Churcher referred could be due to specific mortality processes (see Chapter 4), rather than to preferential calving in the southern ranges. Mammoth calves did not mature for quite a few years, so there would have been young animals present in both north and south ranges whether the animals migrated or not.

Quite possibly the association of bones of different mammoth species in some assemblages, such as that in Medicine Hat, Alberta, as cited by Churcher (1980), is due to wide variations in interspecific dental characters, leading to taxonomic imprecision, as suggested by Churcher (1980:104). In other words, two different species, woolly and columbian mammoths, may not have shared ranges in intermediate latitudes after all.

In spite of the weaknesses in migration models, the possibility that mammoths moved seasonally is still plausible. The carcasses of frozen Siberian mammoths may provide an important clue to the ability of mammoths to travel long distances.

Migrating animals typically increase their stores of body fat before embarking on seasonal movements. Some frozen mammoth carcasses were described as having layers of subcutaneous fat up to 15 cm thick. If those mammoths were indeed as fat as reported, they could have been physiologically prepared to migrate. However, all of the preserved carcasses that have been discovered have been partly decayed, partly eaten by carnivores, and disturbed by erosional exposure. Even the best-preserved carcasses (such as the Beryozovka mammoth) have stunk of decay. It is not clear how the subcutaneous fat could have survived autolysis and decay after skin, viscera, and muscle tissue had begun to rot. Perhaps adipocere was mistaken for natural fat on the bodies.

Nevertheless, the presence of subcutaneous fat does not necessarily indicate a migratory animal. Thick fat is an effective but quite heavy insulation against cold temperatures, although far less efficient than dense hair, which mammoths had. In fact, modern terrestrial mammals of the north (including caribou) do not lay on exceptionally thick subcutaneous fat layers over all the body for insulation (L. Irving 1966; Scholander et al. 1950). What fat is laid on is locally thick, but it serves as metabolic fuel or nutritional reserve rather than as insulation; it is mobilized during the fall and winter months. Elephants in southern Africa do not put on thick fat layers in times of abundant high-quality feed, even though several months of subfreezing nighttime temperatures lie ahead, possibly because during the cold dry season abundant calories are still available as browse, a resource that could have been scarce in late Pleistocene habitats of glacial periods.

It does not seem possible to convincingly falsify the migration hypothesis using the evidence obtainable from the fossil record, nor can it be falsified by calling up ecological models of the behaviors of modern large mammals.

3.9.3. *Male segregation*

Sexual dimorphism among mammoths indicates that there was male–male aggressiveness over access to females. On average, body size differences among antelopes may be greatest in those taxa whose feeding habits and social organization make them type E (Jarman 1974), as modern elephants seem to be, but dimorphic differences need not always correlate positively with sexual segregation. Nonetheless, I think it is probable that male and female mammoths were segregated.

The presence of one or more large mammoths in some fossil assemblages made up mainly of smaller (female?) animals does not necessarily indicate that males consorted with mixed herds for prolonged times. Fossil proboscidean accumulations are most often recovered from old water holes or stream channels; for examples, see Agenbroad (1984, 1990), C. Haynes (1980, 1982, 1984), and Saunders (1977b). Animals that do not live together may visit such areas at different times. Even though modern male and female elephants live apart most of the time, their home ranges often overlap, and they may use (and die in) the same water sources located in the zone of overlap.

However, my impression from the African field studies is that adult male elephants would not be expected to die in the same places where members of mixed herds die in any numbers (see Chapter 4).

3.9.4. *Speculating further about mastodonts*

The mastodont may not have been such an exotic creature as it seems at first glance. Mastodont skeletons outnumber mammoths in the eastern

United States because late Pleistocene habitats were wooded, and mastodonts preferred feeding on the vegetation growing around ponds and lakes in the forest, or backwater bogs and floodplain marshes (Dreimanis 1967; King and Saunders 1984; Saunders 1977b). Mammoths presumably preferred more open habitats, and there were fewer such habitats east of the Mississippi River.

Roughage grazers usually are large, and they eat monocotyledonous plants, which generally are poorly digestible but grow in dense expanses, such as plains or grasslands. Mammoths fit this model. Selective feeders may have either large or small body sizes and may feed on dicotyledonous plants, which often are rapidly fermentable but may grow in clumps that are widely spaced. A large body size would facilitate travel between such patches. Mastodonts fit this model.

When viewed this way, the enormous array of herbivorous animals can be narrowed down to just a few taxa similar in habits to the mastodont. The two taxa of greatest (potential) similarity seem to be the giraffe (*Giraffa camelopardalis*), a browsing ruminant of Africa, and the moose (*Alces alces*, called "elk" in Europe), a browsing ruminant of northern latitudes.

Both giraffe and moose are selective feeders that choose mainly dicotyledonous plants, resources that frequently grow in separated patches. For both taxa, the intestine is relatively short (compared with the intestine of a roughage grazer), and the cecum is large (Hoffman 1985). Moose browse willows, alders, and other deciduous trees and shrubs found along streams or in patches of poorly drained land. Giraffes preferentially browse *Acacia* and other woody plants, especially new growth high up the plant. The leaves, pods, and fruits of such woody plants are rapidly fermentable, with quick passage through the body, but require animals to have high mobility to get to abundant sources (van Hoven and Boomker 1985). As a result, home ranges are nonexclusive for both giraffe and moose, and individual adults show relatively few lasting associations with each other, because feeding competition would otherwise be too risky, especially in winter and during dry seasons. Adult males do not defend territories from other males.

Female giraffes may congregate when they have young at foot. Adult male giraffes are more often loners (Dagg and Foster 1982), although they may or may not feed in company with each other or with groups of females and young. Female moose without young at foot may be solitary or may be found together when "herded" by rutting males (R. L. Peterson 1955). Giraffes may cluster in long-term groups containing ten to forty individuals, including adult females associated with their young and with several adult males. Giraffe "herds" may be widely dispersed and not discernible as single groups, because of tree cover hiding them from human observers, but their great height allows them to maintain visual contact with each other.

This style of dispersed "aggregation" differs from that of dispersed moose – but both are possible analogues for the social spacing of mastodonts.

Among giraffes, young animals drift away from their mothers after the first six weeks of life, but the bond between mother and offspring is strong for over a year; see Dagg and Foster (1982) for references and data. Moose calves are weaned by one year of age and leave the company of their mothers (R. L. Peterson 1955). The leadership of giraffe groups tends to be arbitrary, with adult females or males leading the way to water or forage areas. When moose form groups, a single male attends bunched females, which move under his guidance.

Adult male giraffes spar with each other, and these fights establish dominance hierarchies or rights of access to receptive females (Dagg and Foster 1982). Adult male moose likewise fight each other for access to females.

Undoubtedly there are differences between what *Mammut* behavior would have been and the behaviors of the browsing ruminants *Alces* and *Giraffa*. Socioecologically and behaviorally *Alces* and *Giraffa* are not identical, and so the variable ways in which they feed and interact socially cannot provide lawlike models applicable to all large extinct animals, such as *Mammut*, that might have been browsers.

Moose are boreal cervids; males grow antlers and shed them once a year. Giraffes do not arm themselves with temporary social and physical "weapons" related to reproductive behavior. Moose are incapable of reproductive behavior for most of the year. During seasons of maximum plant growth, moose spend virtually all their time feeding, resting, or traveling in search of food; they have no social interactions to speak of. Their reproductive systems are in effect turned off, as they put on weight, store energy for overwintering, and, if male, grow their massive antlers. When waning sunlight finally triggers their endocrine systems to turn on their reproductive behavior, they behave quite differently, with males and females synchronized to reproduce. Males fight and mate; females bunch up with other females and mate with victorious males. This is the cervid pattern: complete synchrony of the rut, and once-a-year growth of male armament, followed by use of the armament, and finally disposal of it.

Giraffes do not synchronize their mating behavior or herding urges, and they do not go through heavy annual investment periods when males are growing reproductive weaponry. However, like all animals in seasonally variable habitats, they spend part of the year putting on condition and body weight, and part of the year losing it. Giraffes engage in sociable behavior year-round; males and females can be capable of mating at any time during the year, although seasonal birth peaks are typical in most populations.

With moose, all females basically ignore all other females and males most of the year, bunching up to keep company in "harems" only during the autumn. All young moose leave their mothers after a year, when they are weaned and unwelcome in harems.

Giraffe females may form herds, in company with one or several adult males, year-round. Many will have nursing offspring or will be pregnant at any given time, and so only select females will be available for mating.

It seems likely that *Mammut* populations behaved more like giraffes than like moose. Mastodonts probably formed long-lasting social groups made up of adult females and their offspring, with mature males sometimes loosely attached. I would not rule out the possibility that mastodonts formed into harem groups in which a dominant bull attached himself to a female-centered herd. But that is not the modern proboscidean pattern that has been observed in all elephant habitats from savanna grassland to closed forests, and so it seems less likely to have existed during the late Pleistocene.

I think that *Mammut* behaved as do modern *Loxodonta* and *Elephas*. That is, mastodont females would have formed long-lasting associations with other females (including their own mothers), but those social groupings may have fragmented more often than is the case with some modern elephant populations, the result being relatively smaller herds, perhaps averaging fewer than six to ten members most of the time. Adult males may have formed bands in the winter or during dry seasons, but probably were loners much of the time. One major difference between the social behavior of mastodonts and that of modern elephants may have been found in the autumn, when dominant bulls perhaps attached themselves to female/offspring herds in order to mate with estrous members. Male–male aggressive interactions would have occurred during that time because of the thin distribution of females in the range and the urgency felt by bulls to mate with whatever females were encountered.

This is all speculation, of course, lacking empirical support. It is maladaptive for long-lived proboscideans to have no guidance or leadership when growing up, aside from the mother animal. Moss (1988) has reported that among African elephants, new mothers need help from older females in taking care of young calves, and that help comes from living in socially cohesive herds containing old and young animals. Moose succeed in being loners most of the time because their calves mature enough within a year to live independently. But *Mammut* calves needed at least ten years or more to mature, and their mothers therefore required help from other adults for that many years. If the nonpregnant, nonlactating adult female mastodonts were loners, perhaps forming into harem groups under the leadership of individual dominant bulls, then they probably would have been impregnated together in a single breeding season and would have found it necessary to form socially cohesive herds when calves were born. Thus, in the next year there would have been few or no loner females in local populations, and the moose pattern could not have been repeated regularly. Theoretically it is improbable that mastodonts behaved like moose. Surely they acted like elephants.

Perhaps one way to find clues to *Mammut* social behavior is to make a comparative study of bone sites, each of which contains only the accumulated bones of one taxa–giraffe, moose, or mastodon. Many important socioecological characteristics of large-mammal behavior are reflected in bone assemblages produced by populations of these mammals. The tendency for social aggregations or segregated groupings of different ages and sexes in life should result in the creation of bone accumulations that possess different proportions of age classes and sexes. My own studies of elephants and my own and others' studies of giraffe (Dagg and Foster 1982) and moose (G. Haynes 1981:167–206; R. O. Peterson 1977), as well as the annual reports of research of Isle Royale National Park in Michigan, support the general conclusion that mortality rates and trends in the distribution of carcasses of members of different age classes and sexes are dissimilar. For example, the places where adult females and young die rarely contain equal proportions of dead adult males, even when selective predation was not the major cause of death. It would be enlightening to generalize about the bone assemblages produced by giraffes and moose, whose social organizations and behaviors are known, and compare them with the patterns in bone assemblages produced by extinct proboscideans, whose behavior is only inferred.

In northern ranges, where predation accounts for the greatest proportion of carcasses, many moose die in low-lying areas, because moose spend a great deal of time feeding there. Many also die on iced-over lakes, ponds, and sloughs, where they are chased by wolves or are ambushed while feeding. However, moose do feed, travel, and shelter in upland sites some distance from water holes or boggy areas; hence, some die in forested areas that have solid, dry substrates. Such areas are not suitable for bone preservation via submersion or burial (G. Haynes 1981:375–83). Yet even in the low-lying areas, simple submersion in water in areas of mixed coniferous/deciduous forest will not guarantee bone preservation. There must be sediment burial as well, because immersed bone tissue will eventually dissolve in moderately acidic water.

R. O. Peterson (personal communication 1983) analyzed the spatial distribution of dead moose recorded from 1959 to 1983 at Isle Royale National Park, where most mortality results from wolf predation. Calves were found primarily near shorelines, where the most effective means of escape from wolves was to enter the water. Adult females and males were found mainly in the island's interior, but the two sexes usually were found in different places. Of the more than 1,000 carcasses that were found, very few were grouped together in circumscribed "sites" (arbitrarily defined here as an area about 100 m by 100 m).

This kind of segregation of the sexes makes sense when viewed as a reflection of male/female differences in behavior to maximize reproductive success. Females behave in ways that will maximize the survival of their offspring (Geist 1982; R. O. Peterson 1983 personal communication). Sometimes they do so by distributing themselves widely and randomly in suitable ranges in an attempt to reduce their predictability in the eyes of predators. They may end up feeding on lower-quality forage as a result of this type of dispersal, but their young are given better chances for survival. Males, on the other hand, in order to maximize the growth of their bodies and antlers, seek the highest-quality forage, with much less regard for the avoidance of predators, one result being that they aggregate at a limited number of sites, such as patches of preferred feed. This kind of segregating behavior could account for different distributions of dead males, females, and young animals.

Adolph Murie (1934) described Isle Royale in the days before wolves had become established there, although there was some predation on calves by coyotes and hunting by humans. Moose carcasses and skeletons apparently were not uncommon sights, especially near salt licks, bogs, and lakes, the results of deaths caused by winterkill, disease, or accident. Nowadays, when wolves kill and eat most of the moose born on the island, it is rare for travelers to see moose carcasses or bones (G. Haynes 1981:178). Although evidence of moose deaths is difficult to find, except by daily searching from the air, the distribution of carcasses probably is similar to that before the advent of wolf predation.

The giraffe carcasses that have been examined and for which data on distribution are available have been few compared with moose. I have inspected thirty giraffe carcasses or recent skeletal sites; other records, sometimes only superficial, are available in the published literature (e.g., Dagg and Foster 1982).

In present-day African wildlife reserves, many giraffe deaths result from lion predation, including predation on animals already weakened by environmental stress, such as drought. Relatively more giraffes probably die of old age, starvation, trauma, or other nonpredation causes.

About 50 percent of carcasses in my field studies were found within 100 m of a permanent or temporary water source. Some were located directly within or on the edges of water bodies. Interestingly, about 25 percent of the individuals represented in the sites (G. Haynes unpublished data) had one or more fractured limb bones, which apparently had broken before or at the time of death. These traumatic fractures resulted from missteps during pursuit by predators.

Dagg and Foster (1982) saw no sex-specific differences in mortality rates in the recent studies of giraffes. Pienaar (1969) found that the ratio of adults to subadults to suckling young in one death sample was the same as the ratio in the live population of Kruger National Park, South Africa. Predators were making no deliberate selection of certain ages, but were preying at random. However, as is the case with moose, because most adult males seem to prefer feeding in heavily wooded country (maximizing their nutritional intake), whereas adult females and young prefer open woodland and plains (maximizing their safety from predators), it follows that, in general, males, females, and young die in different places (Dagg and Foster 1982:28–9), just as do moose on Isle Royale. Also, in some ranges, predators kill male giraffes more often than females because males usually are alone and easier to ambush. Thus, the sex ratio for predator kills is skewed when compared with the ratio in the live population.

How do these observations of bone sites compare with mastodont bone sites? Numerous bogs in eastern North America contain bones of single mastodonts or small numbers of individuals. Some larger groups have been recovered from artesian spring deposits (King and Saunders 1984; Saunders 1977b). Unfortunately, little in the way of analysis of ages and sexes is available in the published literature, with the few notable exceptions cited. What is needed is a large-scale comparison of late Glacial mastodont ages and sexes in specific localities, such as southern Michigan, where finds dated to 15,000–10,000 years B.P. are common.

Following from the models based on studies of moose and giraffe, adult males and females should be expected to be found in different places, although the ranges of their death sites will overlap. Group death sites should be extremely rare, as they are for moose and giraffe, but if any are found, they should be dominated by females and subadults. Adult females should have died more often in late winter or early spring, as a result of stress caused by lactation, pregnancy, or malnutrition, whereas adult males perhaps should have died in early winter as a result of injuries or weakness following rutting battles. Male moose, giraffe, elephant, and bison (G. Haynes 1981:207–67) suffer high mortality during and following their mating seasons. Adult males would have been found more often where vegetation was thick forest, in similarity to the browsing taxa *Alces* and *Giraffa*, whereas females with young would have preferred open woodland, the shores of lakes or rivers, and other areas where avoidance of predators and chances for escape would have been optimized (if there were any predators targeting mastodonts), but where browse was still available. I would expect that mineral licks would have been heavily used by lactating and pregnant females in spring as their need for certain minerals became acute and that they would have died at such sites more often than would adult males.

Bone sites can tell us much about social groupings and the mortality events that sampled the living population. It is easier to construct models for the social structure of mammoths because data from their bone sites are so much more abundant. Future research no doubt will confirm or refute these simple predictions about mastodonts, but currently the data necessary to test the predictions simply are too few and too difficult to collate from the scattered published reports.

3.10. Afterthoughts and reflections

Clearly, I have based much of my interpretation of mammoth and mastodont behavior on what can be observed of modern elephants. However, modern elephant feeding and reproductive strategies should not be considered directly exemplary models. The behavioral ecology of elephants need not automatically solve mysteries of mammoth and mastodont behavior, even though all the different species are related. I have tried to generalize from knowledge about modern elephants to create referential models. The descriptions of elephants were presented to show how their behavior shapes their fitness. I hope I have provided a conceptual frame for reconstructing past proboscidean behavior; this idea has been discussed by Tooby and DeVore (1986) and Ghiglieri (1987), whose studies were of primates. Such a conceptual model may help substantiate that "a living species is a good parallel to an extinct one" (Tooby and DeVore 1986:186).

Some scholars will claim that modern elephants have developed special adaptations for living in marginal or inhospitable environments, areas where people cannot comfortably live and which are therefore given over to wildlife. If that belief is correct, then modern elephant behavior and ecology will be only superficially analogous with past behavior and ecology.

My reply to that idea is that modern elephant ranges are not universally "marginal," although they are far more restricted than they were 500 years ago. The very earliest descriptions available from travelers in Africa indicated that elephant habitats of the past were similar to elephant habitats of the present; elephants are not so much displaced as they are disappropriated.

It may also be argued by some scholars that because modern and extinct proboscideans are different genera, they therefore have different sets of behaviors, psychological attributes, biology, and so forth. Modern elephants (like modern hunter-gatherers) do not "equal" their past relatives. Of course, I am aware that they are different. But they are clearly related and share common ancestry, and there probably is less time depth in the phylogenetic separation of mammoths and modern elephants than there is between humans and our nearest primate relatives. The two allopatric elephant species alive today share important characteristics of social behavior, ontogeny, feeding styles, and so forth, indicating that much of the modern biology and behavior must be similar to those of the common ancestor. We do not have to invent a long-gone taxon's biology and behavior entirely from zero, using no clues from modern relatives that are members of the same clade and that share derived traits. It would be foolish to spend our time doing so. Even the bone deposits left by the extinct taxa are similar in significant ways to fossil deposits; for example, age profiles and depositional environments are often alike (G. Haynes 1988a); see Chapters 4 and 6. If we believe that the principle of uniformitarianism is valid, we must accept that certain structural similarities between the past and present taxa resulted from similar selective pressures and behavioral responses.

A third type of criticism that might be leveled is that modern elephants have learned many new and different sets of behaviors, such as defensive actions, ways of acting that extinct mammoths and mastodonts would never have learned. A corollary of that idea is that columbian mammoths in North America became extinct because they did not learn how to escape from human hunters. A related corollary is that human contact in modern Africa and Asia has affected all aspects of elephant ecology, especially following the spread of agriculture into formerly wild lands. Such a criticism must always be kept in mind when reconstructing the past based on studies in the present. In fact, in all zoological research, even basic assumptions about animal life must be questioned; for example, what reasons do we now have to think that modern lions behave like lions of only a few millennia ago? The tactics of the hunt, stalking, ambushing, and so forth, may have changed dramatically over the recent past, because prey animals have been subject to evolution and natural selection and have been developing new escape tactics to which the lions must continually be adjusting. Our main clues to animal behavior and biology come from anatomy, depositional environment, faunal associates, and other circumstantial evidence, but once again we must accept uniformitarianism as a guiding principle.

It does not take hundreds of years for proboscideans to learn defensive behaviors. They learn how to survive very quickly, and their behavior is flexible – they are bright and observant animals. Even the earliest European elephant-hunters in Africa were surprised at how rapidly knowledge of the presence of hunters was spread through elephant populations. Elephants communicate caution or fear to each other, and they readily teach each other new lessons in survival. Nowadays, where they are not harassed, elephants can be habituated to close and repeated human presence; but once harassed, even if briefly, they must be rehabituated, a lengthy process.

Alarm or fear of humans can spread through a local elephant population relatively quickly, disseminated through visual cues or infrasound calls that carry kilometers. That the extinct proboscideans were slow learners, easily overexploited by people, seems an improbable idea. Such doubts are difficult to substantiate with cold facts, except by reference to analogous reasoning, using modern elephants as a model.

My arguments addressing the three criticisms may appear to be circular reasoning – I have to assume that I cannot prove, in order to find evidence for the validity of the assumptions. I am not trying to outflank future criticisms; I *know* that modern elephants and extinct mammoths are different animals. But I also know that modern elephant studies may provide lessons that could help us understand the past. If these studies of modern elephants are going to be judged invalid or irrelevant for an understanding of extinct proboscideans, strong arguments of justification will be demanded, arguments that must depend on factually established differences between modern and extinct elephants.

II

ACTUALISTIC STUDIES OF PROBOSCIDEAN MORTALITY

4

Actualistic studies of mass deaths

Wherever we turn we shall find...a singular scarcity of animal remains....
...[an] infrequency of occurrence of skeletons where so many elephants abound....
Animals do not die naturally in crowds when young, and yet we find [fossil] remains of quite young animals abounding in all classes from Mammoths to mice.

—H. Howorth, *The Mammoth and the Flood* (1887)

4.1. Introduction

"Patterns" in mammoth and mastodont bone assemblages must be defined by careful study of the fossils themselves; however, the *meaning* of these patterns can best be determined by actual observations of the proboscidean behavior involved in the creation of analogous bone accumulations. Studies of the dynamic processes of life reflected in death assemblages are "actualistic," involving a type of taphonomic research, which is a more general approach that tracks bone-site disturbances, preservation, and fossilization (Gifford 1981; Gifford-Gonzalez 1989).[1] Binford (1981) has argued vigorously for more of these kinds of studies of the contemporary world, because the ways in which processes in our world work today and the effects that they have on live animals and on bone deposits will in many cases be identical with the end effects of past processes.

A study of elephant bones in nature can provide unexpectedly useful information about the conditions of past animal communities at the time faunal assemblages were forming. A major part of my research on recent elephants is longitudinal study of bone assemblages created by noncultural death processes. My study tactics include recording the modern environ-

---

[1] Taphonomy is the study of all the transitional processes undergone by bones after they have left the biosphere and entered the lithosphere (Yefremov 1940). "Actualism" is here defined as observational study of live animals becoming carcasses, skeletons, or bone scatters (Schäfer 1972). Note that the definitions overlap.

111

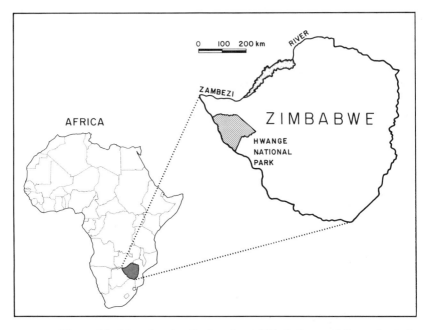

Figure 4.1. Map showing the location of Zimbabwe and the main study area, Hwange National Park, on the border with Botswana.

mental processes affecting populations of live elephants, such as climatic stress and the varying behaviors of predators, and then correlating these with their end effects on bone assemblages. Different mortality events create different kinds of elephant bone sites. The characteristics of the bone sites are very similar to the characteristics of fossil sites, and I propose that the similarities indicate that the sites were formed under essentially parallel conditions.

These studies were carried out mainly around remote water holes in southern African game preserves, in particular Hwange National Park in Zimbabwe (Figure 4.1), as well as Etosha National Park in Namibia and Chobe National Park in Botswana. The animals whose bones were input into the modern assemblages were subjects of major ongoing studies by wildlife ecologists. Bone assemblages in the study underwent processes of formation, preservation, and modification similar to those that Pleistocene fossils may have undergone: Recurring or permanent immersion in shallow water, burial in fine-grained sediments, trampling disturbances, and so forth. Some modern bone accumulations were mainly "attritional," the results of long-term input from common mortality processes affecting elephant populations. Other accumulations originated from "catastrophic" input, a result of very different (and very heavy) mortality episodes; see

Conybeare and Haynes (1984) and G. Haynes (1985b) for discussion of demographic profiles.

I want to emphasize that the elephant die-offs described in this chapter were not the results of recent human influence on wildlife populations. Elephant die-offs during drought years were recorded at least as early as 1935, before artificial water sources were put into the park, and about twenty years before dirt tracks were cut into the most remote wilderness areas. That was also at least thirty years after the cessation of human ivory hunting in the area (R. Gordon 1984; Selous 1881, 1893; Tabler 1955). At the time of the earliest recorded drought-caused die-offs, elephant density was relatively low in the region, and no game fences or other restrictions existed to prevent elephants from moving wherever they desired. Earlier die-offs undoubtedly occurred in the nineteenth century and before. It cannot be argued that compression into the national park precipitated die-offs or that population eruption accounted for mass deaths; die-offs preceded these management problems by three decades or more.

4.2. Study area

Hwange was designated a game reserve in 1927 and was proclaimed a national park on 29 January 1950 (Davison 1967). Its area is about 1,465,100 hectares, or 14,620 km², making it one of the largest national parks in Africa (IUCN 1985).

Less than a third of the park is open to visitors, the remainder being "undeveloped" and a sanctuary for wild game. The park sprawls over the shallow watershed that separates northward water drainage toward the Zambezi River valley from southwestward drainage to the Makgadikgadi Depression in Botswana. About 947,000 hectares are in highly pervious Kalahari Bed substrate, a deep sand possibly 200 m thick (D. Thomas 1988) overlying orthoquartzites, gravels, conglomerates, and Jurassic basalts, Triassic grits and sandstones, and Permian mudstones (Bond 1948; H. Cooke 1957; D. Thomas 1988). Basalts and other rocks outcrop in the northern part of the park, and fertile but shallow soils have developed on them. Very limited sodic soils are found in the Karoo sediments of the northeast; basaltic soils are more common. In the southwest, Kalahari sands have been reworked by wind and water into sandy alluvium/colluvium forming flat expanses. Rather small rivers drain the north and south, of which only the Deka in the north is a reliable source of water year-round. Other northern rivers such as the Inyantue and Lukosi go dry in late winter, although animals can dig in the channels to find water. Only in good years or normal years will the southern rivers (Dzivaninni or Sibaninni, and Nata, which is outside the park) contain permanent pools of water throughout the dry season. Pans (ephemeral ponds) in the shallow-soils areas may hold rainwater for much of the year, but not every year. The Kalahari sand has no permanent rivers or surface water, although numerous pans collect water during the rains and survive for a limited part of the dry season each year. About 4,150 km² of the southwest central part of the park have been designated a wilderness area, the Shakwanki wilderness area (Wankie National Park Management Plans, hereafter WNPMP, 1971).

More than a hundred mammalian species are found in the park (Smithers and Wilson 1979), including large herbivores such as black (browsing) and white (grazing) rhinoceros (*Diceros bicornis* and *Ceratotherium simum*, respectively), giraffe (*Giraffa camelopardalis*), zebra (*Equus burchelli*), and numerous taxa of horned bovids, such as wildebeeste (*Connochaetes taurinus*), kudu (*Tragelaphas strepsiceros*), and buffalo (*Syncerus caffer*). Carnivores are abundant and include lion (*Panthera leo*), leopard (*P. pardus*), cheetah (*Acinonyx jubatus*), wild dog (*Lycaon pictus*), and spotted hyena (*Crocuta crocuta*), among other taxa. Dasmann and Mossman (1962) estimated that the ratio of lions to large ungulates was about 1:110. The abundance of other carnivores has not been well studied. Animal densities vary greatly in different parts of the park, depending on the state of vegetation and the local availability of water, which change through the seasons (B. Williamson 1975a, 1979a, b). Animals are not uniformly distributed throughout Hwange's huge area. B. Williamson (1975a) reported that more than half the elephants counted in an October aerial survey were found in less than 10 percent of the park; at other times of the year elephants were less aggregated and could be found over a larger area.

In 1984, D. Cumming, as cited by Childes (1984), estimated the numbers of the eight largest taxa present in the park (Table 4.1), based on annual aerial censuses. Elephants have been counted once each year and occasionally twice each year since 1964 (D. Cumming 1981; M. Jones 1989 personal communication); numbers for the other taxa are based on frequencies of sightings by park personnel and visitors, as well as observations made during the flights to count elephants.

Presently, during the dry season, Hwange's game animals depend a great deal on artificial water supplies within the nonwilderness parts of the park. A game-proof fence erected in 1960 on the eastern and southern boundaries prevents large animals such as elephant and giraffe from reaching the pumped water sources of farmers and pastoralists; likewise, game animals are prevented or discouraged from moving northeastward out of the park boundaries toward the permanent water of the Gwayi (or Gwaai) River and its tributaries, or north toward the Zambezi River. Discouragement takes the form of shooting. A game fence erected by Botswana in 1981 partially prevents animals from reaching the Nata River southwest of the park. Country to the west of the park is not fenced, but is generally drier than most of Hwange. Because of the presence of the fences and the aridity of the

Table 4.1. *Approximate numbers of some large herbivores in Hwange National Park in the early 1980s*

Taxon	Approximate number
Elephant (*Loxodonta africana*)	22,000
Buffalo (*Syncerus caffer*)	15,000
Sable (*Hippotragus niger*)	2,600
Zebra (*Equus burchelli*)	1,700
Giraffe (*Giraffa camelopardalis*)	2,000
Wildebeeste (*Connochaetes taurinus*)	1,500
Black rhino (*Diceros bicornis*)	150
White rhino (*Ceratotherium simum*)	80

Source: D. Cumming (1984 personal communication).

unfenced lands, during the dry times of the year the game does not move beyond many park boundaries. Additional game often moves into Hwange from the west, which is Botswana, perhaps coming as far as 100 km during the winter dry season. In the wet season, some of these animals then return west to Botswana territory (B. Williamson 1979a, b). Quite possibly there were more perennial pans in the Kalahari sand area 50–100 years ago than in recent times (Campbell and Child 1971; Hodson 1912; Tabler 1955; WNPMP 1971:89), but there are still permanent water sources in this part of the park, most of which are buried and therefore inaccessible to many animal species. These water sources were known to prehistoric and historic humans and to big-game populations in the area for generations (Hodson 1912; Tabler 1955, 1971) and later will be described more fully. Other water used by the game populations in Hwange is provided by pumps at boreholes situated in the eastern and northern parts of the park. Between forty and sixty pans are filled with borehole water during the driest part of the year.

4.3. Method of study

I have visited Hwange National Park every year since 1982. All observations made in the field involved travel on sand or dirt tracks to areas far away from tourist routes and human activities. Much of the time I operated out of temporary camps set up near die-off locales, as described later. I also spent many days exploring other parts of Hwange National Park besides the die-off locales, usually alone and on foot after driving my Landrover into the park. I spent more than eighteen months in the field over a total of nine field seasons. I was present in the park in June, July, and August of nearly every year, and in some years I was also present in September, October, November, December, and January.

Observations of living animals were as important to this study as were detailed examinations of bones. Some individual animals under observation eventually became recognizable because of their distinctive markings, such as torn ears or broken tusks in the case of elephants, or scars and worn horns for antelopes and buffalo, so that their behavior could be monitored during different times in the field.

Written notes were always taken in the field, and longer narratives were written in a separate journal at the end of each day to record sites. I used up a thirty-six-exposure roll of slide film every two days, taken with a variety of different 35-mm cameras and lenses. Quite a few bone specimens were collected each year. Numerous maps and sketches were drawn of bone sites and water holes.

Very remote bone sites look quite different in wet and dry years, but each field season I perservered in searching until I succeeded in finding the sites examined in previous years. Except for the die-off locales, every site selected for detailed long-term ("longitudinal") study was located far from roads, trails, or other areas where people might encounter the specimens. The die-off locales were linked by a track that was impassable except for sturdy four-wheel-drive vehicles; the only traffic on this track during my studies was an occasional antipoaching patrol by rangers and scouts of the national park.

All bone sites described here were minimally affected (or entirely unaffected) by human actions. In many cases the remoteness and difficulty of access to the sites made disturbance virtually impossible or highly improbable. Some sites had originally been spotted from the air during patrols or search flights in small, fixed-wing aircraft and were then sought on the ground under very difficult conditions of bush bashing. In other cases, I sometimes fortuitously stumbled across bone sites while walking about in the bush. At some of these latter sites, the presence of elephant tusks was a certain indicator that no other human had ever visited the site. Ivory is an uncommonly valuable item, and a pair of moderate-sized tusks would be worth more money than the rural people of Zimbabwe or Botswana earn in a year of wage labor.

In the case of the die-off locales situated on the vehicle patrol track, lower jaws and tusks were collected by park rangers and scouts whenever a carcass or skeleton was encountered during the years 1982–90. The tusks were collected because they belong to the government of Zimbabwe; the jaws were collected in order to determine the ages of the dead animals. No other bones or body parts were disturbed.

4.4. Elephant mortality in the study area

Modern elephants are relatively invulnerable to predation (Hanks 1979; Moss 1988), although occasionally young males near or at the age of sexual maturity (eight to twelve years old) wander away from mixed herds and are preyed upon by lions or hyenas (A. Conybeare 1982 personal communication, Moss 1988), as documented by unpublished data on file at Hwange National Park Main Camp. Predation on young elephants occurs much more often during periods of drought, when the animals are already weakened. An accumulation of elephant carcasses lying on land surfaces in one limited geographic locality is invariably the result not of serial predation but of die-offs at water sources, caused by seasonal drought. In the absence of drought-caused die-offs, elephants often die alone some distance from water, although skeletons may begin to accumulate around water holes in the course of several decades, the result of individual mortality due to old age, disease, accident, and so forth. After twenty or more years, many bones on the ground surface around water holes eventually weather away into unidentifiable splinters, or become buried in sediment, never to be seen again. Thus, slowly accumulating assemblages seldom grow visibly large.

The main causes of mortality among modern African elephants may be summarized as follows. The order of presentation is not the order of importance, which varies from year to year.

1   Traumatic fracture involving one or more long bones usually causes a loss of condition and then death in less than a month, although in rare cases slow recovery has been recorded. Although uncommon, traumatic fracturing does occur in nonnegligible numbers in Zimbabwe's wildlife reserves.

2   Accidents such as getting stuck in mud, drowning, or falling from steep slopes may kill outright or slowly. Although uncommon, such mortality nonetheless occurs often enough to be observed in African parks.

3   Injury, such as entanglement in poachers' snares, sometimes causes infection, loss of condition, and death after one to three months, although, again, recovery is possible. Injuries may also result from fighting with other elephants (Hanks 1979), probably in greater numbers than usually recorded.

4   Disease, in some cases, causes sudden death (e.g., as seen with heart ailments, which are thought to be extremely rare in elephants) (McCullagh 1972) or a more gradual loss of condition, followed by death. The diseases to which elephants are vulnerable are not well known; Moss (1988:268–9) discussed the possible types, including anthrax, pneumonia (or pneumonia-like disease), debilitating abscesses, and parasitic infestations.

5   Old age may cause abrupt death due to heart attack or heat prostration or, more commonly, gradual loss of condition due to an inability to feed efficiently, the result of an animal outliving the

useful life span of its teeth (between fifty-five and sixty-five years old) (Laws 1966; G. C. Craig unpublished data). During dry periods, old-age deaths are not especially uncommon and may in fact be frequently encountered in the bush.

6   Drought-related stress may kill numerous animals over short periods of time. Hanks (1979) emphasized that drought is one of the most important factors affecting the sizes of modern African elephant populations. Deaths during drought result from heat prostration, dehydration, and starvation, because the available plant foods provide inadequate nutrition. Starving elephants do not die with empty guts, however, because they eat wood, bark, soils, and whatever else is available to fill themselves.

Hanks (1979) described a drought that killed thirty-one elephants around one Zambian water hole in 1970; most of the animals that he could age were under four years old. Corfield (1973) estimated that nearly 6,000 elephants died in Tsavo National Park in Kenya as a result of the severe 1970–1 drought. Juvenile mortality was high, and adult females were also hard hit. Barker (1953) mentioned an account by Karomojo Bill, an ivory hunter, who in about 1907 found numerous elephant skulls lying about an area where they had died during severe drought. The elephants' deaths were attributed to poisoning by salt springs, although it seems possible that the animals actually had starved to death, the most common cause of death among elephants taking refuge at water sources, as discussed later. Conybeare and Haynes (1984) and G. Haynes (1985b) reviewed the different age-class vulnerabilities to drought in Zimbabwe and found subadults to be overrepresented in death assemblages. Moss (1988:60) observed high juvenile mortality in Amboseli National Park during the drought years 1976 and 1984; adult female mortality was also high, but resulted mainly from spearing or shooting by humans.

7   Human hunting is one important cause of death for modern elephants, inflicted mainly by ivory poachers. The national parks of Zimbabwe are efficiently policed to prevent poaching, although a few isolated incidents occur from time to time, especially near park boundaries. These incidents have never had any statistically significant effect on elephant populations in the national parks and wildlife reserves.

4.5. Kalahari sands: where elephants die

The west central part of Hwange National Park is covered with parallel sand ridges that are oriented east–west. These are the eroded stumps of longitudinal Pleistocene dunes thought to have been formed during very arid times (Flint and Bond 1968), possibly coincidental with the last glacial maximum of about 22,000–18,000 years ago (Heine 1982). The dunes are now 2–35 m in height and average 1,500 m in width, separated by troughs 500–2,000 m wide (Flint and Bond 1968; D. Thomas 1982, 1983, 1984, 1985). Unaligned, braiding channels truncate some of the ridges. The streamways were formed in a humid period following dune formation and trough infilling by erosion (D. Thomas 1981). Vegetation in this region consists of *Baikiaea plurijuga* (Rhodesian "teak") woodland, an *Acacia/Baikiaea* mosaic, and a more open Kalahari sand scrub supporting *Terminalia* and associates. Even though the region is very dry half of the year, some of the woody-shrub leaf material is high-quality browse for herbivores (Rushworth 1975). However, the dominant species, *Baikiaea* and *Terminalia*, are rarely eaten and probably are of low quality. In the more arid west central area, called the Shakwanki[2] Wilderness Block, the mean annual rainfall is 300–500 mm, but rainfall is subject to extreme variability from year to year (Torrance 1981). Drought years are not uncommon. The mean annual temperature is about 22°C, according to the unpublished records of the Hwange National Park Research Station, and day-to-night temperature changes may be on the order of 30°C or greater. The lowest ground temperature recorded in the dry season was − 14.4°C in 1972 at Hwange Main Camp (Rushworth 1978).

Rainfall is strictly seasonal in the park; the wet season is November through March (Torrance 1981), as documented by unpublished records on file at Hwange National Park Main Camp. Between November and the end of January, surface water is widely distributed and plentiful in pans, and new growth is lush. Between February and April, grasses begin drying, rainfall decreases, and nighttime temperatures start to drop. May through July is the cold season; most pans dry up, grasses and herbs continue drying, and ground frosts frequently occur. August through November is a time of increasing heat, the appearance of the first new leaves on some woody taxa, and complete drying of the annual plants (Conybeare 1972). Between March and mid-November, the water deficiency steadily worsens because of high evapotranspiration rates (Childes 1984; Torrance 1981). Neverthe- less, the deep, infertile Kalahari sands support a suprisingly high biomass of woody plants; the grass biomass is relatively low (Rushworth 1978). Mature *Baikiaea* woodland sits atop unconsolidated deep sands, mainly on ridge tops, and *Terminalia* scrub, coppiced *Baikiaea* scrub, or *Burkea* woodland appears where the subsurface sand is compacted into hard layers (Childes 1984; Childes and Walker 1987).

Elephants are nomadic animals, and they seek out different food sources and watering points according to the seasons. Subpopulations of elephants that move to the Shakwanki wilderness are members of a larger regional

---

[2] Also spelled Shakawinki, Tshikwanki, or Schekwanki.

population inhabiting the eastern Kalahari region, including parts of Botswana and Zimbabwe. When the November–March wet-season rainfall has been below normal, elephants that have moved into areas containing remnant water sources may choose not to leave, because of lack of predictable water elsewhere from July through October (the driest part of the year).

An adult elephant requires plentiful moisture each day to digest food and regulate body temperature. During dry seasons of the year, elephants may travel to water every day when possible, although data from radiotelemetric studies (A. Conybeare 1988 personal communication) show female groups remaining without water for forty-eight hours regularly, even when only 5 km from a pan, and male bands going three to four days without water, although never more than 20 km from pans during that time. When there is no water near feeding areas, elephants may travel long distances to water every two to three days, if the available forage contains enough moisture to maintain such spacing. If visits to water are daily, elephants spend a relatively short time drinking, spraying, and wallowing (less than an hour in daytime). When visits are spaced more widely, the time spent at water may be much longer.

The Shakwanki wilderness area is home to few mammals during the November–March wet season. In June, scattered herds of up to thirty to fifty buffalo (*Syncerus caffer*) graze in the shallow drainage-ways, but they abandon such areas when the surface water becomes muddied or is gone and the grasses become quite dry. Impala (*Aepyceros melampus*) have a localized distribution, usually in degraded areas around permanent water points. Zebra (*Equus burchelli*) are present in low density and may eat the grass nearly to the ground surface, while drinking at distant water sources whenever necessary. Much of the grass in the Kalahari sand region is unused by herbivores.

Also observed in low numbers during June are steenbok (*Raphiceros campestris*), kudu (*Tragelaphus strepsiceros*), roan (*Hippotragus equinus*), sable antelope (*Hippotragus niger*), giraffe (*Giraffa camelopardalis*), and gemsbok (*Oryx gazella*). In July through August, single and herding elephants move through on feeding travels that are irregularly spaced. The remaining coarse grass stems and forb stalks in the grassy drainageways (vleis) are pulled up and eaten, or trampled flat. Once grasses have completely dried, they are not eaten by elephants. During drier years, single or herding elephants become resident, and some may die from starvation or dehydration in May through October. After the early rains in October or November, shallow surface water can be found elsewhere in proximity to greening plants that contain abundant moisture and nutrition, therefore allowing the resident elephants to move out of the overutilized water-refuge area, until the next dry season.

Small mammals such as springhare (*Pedetes capensis*) and gerbil (*Tatera leucogaster*), insects, reptiles, terrapins (*Pelomedusa*), and tortoises (*Testudo*) are abundant in the Shakwanki wilderness area year-round.

In a few places the Kalahari sands contain very localized perched aquifers, which Davison (1950) interpreted as resting atop clay layers and covered by redeposited sands. These areas may once have been pans – water holes with clay bottoms that collect and hold rainwater. Based on his random observations, Davison reasoned that the vegetation around these pans had been cleared out by large animals feeding near water, and over time the pans had become covered by drifting sands. Davison (1967) described scooping away the sand cover at one locale called Nehimba, a task requiring several days of work using oxen-drawn scoops. Apparently he reached a "clay" seal, several feet below the ground surface. My own shovel excavations at Nehimba and other seeps failed to reach clay layers, although I did strike oozing water at the distinct upper boundary of a compacted gray sand, 1–2 m below the ground surface. Childes and Walker (1987) reported that soil augers had to be hammered through compacted subsurface sand layers. Little extensive augering has so far been carried out to corroborate Davison's theory; an alternative hypothesis is that subsurface layers of compacted sand, rather than clay, impede drainage.

The areas containing perched aquifers are open, depressed, sandy spaces with no large trees, covered with coarse grass, and usually fringed with scrub acacias, *Baikiaea*, *Combretum*, or *Terminalia*. At least ten such areas are currently used by animals in the Shakwanki wilderness areas and all are called "seeps" (Figure 4.2). Some are located on relatively high ground. The composition of grass species growing in the open areas differs from that of the surrounding land because of herbivore use and, especially, springhare burrowing (Butynski and Mattingly 1979). Long before Hwange National Park was a protected area, before boreholes and artificial watering points (in association with higher temperatures) perhaps lowered the water tables, there were numerous other artesian seepages in the Kalahari sands areas (Hodson 1912; Tabler 1955, 1971). Those areas were well known to San Bushmen "poachers" and white hunters or traders, but many were never mapped, or they dried up before they could be studied; see Austen (1953, 1954a,b) and J. Gordon (1956) for early references to seeps that are now dry more often than not. Those "springs" probably died because of lower precipitation, higher dry-season air temperatures, and a lowered water table. Shrub invasion of what was once open grassland has lowered the water table considerably, a result of cessation of the former frequent practice of burning of vleis by native hunters or pastoralists to encourage greening and more intensified game (or livestock) utilization. Repeated burning destroyed the highly organic upper soil layers, greatly reducing the water-holding capacity of vlei sediments.

During the dry season, elephants visit the seeps to dig shallow wells

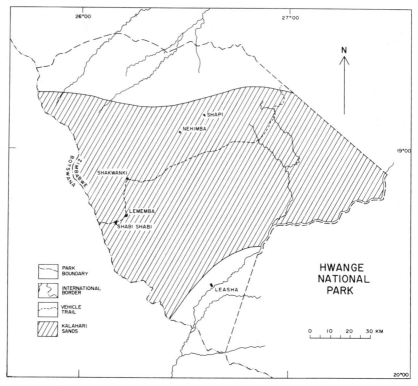

Figure 4.2. Map showing the extent of surficial Kalahari sandbeds in Hwange, and some of the named seeps (Nehimba, Shakwanki, Lememba, and Shabi Shabi). Shapi and Leasha are two major pan complexes with boreholes that usually are pumped in the dry season.

1–3 m deep to the perched water level and wait for water to ooze into the opening; they then draw the water up their trunks and release it into their mouths to drink. This set of behaviors was observed by early explorers such as Chapman, in 1862 (Tabler 1971), Davison (1930–5), and other early workers in the game reserve and was seen mainly during the dry season when pans dried up or were muddied by heavy use (Davison 1930, 1930–5). In very dry years, the water oozes so slowly into the wells that elephants sometimes stand at them waiting for hours to satisfy their thirst, and they ingest a great deal of sand with the water. The sand, when passed with dung, is shaped into boluses containing masticated vegetation.

In a typical year, between August and November the elephant-dug wells at the seeps are the only sources of water for animals in the Shakwanki wilderness area. Water from the seeps is fresh and cool, in contrast to borehole water, which would be so brackish as to be undrinkable by animals. One deep borehole was drilled in the seeps area during the 1970s, but was never pumped because of its high mineral content.

The worst drought year known since records have been kept in the park (regularly since 1951, and occasionally before then) was 1981, when only 335.6 mm of rain fell in the Hwange Main Camp area, which probably receives up to 150 mm more rain than does the Shakwanki wilderness region (M. Jones 1986 personal communication). The following rain season (1982–3) was slightly subnormal in the Main Camp area (receiving about 84 percent of average rainfall, or 475 mm) and also subnormal in the Shakwanki region; in the next year, 1983–4, the rain season was once again very poor (the eleventh worst on record since 1921, with 372 mm of rain in Main Camp, compared with an expected average of about 568 mm), according to unpublished records of the Hwange Research Station. The wet season of 1986–7 was also very poor, and the 1987 dry season was the third worst ever recorded. The early 1987–8 wet season did not shape up well, but late heavy rains prevented the 1988 dry season from being stressful. Rains in 1988–9 and 1989–90 were nearly normal, although they came late in the season.

4.6. Die-offs in the Shakwanki wilderness area

Several of the seeps are situated on long, winding grassy depressions many dozens of kilometers in length. These are extinct stream channels in which surface water no longer flows, even in very wet years, although pans, pools, and mud holes are common in the rainy season. These channel-like features are called "vleis" in Zimbabwe, a word denoting shallow linear depressions that are covered with grass and few or no woody plants and along which drainage is directed (Rattray et al. 1953). "Vlei" is an Afrikaans word loosely translated into English as marsh or swamp; in Botswana, Setswana speakers call the shallow, flat drainage channels "molapo" (Tanaka 1980:17). The Ci-Cewa word "dambo" has been suggested as a more correct designation for such grassy drainageways (Whitlow 1984; Young 1976). The seep called Shabi Shabi[3] is located on a tributary vlei of the Triga[4] vlei system. The long vleis (or dambos) of this system wind through the central part of Hwange National Park and probably are remnants of the headwater drainage of Ntwetwe Pan, the huge pan that makes up the western part of the Makgadikgadi Depression in Botswana (Wellington 1955:421). The main Triga vlei itself terminates at Domshetshe[5] Pan on the border of

---

[3] Perhaps also called Ngobitshani (Collins 1965).
[4] Also spelled Tiriga or Trigo.
[5] Also spelled Domtchetchi, Tamasetje, Tamasatsa, Thamma setchie, Domsetche, or Domshetshu.

Botswana and Zimbabwe (Wellington 1955:483), but smaller vleis continue drainage west and south toward Makgadikgadi.

The earliest written record of Shabi Shabi was made by the assistant game warden J. W. Verney in his unpublished manuscript report of activities during 1938 and 1939, which is on file at Hwange National Park, but the other seeps were given names by the local people (black and white) long before Hwange was a protected game reserve (Tabler 1971). Davison (1930 5) described several seeps along the Botswana border, using their Bushman (not always referring to San) names, which have been preserved. The peculiar behavior of elephants at the seeps was also known early on to Davison and to others working in the area.

During years when rainfall was abundant or well distributed, the seeps were (and still can be) inundated. Davison (1930–5) described some of them as depressions having spring vents, overflow channels, and ponding. The seep called Nehimba can be under several acres of water in the middle of the dry season; that has often been observed following a completely dry appearance during the same season in the previous year, when the only water was 1 m below the ground surface. In wet years, ducks and geese nest at the seeps, and large mammals visit the areas only sporadically. But during the drier years, elephants from as far away as Makgadikgadi to the southwest and Chobe National Park to the northwest make their way to the seeps located near the Botswana-Zimbabwe border (A. Conybeare 1985 personal communication). There almost certainly is two-way movement between Hwange National Park and Chobe National Park, and probably other places in Botswana, but this has never been properly documented (A. Conybeare 1988 personal communication).

During their journey to drought refuges, elephants must pass through scrub vegetation and so-called thornveld, the typical Kalahari vegetation of *Acacia* and widely separated shrubby plants growing among rank grasses. The substrate is loose sand or hard, dry sandy silt and clay. Surface water usually is extremely scarce or nonexistent. Heavily used traditional game trails converge on the seeps from different directions. Many trails follow vlei lines, which are natural routes for travel to the seeps; the region that now includes Hwange National Park once was known as the "land of one thousand vleis" (Tabler 1955).

At the seeps, elephants use their feet to loosen the soil, which they toss away with their trunks. The wells that they dig can reach nearly 2 m in depth (Figure 4.3). Most well-digging apparently is initiated by adult males, which locate the best places to dig by finding damp places on the ground surface. Capillary action brings up moisture in the spots where wells had been dug in earlier years, so that the grasses are lusher and the soil damp. Wells used by elephants collapse after the rains begin, yet they are invariably redug

Figure 4.3. A starving adult female elephant at Shabi Shabi standing in a
well that is about 1 m deep where her forefeet are, and another 50 cm deep
where her trunk is reaching.

within a few feet of where they had been located before, in spite of complete
infilling with compacted sediments.

In the 1980s, according to aerial surveys, 1,000–2,000 elephants were
normally present near the seeps during the dry seasons (D. Cumming
1983–6 personal communications), and competition for access to the
seeping water in wells became intense as the dry season progressed. In spite
of a succession of drought years, hundreds of elephants continued to spend
the dry season in the area of the seeps each year.

During the hot, dry season, many elephants went to the wells in the dark
of night or very early in the morning, thus avoiding the heat of day (Guy
1974; Weir and Davison 1965). One other reason for that timing was that
food sources were located farther and farther away as the season
progressed, and feeding by elephants had exhausted nearby forage, forcing
animals to walk much of the day to food and back to water. The hottest
hours of the day were spent standing or lying down in the shade to rest,
whenever possible. Elephants began to feed again as they made their way
back to water at the seeps. As the heat increased in August and September,
the walk back to the seeps was begun later and later in the day. Judging
from the abundant evidence of elephant dung at the seeps, their dry-season
diet consisted mainly of bark and twigs, some leaves, and acacia pods.
Many trees are not browsed in the dry season, and the dry grasses are also

avoided; it appears that acacia pods are preferred food, although abundant sources are located at increasing distances from the seeps as the season progresses.

Maximum walking speed is around 8 km per hour for adult elephants (estimated from field observations, as well as mathematical formulas using limb lengths), and they will walk for hours, if necessary, to reach food or water. Elephants travel at sustained rates of about 5 km per hour or faster over long distances [Oberjohann (1952) estimated speeds of about 2.5 km per hour for feeding search, about 6.5 km per hour for brisk march between feeding spots, and about 15 km per hour for forced escape], but they become heat-stressed when forced to travel during the hottest part of the day at the height of the dry season. The maximum daily foraging radius for an adult in this season probably is around 40 km, but in actuality the utilized range is much smaller, because several hours are required for feeding during the day, and walking is restricted to cool nighttime hours.

Big-game taxa other than elephants also utilized the seeps, especially at the beginning of the dry season when the available vegetation was still moist. Competition for food did not affect the movements of these taxa as much as did competition for water, which led to abandonment of areas without abundant water (Duckworth 1972). Buffalo herds abandoned the seeps by the middle of the dry season, unless there were pans nearby that still had water remaining in them. Antelope such as impala (*Aepyceros melampus*), kudu (*Tragelaphus strepsiceros*), roan (*Hippotragus equinus*), and sable (*Hippotragus niger*) also abandoned the area, although sometimes individuals or very small groups remained throughout the dry season, subsisting on moisture from vegetation, or crawling down into the larger elephant-dug wells to drink, if possible. Some large carnivores such as spotted hyena (*Crocuta crocuta*) and leopard (*Panthera pardus*) remained in their home ranges near the seeps year-round, whereas other taxa such as cheetah (*Acinonyx jubatus*) and lion (*P. leo*) increased their wanderings in the dry season and visited the seeps, where their abundance in the wet season is poorly known.

At the seeps, elephants of all ages frequently were aggressive over access to the limited numbers of wells that had already been dug. Interactions between individual elephants usually involved a rapid acknowledgment of dominance or submission when two animals approached the same excavated well. However, some encountering animals did act violently and clashed over rights of access, although such encounters were brief. The usual "fight" was a violent shove or butting against one animal's flank, rear, or head by the dominant animal using its tusks and forehead. Aggressive encounters were rarely reciprocal – only one animal struck the other. These shoves created loud cracking noises and resounding thuds, as tusks and skulls contacted skin, attesting to a great deal of applied force. Occasionally tusks broke during fights (Deraniyagala 1955:66; Moss 1988; Sutherland 1912:95–9). Although broken tusk fragments can be found at virtually any water source visited by elephants in Hwange National Park in any year, great quantities of fragmented ivory are to be found during the especially dry years.

The earliest mention of broken tusks resulting from drought-year fights was by the assistant game warden J. W. Verney (1938–9), observed at the seep called Nqwasha (several spelling variations exist). Davison (1950:24) collected 200 pounds (ca. 91 kg) of ivory chips at various water sources over a short period of time, presumably during a drought period. The seep called Shabi Shabi yielded an estimated 100 kg of ivory chunks, tips, and flakes between 1981 and 1988 (Figures 4.4–4.7). Those fragments were collected in an area measuring 150 m by 50 m, located within a meandering depression of the vlei. That part of the vlei also contained the skeletons or carcasses of over fifty elephants that had died during the severe stress of the 1981–4 and 1986–7 dry seasons (Figure 4.8). All whole tusks and tusk fragments were collected by personnel of the Zimbabwe Department of National Parks and Wild Life Management.

As a rule, tusks were not used to dig out wells at the seeps, although very rarely an elephant would loosen compacted sediments with its tusk tips. However, tusks frequently were used in other parts of Hwange as aids in breaking up sediments to be eaten for their mineral content. Termite mounds sometimes were broken apart and eaten, and so-called salt licks often were visited at seasonal pans where the high clay content and carbonates had cemented the sediments into brick-hard layers valued for their minerals.

Broken tusks were never found within elephant-dug excavations at the seeps and in fact were always found on the ground surfaces between wells, where aggressive encounters occurred. Tusks did not break because of well-digging.

In the early dry season of 1987, an average one tusk tip and one tusk-shaft segment were broken off every day at each seep, a relatively low rate that reflects the smaller size of the local subpopulation of elephants then utilizing the seeps, as compared with earlier years. The decrease in numbers may have been partly a result of large-scale culling in other parts of the park from 1983 to 1986. Culling has not been carried out in the Shakwanki wilderness area.

The earliest written record of elephant die-offs at the seeps was provided by Davison in his unpublished report of activities during 1935. Later reports by the game wardens Davison, Till, and Verney also mentioned dry-season deaths during the 1930s. Deaths during the 1960s, 1970s, and 1980s were witnessed by the senior research scouts Million Sibanda, Morven Mdondo,

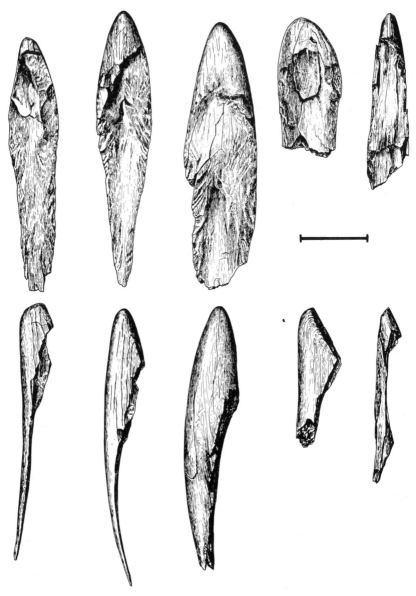

Figure 4.4. A sample of tusk tips broken by elephants pushing and fighting each other over access to water in the wells. The scale bar represents 5 cm. Drawn by M. Bakry.

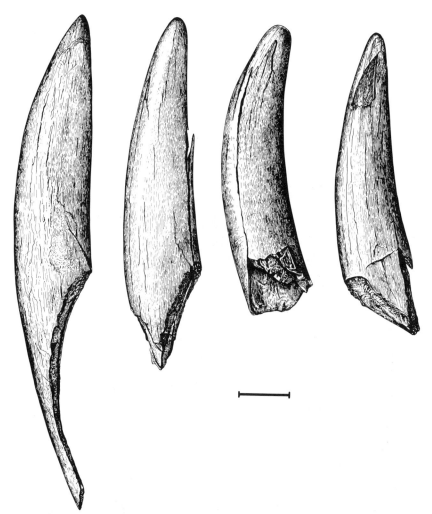

Figure 4.5. Larger tusk tips showing the variation possible in breakage patterns when elephants fight. The scale bar represents 5 cm. Drawn by M. Bakry.

Peter Ngwenya, and William Deviyas, who described the die-offs in terms similar to those used by the earlier workers such as Davison.

4.7. General descriptions of bone sites

General descriptions of the die-off sites have already been summarized (G. Haynes 1985a, b, 1987a, c, 1988a–c). These sites can be enormous in area,

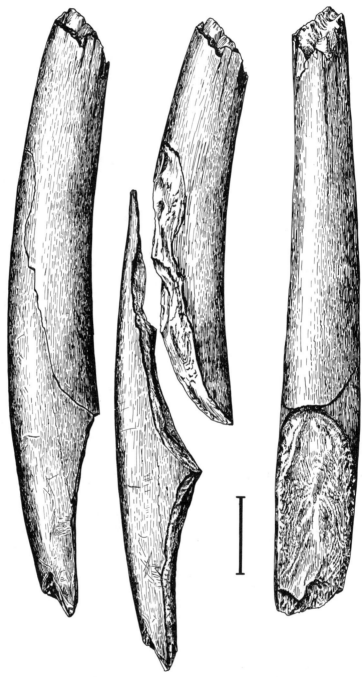

Figure 4.6. Three views of a tusk midsection broken during fighting over water. The scale bar represents 5 cm. Drawn by M. Bakry.

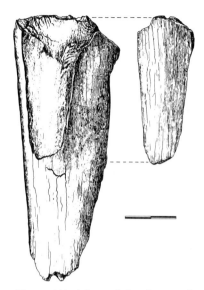

Figure 4.7. A large flake that can be refitted to an even larger flake of ivory, broken by a bull fighting another bull. The scale bar represents 5 cm. Drawn by M. Bakry.

although many are rather compact, with sizes ranging from 1 km by 1 km to 40 m by 40 m. In the largest site, Shabi Shabi, a wooded "outer" region contains bones and carcasses of dozens of animals that died in patches of trees and shrubs, whereas the "central" area contains the remains of dozens of animals that died while seeking water. The central areas of all the major seeps (Shabi Shabi, Lememba, Shakwanki, Nehimba, Tamafupa, Tamasanka, Ngerewa, Nqwasha, and Domshetshe) are open and treeless, surrounded by a fringe of trees and shrubs (Figure 4.9). However, large sections of tree trunks lying on the ground or partly buried are common within the central, treeless areas of all seeps. These tree trunks often are charred at one or both ends and sometimes are rounded from weathering and trampling wear (Figure 4.10). Lightning-caused fires have burned through the seeps many times in the past, although thick stands of large trees survive in nonuniform distributions around the different seeps.

The youngest elephants sometimes died much nearer the inner areas than did older animals. In many cases elephants under two years of age died within wells dug out earlier by other elephants (Figure 4.11). In contrast, the larger carcasses were located several meters from any well. One reason for this distribution of ages probably is that the smallest and youngest animals competed much less successfully for access to the dug-out wells, and thus stayed longer in the open central areas awaiting a chance to drink. These

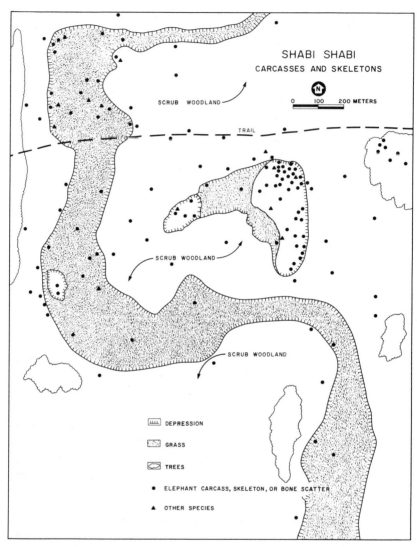

Figure 4.8. Map of the central meanders of the fossil channel at Shabi Shabi, showing known locations of most carcasses and skeletons of mammals that died between 1980 and 1989.

smaller animals also could not feed well away from the seeps where the vegetation was being stripped by heavy feeding pressure from larger animals.

The smaller elephants were too weakened by thirst and starvation to walk the long distances necessary to find food, and so they stayed near the

Figure 4.9. Oblique aerial view of elephants and two zebras (center) congregated at Shakwanki seeking water in elephant-dug wells. This main area of Shakwanki is a basin fringed with shrubs (lower part of the photograph) and trees (not shown).

Figure 4.10. A branch or tree trunk (foreground) about 15 cm in diameter and 60 cm long lying in elephant spoor in the central part of Nehimba, photographed in 1987. No trees this size grow nearer than about 500 m to the seeps area.

Figure 4.11. The carcass of a small calf lying in the opening of an elephant-dug well at Shabi Shabi, 1982, about two weeks after death.

water. The water itself was out of reach to many of them, because heavy elephant pressure on the supply from the wells lowered the level of the oozing moisture beyond young trunks. Very compacted sediments around caved-in wells could not be excavated by weakened and small animals. The smaller elephants were shoved (or threatened) out of the way by thirsty older competitors.

A final factor suggested to account for very young elephants dying closest to the wells in the open areas is the air temperature at the seeps – nighttime temperatures in the Kalahari sand areas often dropped below freezing during June and July, and the smaller animals may have succumbed to this additional stress. Ground and air temperatures under the trees were several degrees warmer.

Wells often were redug, cleaned out, and enlarged by elephants capable of doing so, especially after traffic to the seeps and their use as water sources increased dramatically in the middle of the dry season. As the ground around wells was trampled bare, the sediments dried out, and wells often collapsed. If a carcass or skeleton lay within the depression created by digging or trampling around a well opening, or within the well itself, elephants would kick it, shove it with their feet, and move it using their trunks to get it out of the way. Elephants dug through buried carcasses as well, after young animals had died within wells and caused the holes to slump or collapse. Bones or body parts subjected to well-digging were

Figure 4.12. An early stage in the controlled excavation of a small elephant skeleton that had been buried by the collapse of an elephant-dug well at Shabi Shabi. To the right are several ribs in the articulated unit of thorax, spine, neck, and head. Note that limb bones are disarticulated and were only very shallowly buried or were exposed atop the ground surface.

kicked, stepped on, pushed, and lifted, although sometimes carcasses and skeletons were removed in sections, not all of which were subjected to identical disturbances. For example, eight months after the 1983 die-off, I excavated a collapsed well at Shabi Shabi (Figures 4.12 and 4.13) and recovered the complete ribcage, spine, neck, skull, and mandible of a young elephant. The bones were unbroken and in complete articulation, buried under about 1 m of sediments. However, all limb bones were out of articulation, and a few were only shallowly buried; some were located atop the ground surface, where they had been trampled. Breakage and trample marking of the bone assemblages are described later.

In 1987 I observed that spotted hyenas had dug burrows into two collapsed well depressions and were living within the cool, damp sediments about 2 m below the ground surface. No doubt they had dug passageways through buried bones and skeletons, although the unmistakable presence of adult hyenas in the burrows during the daylight hours prevented me from crawling below to investigate. Mounds of loose, dug-out sediments at the openings of the burrows contained several elephant bones, some of which had been carried there by hyenas, but some of which appeared to have been

Figure 4.13. The final stage in the controlled excavation of the buried skeleton shown in Figure 4.12. The skull and lower jaw are articulated, with the face directed downward, and the right side of the rib cage is now exposed, following removal of the articulated left ribs and the cervical and thoracic vertebrae.

thrown up out of the excavations (Figure 4.14). Elephant mandibles were favored as elements to be transported to the burrows. Numerous smaller burrows were abundant in the seeps areas, dug by springhares or gerbils, but those never showed evidence of bones thrown up from below. Springhares excavate rather large underground burrow networks, generally no deeper than 70–100 cm below the surface. Each burrow has several openings. Snakes (such as puff adder and cobras), other reptiles (such as lizards), and small rodents (such as gerbils) also utilize these burrows, which springhares reuse every year, and occasionally enlarge (Butynski and Mattingly 1979).

One hyena burrow had been dug in the same well depression where two years earlier I had placed a test excavation. The hyena burrow was angled downward so that the opening faced west; elephants, on the other hand, had always dug out their well at that spot such that its opening would face east. Thus, the hyenas had selected the path of their burrow through compacted sediments that had not been dug out before by elephants.

The test pit I had dug in 1985 had no bones in it, but I did find one plain body-sherd of sand-tempered pottery (Figure 4.15). The sherd was found at the transitional zone between light gray sand and the overlying sandy layers

Figure 4.14. The entrance to a spotted hyena burrow at Shabi Shabi. The bones visible in this photograph were disturbed when the burrow was dug into an infilled well containing a small elephant skeleton. Other bones (not shown) were carried to the burrow later by hyenas. Mandibles seem to be slightly favored for transport to burrows such as this one.

Figure 4.15. Plain body sherd recovered about 30 cm below the ground surface at Shabi Shabi.

containing abundant fragments of vegetation derived from elephant droppings. The light gray sand is the level at which water oozes into an excavation. Although the seeps are located in a part of southern Africa that is considered traditional Bushman country, and deeply stratified sites provide evidence for thousands of years of hunting and gathering by cultures very much like the modern San (Bushmen) (e.g., A. Brooks and Yellen 1977; C. Cooke 1967; Helgren and Brooks 1983), there is strong

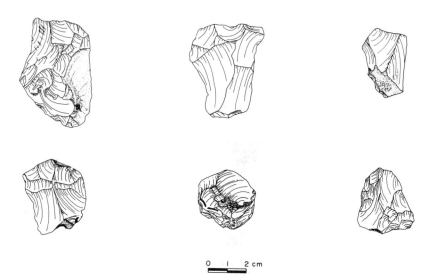

0 | 2 cm

Figure 4.16. Flakes and flake tools from Nehimba. Most are made of silcrete or chalcedony that has become heavily patinated. The specimen on the lower right is quartz. The specimen on the lower left has been flaked bipolarly.

evidence (Denbow 1984; Schrire 1980) that pastoralists and farmers inhabited the eastern Kalahari almost 1,500 years ago, and foraging people had adopted some of the more "settled" technology such as pottery making (Denbow 1984:184; Sampson 1988; Wilmsen 1989). Also, stone ruins (lesser "zimbabwes") can be found in the northern part of Hwange National Park; these and other Iron Age sites are dated to 200–1,000 years old and contain abundant pottery similar to the sherds found in the seeps (Fothergill 1953; Kearney 1907; K. Robinson 1966; Summers 1971).

Potsherds are not the only prehistoric artifacts found at the seeps. Another potsherd was found at Shakwanki, and the seep called Nehimba has yielded dozens of stone flakes, some from the ground surface and some in test pits dug to the water level (Figure 4.16). The stone artifacts were made of silcrete, chalcedony, or quartz, originally found in quarry sites, usually as nodules within calcrete deposits; Nehimba is much closer to large quarries than are the other important seeps, where flakes are rarer. It is difficult to determine if the other seeps contain stone tools in nonnegligible quantities, because ground surfaces usually are well vegetated or covered with trampled elephant dung. I have made numerous finds of single flakes and cores (some bipolar) at many seeps and seasonal pans in the park; some of the artifacts may be from the Middle Stone Age (MSA) (Sleigh 1970).

Thick bifacial points are relatively abundant at several calcrete pans (Figure 4.17).

These artifacts often were directly associated with elephant bones or carcasses, and even when not located within skeletal scatters they were still bedded in sediments that also contained bones. The upper 50 cm to 1 m of sediments at the seeps has been reworked by elephants, burrowing mammals and reptiles, water, wind, and other agencies for at least a century, and almost certainly longer. The typical stratigraphic column at a seep consists of a surface layer of trampled dung (masticated and clipped twigs and other vegetative parts) overlying a 5–10-cm-thick layer of dry, gray, fine to medium sand, below which is 50 cm to 1 m of a dark brown, nearly gray-black, fine to medium silty sand containing scattered lenses of darker and lighter sediments and numerous woody plant fragments. Below that is a moist, dark brown sand containing less silt, few plant fragments, and often stone tools. This layer's upper boundary usually is distinct and horizontal, and its lower boundary is always abrupt. The layer below is a waterlogged, pale gray sand. Water seeps into excavations at the upper boundary of this very clean sand. The textures of the various layers are similar – usually a fine to medium silty sand containing some organic matter. Because of frequent excavations and refilling, as well as burrowing and seasonal trampling, the seep areas at wells have complex cut-and-fill stratigraphy, giving the appearance of marble-cake swirls and patterning on both vertical and horizontal exposures.

The existence of one potsherd at Shabi Shabi and another at Shakwanki is not proof that pottery-using people visited the sites. Elephants sometimes ingest stones when they feed from the ground, such as at salt licks or in dry grasslands, and then pass those stones many kilometers away. Several times I have found a particular kind of crushed gravel (which is used in the construction of concrete troughs at some pans in the park's eastern area) over 50 km from the nearest possible source, the result of long-distance movements of elephants between one point where the rock was ingested and another where it was later passed.

The stone tools, on the other hand, are abundant at some seeps, especially Nehimba, and they certainly have resulted from human occu-pation, rather than elephant deposition. Darkly stained and spirally fractured mammalian limb bones have been found buried at the seeps, although the age of these old-appearing specimens cannot be determined short of dating all of them by radiometric methods. Field studies of burrowing rodents (such as springhares in the Kalahari) have shown that underground microenvironments possess stable and moderate tempera-tures, with relatively high humidity levels – ideal conditions for bone preservation (Butynski and Mattingly 1979). The buried bones may be as old as the stone tools, which could be hundreds or thousands of years, or

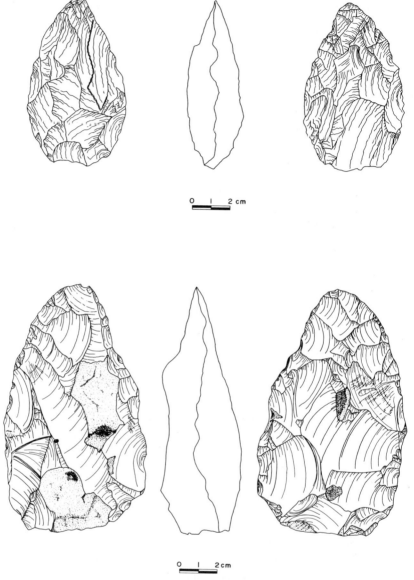

Figure 4.17. Two bifacial points from Hwange National Park. At the top is a white quartz hand ax from Dopi Pan; at the bottom is a silcrete hand ax recovered in road metal quarried between Manga Pan and Ngwahla Pan.

they may be much more recent. The stone artifacts often are patinated and bear many morphological similarities to MSA and Late Stone Age (LSA) assemblages discovered elsewhere in south central Africa (A. Brooks and Yellen 1977; C. Cooke 1967, 1976, 1979; Goodwin 1929; Volman 1984). Many "hand axes" have been found in Hwange near quarry sites where chalcedony or white quartz was obtained to make the flake tools, but the age of this type of implement need not be Acheulean or mid-Pleistocene, because thick bifaces were ubiquitous in several later stages or periods.

Regardless of the age of the flake tools, the point emphasized here is that they are now associated with massive bone accumulations that developed nonculturally. Kalahari sandbeds are known to "mix" stone artifacts by redistributing them at differential descending rates in the sand (Cahen and Moeyersons 1977; Moeyersons 1978); in other words, artifacts migrate downward at different velocities, depending on size, biogenic activity in the sediment column, and the frequency of wetting and drying cycles. Thus, archeological or nonarcheological materials such as wood, bone, charcoal, and stone tools of different ages may eventually end up redistributed vertically, perhaps associated with each other, within Kalahari beds (which cover an enormous part of central Africa) (H. Cooke 1957:Figure 10).

4.8. Bone fracturing, trample marking, weathering, and other modifications

Table 4.2 shows numbers and percentages of spirally fractured elements counted at two seeps in 1985, a year of rather low bone visibility in the grass, but no die-offs; in that year, elephant trampling was minimal because water was abundant throughout the park. In 1987 I made a partial count of bones in some parts of Shabi Shabi, and I found that the proportion of fractured elements had increased since the earlier survey. I could not completely count the bones because numerous unhappy elephants were present.

4.8.1. *Fracturing*

Fractured elephant limb bones were abundant in all die-off locales. Many fractures were of the so-called spiral type, characterized by helical fracture fronts, an absence of large right-angle offsets in fracture outlines (G. Haynes 1983b; Morlan 1980), and several other attributes described by E. Johnson (1985) (Figure 4.18). Most fractures were created by elephants trampling on elements partly buried in the surface sediments or lying atop the ground. In a few cases, hyenas had first gnawed off epiphyses, a finding especially common with bones from elephants whose epiphyses had not synostosed with shafts at the time of death.

In 1987, hyenas were denning at Shabi Shabi for the first time in a dry

Table 4.2.   *Elephant limb bones and tusk fragments
counted in 1985 at Shabi Shabi and Shakwanki seeps[a]*

Element	Unbroken	Broken	Total[b]
Shabi Shabi[c]			
Humerus	26	4	30
Femur	26	13	39
Tibia	18	0	18
Innominate	14	19	33
Scapula	6	18	24
Radius	4	4	8
Ulna	18	4	22
Tusk	—[d]	26	26
Total	112	88	200
Shakwanki[e]			
Long bone	8	13	21
Tusk	—	19	19
Total	8	32	40

[a]Note that tusks were collected from these sites before the counts were made.
[b]Total number of elephant bones (including elements not in this table) = 624.
[c]Shabi Shabi: surface area surveyed = 150 m × 50 m (7,500 m^2); MNI (minimum number of individuals) represented by bones = 19; actual MNI (from counted carcasses, 1982–5) = 31; percentage of long bones broken = 36% (62 of 174).
[d]None counted.
[e]Shakwanki: surface area = 40 m × 40 m (1,600 m^2); MNI represented = 3; actual MNI = 10; percentage of long bones broken = 62% (13 of 21).

season since I had begun visiting the site (Figure 4.14). Some fresh limb bones of subadult elephants had been gnawed into open-ended cylinders, a type of breakage that I had only rarely before seen in elephant bone assemblages (Figure 4.19). Cylinders were more often seen in bones from animals smaller than elephants, such as buffalo, both adults and subadults. The epiphyses of adult elephant bones, when they had fused to diaphyses before death, sometimes showed hyena tooth marks, but as a rule gnaw marking was uncommon in the die-off assemblages.

In the elephant die-off assemblages, hyena gnawing accounts for the presence of cylinders virtually 100 percent of the time, whereas spirally fractured limb bones are created by trampling far more often than by carnivore gnawing. For comparison, Figure 4.20 shows the percentages of bison bones broken by trampling and gnawing in modern bone

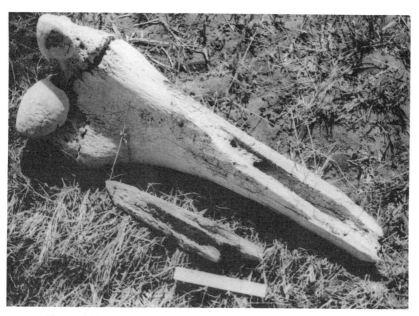

Figure 4.18. A spirally fractured elephant femur, broken by trampling at Shabi Shabi. The scale bar at bottom is 15 cm long.

Figure 4.19. Limb-bone shaft from a very young elephant, broken by spotted hyena into a cylinder with open ends.

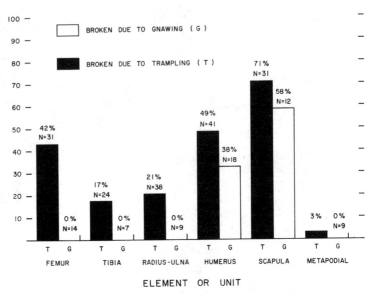

Figure 4.20. Comparison of the proportions of fresh *Bison bison* bones in Canada broken by wolves and by bison trampling. Sample sizes per element differ because of the scattering or removal of some from different sites. Drawn by R. Lewis.

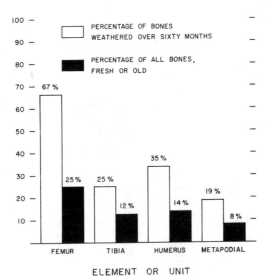

Figure 4.21. The proportions of *Bison* long bones that were broken five or more years after death compared with the proportions of any bones, fresh or weathered, that were found broken. Drawn by R. Lewis.

assemblages examined in field studies in Canada (G. Haynes 1981, 1982, 1983a, b).

Although the findings have not been quantified, it seems apparent that elephant bones lying on ground surfaces in the die-off sites are broken in increasing proportions over time. I have found this generalization to be true for bones in many different habitats and environments. Figure 4.21 shows percentages of bison bones in North America that were fractured, comparing assemblages that had been deposited for different lengths of time. In the case of the elephant bones at the die-off locales, the fracturing was mainly spiral, because the bones in general had been well protected from weathering by burial in thick vegetation or moist sediments. Many bones were buried in wet sediments that remained cool nearly year-round, thus preserving them in a state that allowed spiral (rather than dry-bone) fracturing.

A simple experiment will confirm that naturally degreased dry bone that has not weathered past stage 2 or a stage-1–2 phase (Behrensmeyer 1978) may be capable of breaking in a form similar to that of very fresh bone – in other words, with helical fracture fronts ("spiral fracturing") creating relatively smooth, pebbly fracture surfaces indicative of cleaved (not longitudinally split) collagen bundles.

If such degreased bone has been only somewhat degraded by partial decay of the organic component, the addition of moisture to the remaining organic and inorganic structure will restore enough flexibility to permit spiral fracturing, as opposed to the linear or longitudinal fracturing that often occurs when bones are very brittle because of greater loss of organic material and moisture. Apparently, added moisture causes incipient microscopic split-line cracks to close up solidly, thus allowing fracture fronts to move *across* them and not be redirected *along* them, thereby resulting in spiral rather than linear fractures.

This microscopic effect of rewetting has not been empirically verified. Rewetted bone increases in denseness, as shown by its increased weight. Rewetting may also cause whole long bones that floated when dry to sink when soaked in water for several minutes. The sinking is not simply due to water flooding into narrow cavities, because such bones may sink immediately even after cavities are drained of water through their nutrient foramina.

Because the initial conditions of postmorten weathering undergone by a single bone or parts of one bone can vary tremendously within a site, it is not yet possible to isolate the specific ranges of temperature and humidity, length of sunlight exposure (hence ultraviolet radiation), and cycling times for changes in these variables that correlate with spiral-fracturing capabilities. Many bones in the African deposits under study had been subjected to occasional freeze–thaw cycles, wet–dry cycles, and strong-sunlight–

Figure 4.22. Two elephant bone flakes that refit onto the shaft specimen in the center, broken off by spotted hyenas.

deep-shade cycles for up to three years, and when trampled, chewed, or struck, they still fractured with spiral morphology. Many bones in north central Canada (G. Haynes 1982, 1988c) have been spirally broken ten years after original deposition because they had lain only in thick vegetation and mud that had not leached out sufficient organic material to turn them brittle, or sufficient inorganic material to turn them chalky and too soft to split apart when impacted by trampling hooves.

Spirally fractured elephant limb bones can be trampled by animals visiting the sites, thus further fracturing them. In some cases, these trampled fragments bear notched edges that appear similar to the impact damage created by humans who break open long bones for the marrow within.

I have recovered a number of fragments of elephant limb bones that appear similar in many ways to flake cores (Figure 4.22). These kinds of

Figure 4.23. Traumatically fractured tibia from a giraffe (distal end to the right), showing a transverse fracture and the presence of spiral fracture-front cracks and rounded fracture edges. The rounding probably resulted from rubbing of the edges against other tissue before death.

specimens were created both by carnivores gnawing bones and by trampling. Carnivores seem able to create the core mimics out of angular bones such as scapulae and ulnae, as well as out of cylindrical bones. The aim of breaking back heavy elephant bones is to obtain greasy fragments that can then be swallowed, as well as to reach farther down into the interior of bones in order to eat oily cancellous tissue.

With cylindrical bones such as femora, carnivores first gnaw or bite off an epiphysis, then grasp a part of the remaining edge in the teeth and pull back until a part of the edge or a flake from the bone-shaft surface detaches. Bone flakes and spirally fractured fragments may be 30 cm long or more. With angular bones, such as the proximal ulna, a protuberance is grasped in the teeth and pulled strongly until it detaches. The bones flaked into core mimics most often are limb bones from elephants whose epiphyses had not fused before death; note that some epiphyses in elephants remain unfused well into the fifth decade of life (see the Appendix).

Traumatic fracturing during life may contribute a number of spirally fractured limb bones and other elements to modern deposits. Giraffe bones are especially prone to traumatic fracture. In Hwange, giraffes are plentiful, and during certain seasons lions select them as prey. Of the thirty giraffe

Figure 4.24. Fragments of the shaft of a traumatically fractured elephant femur, showing several rounded edges. The scale is 15 cm long.

carcasses or skeletons that I have examined in Hwange National Park, eight had suffered a broken limb bone prior to death; in some cases the fracture may have occurred a considerable time before death, because fracture edges either had been rubbed smooth during attempts to use the limb or had been resorbed and had become smooth (Figure 4.23). I have found four elephant skeletons or carcasses with premortem limb fractures (Figures 4.24 and 4.25) and a few others whose fractures had begun the process of healing before death.

4.8.2. Gnaw damage

The degree of damage attributable to carnivore teeth is extremely variable in the die-off sites from year to year and from location to location. It varies from very light to very heavy. For example, in 1987 an adult male sable antelope (*Hippotragus niger*) died at Shabi Shabi, and its flesh, viscera, and skin were consumed by jackals (*Canis* spp.) and spotted hyenas (*Crocuta crocuta*). Its horn tips had been chewed off, and yet greasy limb bones remained nearby, as did the nearly complete fresh carcass of an elephant cow and at least six more partly eaten fresh carcasses of smaller elephants. The chewing of the antelope horns seems to have been an inefficient way to

Figure 4.25. View of part of a traumatically fractured elephant ulna. The left margin of the shaft probably was rounded by rubbing against other tissue before death.

Figure 4.26. Carcass of an old female lying between elephant-dug wells at Shabi Shabi, about five days after death in 1983. Note how the head is partially buried by sediments thrown out during elephant excavations to enlarge well openings. The carcass has been nearly ignored by carnivorous scavengers.

get food, especially in light of the abundance of other food within sight. I assume that carnivores competed for access to this sable's carcass and utilized it rather fully until the elephants died, at which time it was abandoned in favor of the more abundant and fresher sources of food. Yet the well-utilized sable's bones are penecontemporaneous with the poorly utilized elephants' bones in Shabi Shabi, a potential source of interpretive confusion to some future analyst.

It is important to note that although the degrees of gnaw damage on bones in the assemblage were variable, the sequence of damage done by carnivores was very predictable. In other words, the order in which specific elements were damaged or eaten was regular and nonrandom, much as it is with smaller animals such as bison or moose in North America (G. Haynes 1982). In fact, the sequences of damage stages on elephant and bison bones are much alike in spite of the differences in bone sizes and the types of carnivores inflicting the damage.

White dung from hyenas was abundant at Shabi Shabi in 1987; it had always been rare or nonexistent before. White dung appears to be made up mainly of digested bone, indicating very full feeding utilization of carcasses. Several latrine areas had been created at Shabi Shabi, and regurgitated hairballs were plentiful. In previous years I had rarely been able to track down such hyena-deposited materials, in spite of a strong effort to do so. During earlier dry seasons, so many elephant carcasses had been available

SKIN

Figure 4.27. Skeletons of two elephants, including the old female seen in Figure 4.26 (left) and a very small calf (center), mapped in Shabi Shabi, 1985, two years after death. The tusks and the lower jaw of the old female were collected in 1983. Note that stomach contents are preserved within the bones, and the skeletons are still in close anatomical order, a result of very poor carcass use by scavengers. Map drawn by E. Paige.

to the hyenas of the region that they had not eaten bone and fur or fully utilized the dead animal resources (Figures 4.26–4.29). However, in 1985 and 1986 there were no die-offs, and it is possible that a population eruption of hyenas that had benefited from the earlier mass deaths was in 1985–6 forcing the hyenas to compete fiercely for food.

Figure 4.28. Two elephant skeletons lying among trees near the Shabi Shabi channel, mapped in 1985. The elephant lying more to the right died in 1983, and the one more to the left died in 1982. Tusks and lower jaws were collected by park officials. Drawn by E. Paige.

4.8.3.   *Miscellaneous agencies of bone modification*

Bone chewing by herbivores has been observed several times in Hwange National Park and other places in Africa and North America. Giraffes and other ungulates in the Shakwanki wilderness area often search among skeletons of elephants and other animals for bones to chew. The searching takes the form of standing within bone scatters and moving elements around with the feet, or taking elements in the mouth and picking them up

Figure 4.29. Map of bone scatter in the central part of Nehimba, 1985. Bones at the lower left are mostly from an old cow that died in 1983. Other bones came from several subadults that died in 1982 and 1983. The "seep area" is a pool of water oozing up from underground. Note how the bone scatter has been limited because of light scavenger activity and elephant trampling. Drawn by E. Paige.

Figure 4.30. Four fragments of bovid bones chewed by giraffes in Hwange National Park. Note the grooving and crushing.

while trying to place them between the teeth for chewing . Aside from very heavy and large elements such as skulls, the bones selected for chewing may be practically any size and shape; eventually they are reduced to small fragments or are fully eaten. Well-chewed bones have crushed or extremely abraded and polished surfaces pitted with teeth marks, as well as rounded edges, and often they are bright white from cleaning in the mouth (Figure 4.30). Less well chewed bones are less obviously affected, except for deep grooves cut into the surfaces (Figure 4.31). Very fresh, greasy bones are rarely, if ever, chewed by herbivores, but defleshed and slightly greasy elements may be selected, as are weathered and chalky elements. Tusks may also be chewed, and in some cases eaten entirely. For comparison,

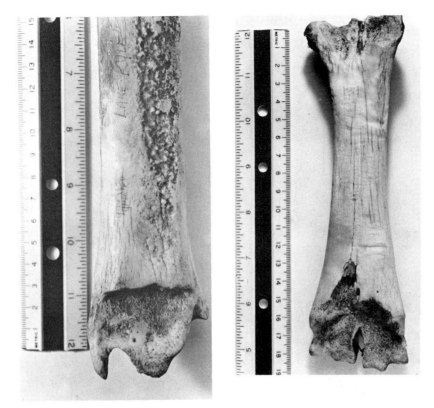

Figure 4.31. (a). Distal end of a tibia of wildebeeste (*Connochaetes taurinus*) from Lake Kyle, Zimbabwe, deeply grooved by giraffe chewing. The shaft also shows the effects of dissolution in water. (b). Wildebeeste metacarpal from Lake Kyle, grooved, crushed, incised, and ground by giraffe chewing.

Figure 4.32 shows *Camelus* and *Bos* bones eaten by feral *Camelus* in central Australia. Note the similarities to bones chewed by giraffes.

Some insects, especially termites, eat bones (Behrensmeyer 1978; Derry 1911; Thorne and Kimsey 1983; Watson and Abbey 1986), although termite damage has not been documented on bones specifically in the die-off locales. Termites lay down thin layers of earth or sand over bone elements, cementing the particles with body secretions. The earthen layers may be continuous sheets (called cartons) or, more often in the case of bones, hollow networking tunnels (called galleries). Termite damage appears as numerous roughened pits and grooves of varying widths and depths, probably depending on the taxon of termite. This damage in some ways is similar to the tooth pitting produced by large mammalian carnivores.

Figure 4.32. Fragments of bones of *Bos* (cattle) and *Camelus* (camel) chewed by wild-roaming camels in central Australia. Note the grooving and crushing of surfaces.

Figures 4.33 and 4.34 show elephant bones with cortical surfaces partly eaten by insects (possibly *Pseudacanthotermes militaris*) in Hwange National Park. Insects such as termites seem to prefer relatively fresh bones that are somewhat degreased but not badly weathered. The attraction of bone probably is related to seasonal nutritional needs, such as during the season when progeny are being fed, prior to their nuptial-flight dispersal (Thorne and Kimsey 1983).

Grass fires can create localized damage on bone surfaces where grease, cartilage, or ligaments have smoldered. At first the burned places are

Figure 4.33. Ventral surface of an elephant patella showing termite damage. Note similarities to carnivore tooth impressions.

blackened and crumbly, but this brittle surface breaks off, leaving a white or gray patch of bone with the characteristics of tissue in an extremely advanced state of weathering. Thus, it is possible to find single bone elements showing weathering stages from 0 to 5 (Behrensmeyer 1978) (Figure 4.35). One entire side of a bone or one very localized patch may be affected by fires in this way.

Another highly localized type of surface modification produced in die-off locales is trample marking; many bones,when trampled, are scratched with deceptive cut-mark mimics. These marks may be single, sharply incised scratches or sets of scratches that sometimes appear preferentially (not necessarily randomly) oriented (Figures 4.36–4.38), and therefore easily mistaken for butchering cut marks (E. Johnson and Shipman 1986; Potts and Shipman 1981; Shipman and Rose 1983). Figure 4.39 shows photomicrographs made with a scanning electron microscope (SEM). Edge rounding may also be highly localized on fractured bones, resulting from partial protection of fracture edges by burial or resulting from trampling that impacts only parts of edges.

Elephants are inquisitive and exploratory animals, and they use their feet to manipulate objects on the ground. At die-off locales the most numerous objects that invite curiosity are the bones of other elephants; see the photos

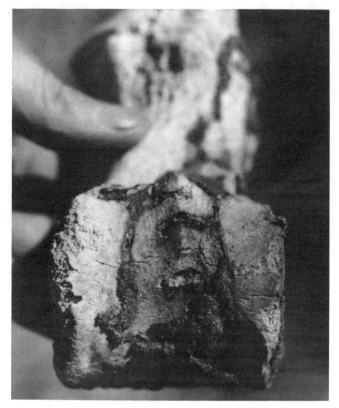

Figure 4.34. Distal end of an elephant metapodial showing termite damage and adhering galleries formed of sediments cemented by termite secretions.

of elephants investigating bones in the publications by Douglas-Hamilton and Douglas-Hamilton (1975) and Moss (1988). Several bones at the seeps had been kicked and pushed by elephants using their feet, and those bones possessed very localized "trample marks" on surfaces that could not have been walked on or actually trampled. For example, the back of one elephant cranium had been scratched above the condyles (Figure 4.40); in another example, a skull's tusk alveoli had been scratched. Such manipulation marks may follow the contours of the bone, or they may skip over depressions on bone surfaces, just as do cut marks made by stone tools.

4.9.   A word on "experimental" studies

Olsen and Shipman (1988) reported on several "experiments" to see if true cut marks made by stone tools could be distinguished from marks made by

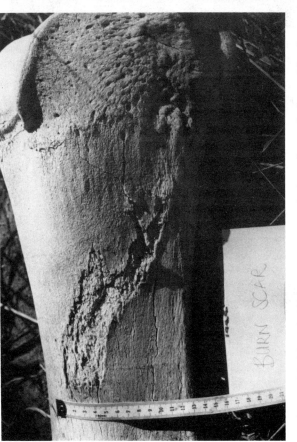

Figure 4.35. (a). Isolated burn scar on an elephant femur shaft, resulting from a patch of dried periosteal soft tissue smoldering during a low-heat grass fire. (b). Posterior surface of an elephant tibia showing a very crumbly patch left after a grass fire ignited grease soaked into the compacta.

Figure 4.36. Elephant stylohyoid bone recovered from a buried context at Nehimba. Note the parallel incisions on the main body (just left of center), the results of trampling by elephants.

Figure 4.37. The head of a femur of a lion (*Panthera leo*) found at Shabi Shabi. Note the fine parallel incisions, which are trample marks.

Figure 4.38. An elephant's cervical vertebra found partly buried at Nehimba. Note the two sharply incised trample marks on the centrum plate.

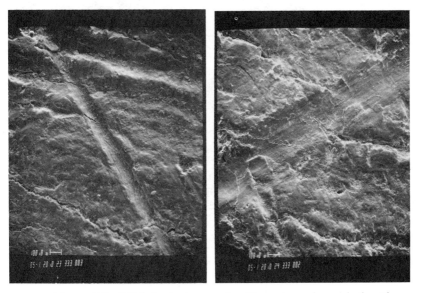

Figure 4.39. SEM photomicrographs of trample marks on elephant bone from Nehimba. Note the parallel, longitudinal striations within the single marks, and the raised edge on one side of each mark.

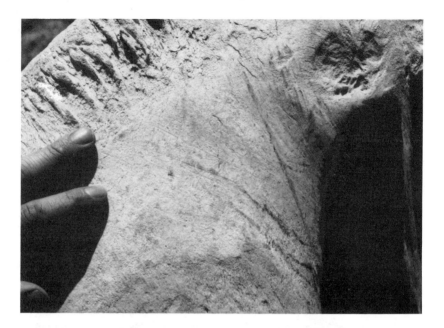

Figure 4.40. (a). The rear of an elephant skull from Makololo 2, in Hwange National Park. The marks resulted from an elephant manipulating the skull using its feet. Note the sharply incised parallel marks on the surface. (b). Closer view of some of the manipulation marks on the rear of the elephant skull.

trampling or sedimentary abrasion, and they concluded that there were significant differences between surface modifications created by the different agencies. However, although their conclusions may prove to be valid, their experiments seem to have been biased in several ways. First, they apparently assumed that all trampled bone assemblages share the same characteristics, such as having (1) uncountably high numbers of scratch marks on each bone, (2) very fine and shallow scratch marks, (3) randomly oriented scratch marks, and (4) polish on bones in association with the scratch marks. Trample-marked bones in the modern elephant assemblages described here infrequently conformed to these expectations.

A second bias in Olsen and Shipman's experiments was that they compared a sample of trample-marked bones to a sample of "butchered" bones that were in fact *not* butchered, but were deliberately cut with stone tools in order to make visible marks. The cut marking was not created as a byproduct of normal butchering procedures and hence may not have produced valid or realistic examples for use as norms in describing actual cut marks on bones. The experimentally created cut marks may have been much larger and deeper and more sharply incised than are many true cut marks made by humans who habitually butcher game animals. Olsen and Shipman have not demonstrated how their standardized cut marks can be replicated by anyone else desiring to create such marks – that is, they have not described how their experimental tools were shaped, applied to the bone surface, and moved – nor how those actions differ from the actions of experienced butchers more interested in extracting resources from an animal carcass (and perhaps saving tool edges, as well) than in marking up bone surfaces. Their conclusions about the differences between cut marks and trample marks should perhaps be cautiously reexamined under more stringent control of the potential variables.

Their study also suffered from inadequate reportage about how much searching was involved in finding SEM photo images that differentiated their experimentally created cut marks from experimentally created trample marks. How many hours were spent in examination of all marks on all specimens? Were all marks examined equally carefully? Were *similarities* between marks inadvertently ignored in favor of the *differences* that were being sought?

It should also be pointed out that the examination of such a small number of actual cut marks probably has created an unusually leptokurtic sample, where central-range values are very narrowly defined and most of the observed attributes are on one end or the other of a frequency graph (i.e., they are either very rare or very common in larger samples). Small samples may be useful for gross differentiation, but not for fine distinctions between agencies.

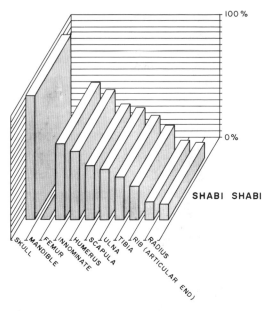

Figure 4.41. Bone representation at Shabi Shabi, determined by dividing the expected number of each element by the actual number found.

4.10. Bone representation

During 1982 and 1983, all elephant mandibles and tusks were gathered from death sites and placed in my field campsite in order to determine the ages of dead animals. After 1983 I left all mandibles and tusks in the original death sites. No other bones were removed from any carcass or skeleton, except by noncultural agencies such as hyenas and elephants.

I made several 100 percent walkover counts of all surface bones at the seeps (see Table 4.2 for counts at Shabi Shabi and Shakwanki). Bone representations at all seeps were closely similar – skulls represented MNIs (minimum numbers of individuals) most fairly, followed by femora and innominates, humeri, scapulae, and so forth (tusks and mandibles would also have been relatively well represented had they not been collected prior to 1983) (Figure 4.41).

Bones were subtracted from these assemblages by burial, by scavenging carnivores taking them away, or by elephants carrying them off. In a very few instances, spotted hyenas ate nearly entire skeletons of certain individual animals that died early in die-off periods, before carcasses became plentiful.

Natural burial in the channel bottom was fairly common, in some cases involving articulated skeletons or body parts (Figures 4.11 and 4.12). The buried bones, unless fortuitously found during my sampling and surveying, were, of course, uncounted.

4.11. Demography of the die-offs

Table 4.3 lists the numbers of elephants that died at the main seeps during drought years, as counted in the field at the end of each dry season by park officers exploring the areas.

Table 4.4 shows the minimum numbers of individuls of all taxa found dead at the four main seeps. These totals are not based on bone representation at the seeps, but on actual counts of carcasses made at the end of each dry season between 1981 and 1987.

Table 4.5 shows the representation of taxa and individuals based on counts of surviving bones remaining at the central parts of four sites in 1986, 1987, and 1988.

Table 4.6 illustrates 1987 elephant bone counts at the main loci of the two largest seep areas, Shabi Shabi and Lememba, which are about 4 km apart.

Table 4.7 shows the "simplified" age structure of Hwange's drought die-offs at the main seeps. This type of information was provided in much more detail by G. Haynes (1985b) and Conybeare and Haynes (1984); however, I have found that the two-year age classes reported in earlier studies are difficult to assign to some skeletons, and that in fact reporting only by twelve-year classes is adequate. These twelve-year age classes separate important intervals in the life history of the elephant (see Chapter 3).

Figure 4.42 presents a comparison of the age profile created in the Shabi Shabi die-off (central area) and the age profile of a stable elephant population derived from data reported by Pilgram and Western (1986), T. Pilgram (1986 personal communication), and G. Haynes (1987c) (see also Chapter 3).

4.12. Discussion

I have chosen to discuss more fully only those bones counted in the main die-off locales (Tables 4.5–4.7) because this subassemblage has the highest chance of fossilization and future discovery. Also note that I discuss only bones found on the ground surface; many bones are buried at the seeps, as I have discovered from test-pit excavations at several loci within the main die-off areas. The percentage of excavated areas is negligible compared with unexcavated areas, and it is therefore impossible to predict numbers of bones (and MNIs) buried in the deposits, because carcasses and skeletons are nonrandomly distributed.

Table 4.3.   *Dead elephants counted at main seeps in Hwange National Park*

Year of death	Shabi Shabi	Lememba	Nehimba	Shakwanki	Border seeps[a]	Total
1968	—[b]	—	—	—	—	110[b]
1970	—	—	—	—	5	5(?)
1971	1	—	3	1	2	7(?)
1981	1	—	—	1	0	18(?)
1982	147	10	16	5	0	178
1983	44	10	6	3	0	63
1984	1	1	0	1	0	3
1985	1	0	0	0	0	1
1986	0	0	0	0	0	0
1987	21	0	29	1	0	51
Total (minimum)	216	21	54	12	7	436

[a] Border seeps were Tamafupa, Tamasanka, Ngerewa, Nqwasha, and Domshetshe.
[b] Dashes do not indicate that no animals died that year, only that no counts were made or that totals from different seeps were not recorded separately.

Table 4.4.   *Minimum numbers of individual animals that died at four seeps (1981–8)*

Taxon	Shabi Shabi	Lememba	Shakwanki	Nehimba
Loxodonta africana (elephant)	215	21	12	51
Ceratotherium simum (white rhino)	—[a]	—	—	1
Syncerus caffer (buffalo)	8	4	1	2
Equus burchelli (zebra)	2	2	—	—
Tragelaphus strepsiceros (kudu)	2	4	—	1
Giraffa camelopardalis (giraffe)	—	3	1	—
Phacochoerus aethiopicus (warthog)	—	—	1	1
Panthera leo (lion)	1	—	—	—
Taurotragus oryx (eland)	1	—	—	—
Hippotragus niger (sable)	2	—	1	1
Raphicerus campestris (steenbok)	1	—	—	—
Crocuta crocuta (hyena)	1	1	—	1
Canis mesomelas (jackal)	—	3	—	—
Orycteropus afer (antbear)	—	—	—	1
Manis gigantea (pangolin)	1	—	—	—
Large bird	5	2	—	—
Pelomedusa subrufa (terrapin)	4	2	1	1
Testudo (tortoise)	1	2	1	1
Total MNI	244	45	18	61
Taxa represented	13	10	7	10

[a] No deaths.

Table 4.5. *Bone counts and MNIs in the central parts of the four seeps (1986–8)*

Site	Area of highest carcass density (m)	Causes of deaths, dates of input	Number of taxa represented	Number of bones	MNI (all taxa) (from bone counts)
Shabi Shabi, main locus, (east)	150 × 50	Drought & starvation, 1981–3, 1987 (very little predation)	6	> 1,000[a]	43
Shakwanki (main basin)	20 × 20	Drought & starvation, 1981–3, 1987	3	49[a]	8
Lememba (NW pothole)	12 × 12	Drought & starvation, 1981–3	2	48[a]	5
Nehimba (main basin)	20 × 20	Drought & starvation, 1981–3	8	171[b]	19

[a] Many elephant tusks and mandibles were not counted, because they had been collected earlier.
[b] Area was flooded after deaths, so the total number of bones counted was very low compared with the total actually present.

Table 4.6.   *Minimum numbers of elephants at main parts of Shabi Shabi and Lememba, determined from bone counts during 1987*

Element	Shabi Shabi[a]		Lememba[b]	
	No. bones	MNI	No. bones	MNI
Skull	61	23	5	2
Mandible	5[c]			
Vertebra	207	?	—[d]	—
Hyoid	5	3	—	—
Rib	250	?	—	—
Scapula	18	9	2	1
Humerus	31	16	7	4
Radius	11	6	3	2
Ulna	25	13	2	2
Innominate	39	23	1	1
Femur	41	22	2	2
Tibia	20	11	1	1
Lower-leg/foot bones	88	27	—	—

[a] Area surveyed = 150 × 50 m.
[b] Area surveyed = 12 × 12 m.
[c] All lower jaws were collected from carcasses when they were discovered, in order to determine ages.
[d] None counted.

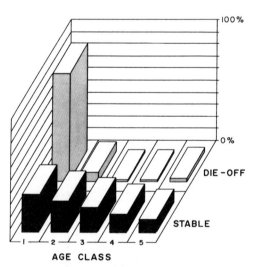

Figure 4.42. Comparison of the age profile from the Shabi Shabi 1980s die-off and the age profile expected in a stable elephant population.

Table 4.7. *Age structure of the death sample, 1982–7 die-offs*

Site and age interval (years)	1987	1985	1984	1983	1982	Total by age interval
Shabi Shabi						
0–12	19	0	0	34	138	191
13–24	0	1	0	8	5	14
25–36	0	0	0	0	3	3
37–48	0	0	1	1	1	3
49–60	2	0	0	1	0	3
Lememba						
0–12	0	0	1	7	10	18
13–24	0	0	0	2	0	2
25–36	0	0	0	0	0	0
37–48	0	0	0	0	0	0
49–60	0	0	0	1	0	1
Nehimba						
0–12	28	0	0	4	13	45
13–24	0	0	0	1	2	3
25–36	0	0	0	0	0	0
37–48	0	0	0	0	1	1
49–60	1	0	0	1	0	2
Shakwanki						
0–12	1	0	0	3	5	9
13–24	0	0	0	0	0	0
25–36	0	0	0	0	0	0
37–48	0	0	0	0	0	0
49–60	0	0	1	0	0	1
Totals	51	1	3	63	178	296

4.12.1. Demography

Figure 4.43 is a visual presentation of the age profile for the central Shabi Shabi death assemblage (based on carcass counts at the time of the die-offs, not on later bone counts) compared with the age profile for the entire death sample. The proportions of animals in each age class are closely similar in the two cases, so the central die-off subsample (recorded at the time of death) is fairly representative of the entire sample.

Comparison of Tables 4.5 and 4.6 shows that the numbers of dead animals represented by bones remaining on ground surface at Shabi Shabi and Lememba may have been far lower than the numbers that actually died in the main parts of these seeps. For example, there are only six elephants represented by radii at Shabi Shabi, and yet twenty-four elephants died there between 1981 and 1988. Skulls and skull fragments, innominates, and femora represent most or all animals that died, but no other elements fairly

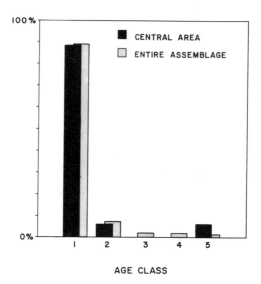

Figure 4.43. The age profile from only the central area of Shabi Shabi compared with the age profile from the entire die-off assemblage (central area plus all surrounding areas).

represent the true number of deaths. Figure 4.41 illustrates "representation" at the site, or the numbers of each element surviving compared with the numbers expected.

Currently there are no reliable methods for determining age profiles from analysis of postcranial *Loxodonta* elements. The category "Skull" in Table 4.6 includes fragments that were not teeth-bearing, so even this element is not adequate for deriving a meaningful age profile. If the scheduling of epiphyseal fusion were known in detail for all postcranial and cranial elements of the proboscidean skeleton, it would be possible to use bones other than teeth to determine age profiles. Roth (1982, 1984) has contributed to our knowledge of skeletal maturation rates, and the Appendix describes a limited sample and first approximation of fusion scheduling for Hwange elephants. Using these data, it is possible to assign a vague age profile to the Shabi Shabi bones: Of the twenty-three individuals represented, two were adults over about thirty-two years of age, and the rest were under eighteen years of age.

This "age profile" is information-poor, but it does indicate that relatively young animals predominate in the surviving bone sample. The most useful and most detailed age profiles must be determined on the basis of dentition; however, because most lower jaws had been collected from the seeps at the time of the die-offs, it is difficult to judge whether or not the age profile of the bone assemblage is fairly representative of the actual die-off sample. In the absence of lower jaws, the upper teeth of at least nineteen individual elephants remaining on the ground surfaces represent two animals over forty-eight years old (11 percent of the total), one animal under eighteen but over twelve years old (5 percent of the total), and sixteen animals under twelve years old (84 percent of the total). This age profile, though lacking detail, is indeed similar to the age profile for the actual die-off assemblage.

The oldest female that died in the main part of Shabi Shabi was aged about fifty-five years, based on criteria of tooth eruption and wear; in fact, there are bones of two fifty-five-year-old females within the central die-off area. If we consider the "group" of dead elephants represented by bones within this area to have been members of a single herd that died together, then the existence of two very old females should indicate that the original herd visiting the seeps may have been larger than the death sample that was found. In Chapter 3 I described age structures from a large sample of culled elephants and pointed out that whenever groups with fewer than thirty members were led by very old matriarchs (i.e., matriarchs over about fifty years of age), there were no recorded cases where a second very old female was also present in the herd. In those cases in the culling when very old matriarchs were leading medium or small groups, the next oldest female was about thirteen to twenty years younger than the matriarch. On the other hand, in cases where more than one very old female was present in a mixed herd, the group size was greater than thirty members, and usually greater than forty members.

Hence, if we view the Shabi Shabi assemblage as we would view a fossil assemblage, the presence of two fifty-five-year-old females in a death assemblage of twenty-three elephants is an anomaly.

The "group" that did visit Shabi Shabi was much larger than twenty-three members, of course. In fact, over 200 elephants and other animals died around this seep (Tables 4.3 and 4.4), although the bones of many were not deposited in the central part, where the chances of burial and preservation were optimal. If the central part of Shabi Shabi were ever discovered as a fossil deposit, the ages of the oldest females should alert analysts to the possibilities (1) that far more elephants died at this site than were preserved, (2) that a large number of herd members survived the death event that created the assemblage, or (3) that the assemblage has been time-averaged and reflects deaths from different causes.

The oldest female in the Nehimba bone assemblage was also in her middle fifties when she died; the assemblage of bones remaining at the center of Nehimba represents a total of only six elephants. The age difference between this female and the next oldest one at the site was about forty years (G. Haynes unpublished data). This age difference seems somewhat excessive, but in theory it does fit the model derived from analyses of the Hwange cull data (Chapter 3): When very old matriarchs lead small herds, the next oldest female is young enough to be a daughter or granddaughter.

However, further analysis of the Nehimba age structure in Table 4.7 (G. Haynes unpublished data) shows that this bone assemblage cannot possibly represent a complete herd, because there are so many sexually immature animals represented and not enough adults that could have borne or nursed them. Therefore, this assemblage's age structure also should alert future analysts that not all the elephants present at this site died during the death events that created the bone assemblage.

4.12.2. *Bulls at the seeps and differential vulnerability of the sexes*

There was only one adult male elephant in all the seeps die-off assemblages, and that animal had been shot in Botswana at the end of the 1982 dry season. It walked to Shabi Shabi and died there, about 300 m from the main die-off area. Although no other adult males died at the seeps, substantial numbers were present in the Shakwanki wilderness area each dry season during the die-offs. Adult males initiated well-digging at the seeps in the middle of the dry season, even when surface water still remained in nearby seasonal pans.

In 1985 and 1986, no seeps were dug out for water because the vegetation was so lush and surface water could be found up until September. In 1984, bulls dug out wells at Lememba in July, although two large seasonal pans still held a great deal of water about 200 m away. Pan water was muddied and fouled by animals bathing, walking, and defecating in it, and it was also quite warmed by the sun each day. Elephant bulls would wallow and bathe in these pans, but then would walk to wells that they had dug in order to drink cool, clean water. There is a lesson here for interpreting fossil assemblages: Elephants will dig holes for water even when it is available in ponds or streams. Elephants prefer drinking clean water at springs, although they may have to make an effort to do so. Deraniyagala (1955:65) also noted Asian elephants drinking from pits they had taken the trouble to dig in sandbanks adjacent to flowing streams.

Between 1982 and 1987, Shakwanki seemed to be visited by bull herds more often than by mixed herds, although mixed herds watered there and lost members that died during the drought seasons. Shakwanki is located relatively nearer to pans that are supplied from pumped boreholes than are most other seeps. The nearness of Shakwanki to other water and the preponderance of less vulnerable bulls in the subpopulation using the seep perhaps account for the smaller die-off assemblage at Shakwanki than elsewhere in the wilderness area.

Nehimba is also located relatively near to pans that are pumped in the dry season, although mixed herds predominate in the subpopulation depending on this seep. The death sample at Nehimba is unexpectedly large

Figure 4.44. Adult female with a broken or dislocated right wrist and a
starving calf trying to nurse, Shabi Shabi, 1987.

when its nearness to other water sources (about 30 km) is considered.
Clearly, more than a simple lack of water was contributing to the deaths of
elephants. The main cause of death was starvation. The high-quality
browse found in the Kalahari sand region around the seeps is preferred feed
during the dry season, in spite of the scarcity of water, but it is a limited
resource that cannot sustain heavy elephant pressure as the season
progresses.

Being larger, adult males can compete successfully against adult females
and subadults for access to water. They are also capable of traveling farther
and faster to seek food, unlike mixed herds, which are hampered by the
presence of young and very small elephants unable to travel long distances
quickly. The young elephants in mixed herds must nurse from time to time,
an activity that is not possible while cows are walking, because an
elephant's two mammaries are located under the forelegs, in a position
analogous to human breasts. Cows that must walk long distances to find
food create additional hardship for their unweaned calves (Figure 4.44).
Bodenheimer (1957) reported that reduced forage in semiarid Namibia
resulted in reduced lactation by elephants and other animals. Starving
adult females thus might be inadvertently starving their nursing calves. The
nutrient quality of milk drops significantly during drought.

Elephant cows with nursing young often are separated from the rest of

Figure 4.45. An adult bull standing in a well at Shabi Shabi, with the fresh carcass of a calf lying in front of it, 1987. Note the slack skin, especially behind the foreleg. Two other bulls are drinking in wells in the background.

the herd, and it is quite common to find single cow-calf units at the seeps searching for water after the middle of the dry season. This kind of fissioning of familial social groups probably adds greatly to the stress and trauma of drought survival. Moss (1988:56–7) observed separation of cows with youngest calves during drought in Amboseli National Park in Kenya. Old cows that are nursing may die during drought, leaving their calves to die after lingering a few days in terror, if predators do not kill them at once. In some cases I have seen lactating cows without calves in tow, resulting from the calves' death or abandonment by the mother.

Even though adult males survive the drought die-offs at the seeps, they show clear signs of stress during severe drought; usually they lose condition, shown by slackening of flesh and sharply defined protuberance of bones under the skin (Figure 4.45). Perhaps the absence of bulls in the death assemblages is related to their ability to travel faster and farther and to win access to water after competitive encounters with mixed herds; there may also be a much lower density of adult males present in the die-off areas, with the result that far fewer adult males die there. Thus, a sampling error could account for the absence of bulls in the die-off sample.

The carcasses of a few old males have been found elsewhere in Hwange National Park, many of which had died during unusually dry years. Yet all of them had died near permanent water. Perhaps weakened males abandon the seeps before death and leave their carcasses in different regions than do adult females and subadults.

R. Martin (1978) found that adult male elephants in parts of Zimbabwe maintained home ranges that overlapped the ranges of mixed herds, but that were distinct. A. Conybeare (1987 personal communication) has suggested that bulls spend the dry season in different parts of ranges than do mixed herds. Based on these generalizations, it can be postulated that because bulls feed and live in different areas than do mixed herds, at least for part of the year, there may exist separate and distinctive all-bull death assemblages in places apart from the sites where mixed herds produce death assemblages. This possibility is being explored through surveys of more Kalahari sandveld and surrounding regions in and out of Hwange National Park.

Like male elephants, the adult males of other large-mammal taxa are relatively less vulnerable to environmental stresses such as drought and progressive loss of habitat. Child (1965) described the effects of flooding on a buffalo population inhabiting an island in Lake Kariba in Zimbabwe. He found that the younger animals were the first to succumb to the decreasing availability of food, followed in death by adult females, then by old males, and finally by prime-age males. In other studies, Child (1968, 1972) described drought-caused die-offs of buffalo and wildebeeste in Botswana, near the Shakwanki wilderness in Hwange. The sequence of age- and sex-related vulnerability was similar in many respects.

My studies of North American bison in north central Canada (G. Haynes 1981, 1982, unpublished data) have demonstrated that when faced with very harsh environmental conditions, bison show the same differential vulnera- bility as African buffalo. The Canadian bison were affected by wet weather that had caused much of their summer and autumn feeding range to be flooded, and they were then subjected to ice crusting atop the wet ranges; finally, they were subjected to a large spring flood in these ranges (G. Haynes 1981, 1982). Even after those extremely severe conditions, prime- age adult males were infrequently represented in bone assemblages, not because they had been absent from the ranges but because they had been able to survive. Younger animals of both sexes and adult females were abundant in the death sample.

Thus, there is good reason to expect adult elephant males to be much less vulnerable to drought, accounting for their absence from the seeps bone assemblages.

4.13. Conclusion

Tons of elephant bones are lying on ground surfaces or are buried at the African die-off locales, and my analyses should be considered in an initial stage; in fact, ideally this study will continue for decades, to follow the remains of Hwange elephants during their early transition from biosphere to lithosphere. The information about social behavior and ecology clearly reflected in some aspects of these bone assemblages may be preserved or made ambiguous in the future as the bones are trampled, weathered, diagenetically modified, or protected by deep burial in gentle micro-environments.

At this time I can make certain detailed and complex statements about elephant behavior, based on the characteristics of the bone accumulations, because I have seen and studied the behavior at the same time that I have watched the bones accumulate. But as the behavior that contributed to the different characteristics of the assemblages becomes more and more removed in time, and as the bones change during their future career as fossils, the kinds of statements that the analyses will warrant may be quite different.

In the next chapter I describe another kind of bone accumulation, one that resulted from mass kills by riflemen working for the Zimbabwe Department of National Parks and Wild Life Management. As in the mass die-off sites, the bones in these kill sites preserve some rather unambiguous reflections of elephant social behavior, seen especially in sample age profiles. However, like the bones in the mass die-off sites, the bones in these sites may be altered in some ways over the next decades, and the kind of information they can give us about elephants may not be the same in the future. I do not present the analysis in this and the next chapter as the final word on elephant bone accumulations, but more as a starting point for teasing out the meaning that may hide in fossil bone assemblages.

5

Actualistic studies
of mass kills

*I should not greatly care to kill any more elephants. They are
too big, too old, and too wise to be classed as mere game.*

—W. S. Rainsford, quoted by Kunz (1916)

5.1. Introduction

As mentioned in Chapter 3, the Zimbabwe Department of National Parks
and Wild Life Management found it necessary to kill a substantial number
of elephants in Hwange National Park between 1983 and 1986 to prevent
overuse of the park's vegetation. The population was reduced to the level of
about twenty years earlier. Over 9,000 elephants were shot.

The culling was a highly organized operation in effect during the cooler
months of each year, and it involved daily kills of individual herds until the
annual take-off quota had been reached. The culling required the labor of
hundreds of people, mainly as butchers and processors of meat and skins.
Ecologists or technicians were also in the field at all times to gather
scientific data. Specialists employed by the Department of National Parks
were in charge of locating herds, stalking and shooting them, and ensuring
that skins, ivory, meat, and other by-products were properly salvaged and
prepared.

During my association with the culling (1983–6) I stalked herds on foot
with riflemen, and I made a special effort to record elephant behavior when
the herd discovered us, just before the shooting started. I photographed as
much as possible and took detailed notes at the end of the day.

It was a disturbing experience to participate in the killing of elephant
herds day after day, but it was instructive. There was a great deal of
sameness to the violence, but also sporadic departures from the unformity,
such as when unexpected bulls in mixed herds fled in all directions, running
near enough to knock people down, or when angry cows put their heads
down and charged. I did not see every elephant in every herd make its last
stand, but I did witness the deaths of several hundred animals in dozens of
different herds. These experiences may make good anecdotes, but they
cannot be quantified.

177

In this chapter I present detailed descriptions of elephant butchering in the field, because there are lessons to be learned about carcass utilization, and also because I think the cull sites created by the butchering make an appropriate sample to compare with supposed mass kill sites in the fossil record.

After explaining the butchering work, I describe the sites themselves following abandonment. The sites are under long-term study and will be reexamined periodically (Figure 5.1) in order to inventory the remaining elements. After some years have passed, I plan to compare the characteristics of single-elephant kill/butchering sites with those of mass-death sites.

5.2. Hwange cull sites

Each cull site was abandoned after the animals were fully butchered and studied. The skeletons remain in the bush. The only elements removed from the sites were mandibles, tusks still rooted in the alveolar bone (chopped off the skulls), and an occasional limb bone that I took for other studies.

The age profile of the culled sample is thought to be fairly representative of the overall Hwange population (see Figure 3.3). The elephant culling sites are true culture assemblages whose age distributions are "catastrophic." However, not all findings from these sites can be considered analogues for interpretation of prehistoric sites, because the Hwange herds were destroyed in seconds by high-caliber rifle fire. All animals in all age classes were equally vulnerable to that death process. These sites are described for purposes of comparison with the noncultural sites, where members of certain age classes in local populations were more successful than others in their attempts to escape death.

The labor gangs in Hwange National Park butchered well over 10,000 elephants in the 1980s. About 200 local men were hired each culling season, of which about 50–100 accompanied the shooting team daily and butchered the animals in the field. The remaining laborers worked in the base camp cutting meat into strips for air-drying into biltong, or cleaning and salting skins for drying.

Most of the field laborers worked in the culling each year in Hwange and were thoroughly experienced. The gangs were divided into skinners, meat cutters, tusk choppers, and carriers (called "waiters"). The skinners and meat cutters were issued knives and sharpening steels at the beginning of the cull season and maintained their own tools for the duration of the cull. Tusk choppers carried steel axes with long wooden handles. These men chopped off the front of the head of each tusk-bearing elephant, without damaging the ivory, so that the tusks themselves could later be carefully chopped free of bone. Tusks are tightly rooted, especially in adults, and even in the case of degreased and weathered skulls the tusks are difficult to remove.

Figure 5.1. Site where seventy-three elephants were shot in 1983, photographed three years later. Clustered bones represent individual or clustered carcasses.

Axes were never used for any other field butchering. The field butchers used only steel knives that began their use-life with blades six to eight inches long, mounted in plastic or wooden handles. Repeated resharpening of the relatively soft steel blades shortened and narrowed them considerably over the course of the culling season.

5.3. Kills

Each herd selected was destroyed as a unit by a skilled team of riflemen. The elephants' initial perception of the riflemen typically occurred at about 20 m distance, sometimes nearer. A small, fixed-wing airplane first located each mixed herd and radioed its location to members of a ground team, who carefully approached on foot from downwind. The elephants in the chosen herd usually were preoccupied with the noise of the small plane or with feeding and did not notice the approaching riflemen and gun-bearers.

Three or more experienced marksmen shot the animals. One man was in a central position in respect to the confronted herd, with right and left flankers. The center man attempted to initiate the action by waiting for the herd's matriarch to step forward to investigate; on occasion the matriarch would scream, spread her ears, and make a bluff rush, then flee,

Figure 5.2. Moments after initiating a cull, 1984. To the right is the matriarch's body collapsing in death.

with the rest of the herd massing to follow. On an ideal day the central rifleman immediately shot the matriarch, and then he and the flankers killed all the rest of the larger animals as they milled about or stumbled into each other (Figure 5.2). Brain shots dropped many animals straight down, instantly dead. Others had to be shot through the lungs and heart or spine if a clean shot to the head was impossible. Many animals were trapped within a circle of dead adults (Figure 5.3) and thus could not escape after the first shots were fired. Younger animals such as calves often bolted as they were cornered, but many stayed near their dead mothers and were shot after the adults were dead.

A large group of thirty to fifty animals could be shot dead in less than one minute if they did not break or try to escape into thick vegetation. If breaks occurred, every animal had to be tracked, no matter how far the pursuit, and finished off. No animals in a herd were allowed to live, because escaping animals would have been severely traumatized by the violence. Young calves were tranquilized and captured alive if there was demand from other parks, animal collectors, or zoos.

After the animals were down, they were quickly checked to be sure that all were dead. A few required a close-range brain shot to finish them off. At once a team of research technicians began measuring the shoulder heights

Figure 5.3. Elephants lying as they died, unable to escape through the enclosing circle of carcasses.

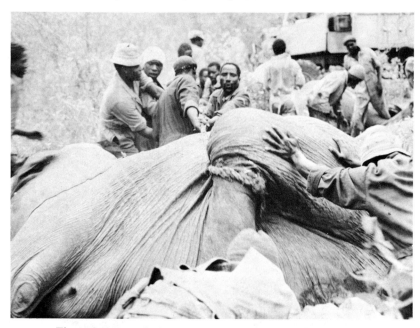

Figure 5.4. Part of a butchering team turning over the carcass of an adult cow elephant shot in the cull. The rope has been looped around the left rear leg, and about ten men are pulling.

of all animals (by straightening forelimbs, if possible), assigning numbers to all carcasses (written in chalk on the skins), and attaching aluminum number tags to lower jaws, which were later collected for age determination.

The butchering team then began pulling carcasses onto their sides (Figure 5.4), if necessary, or separating tightly bunched groups. The heavy lorries that carried meat and skins were driven into the sites, often a slow trip of several kilometers from the nearest track. These lorries sometimes were used to separate carcasses when crowding was a problem.

To turn or separate carcasses, a heavy hemp rope was tied around one leg of an animal, and about a dozen laborers grabbed it in line and pulled, while a few other men pushed. A rhythmic chant set the pace, and the heavy carcasses were slowly turned over; at each pause, the carcass was held in position and a new hold taken on the rope, until the carcass finally fell to a stable position. A small number of prehistoric hunters would have had no trouble turning over any adult mammoth or mastodont carcass, especially if heavy cord were available.

The laborers then processed each carcass in waves. First, several tusk removers walked through the carcass site chopping off the fronts of the skulls, which were carried to a special lorry. At the same time, several other

Figure 5.5. Part of a butchering team making skinning cuts in the side of an adult cow elephant shot in the cull.

men removed tails (if they were to be salvaged for tourist-trade bracelets), ears, and trunks. The skin from ears and trunks was then removed by the skinners. Meanwhile, a number of carcass skinners made the first cut in one side of the hide of each carcass (Figure 5.5). Waiters took the pieces of the hide to the skin lorry.

Meat cutters then processed each carcass in turn, removing meat from limbs and trunk on the upward-facing side. The meat was carried to the lorries by the waiters. The carcass was then flipped over with the cooperative effort of nearby laborers, and the process was repeated on the other side.

Skinning and meat removal were carried out in patterned ways. Every carcass was treated the same and went through the same sequence of skinning, butchering, and meat removal. The butchering team performed the most efficient, quickest sets of steps to remove nearly all the skin in panels and most meat in single masses without meticulous filleting. As a result, a relatively small amount of meat was left unsalvaged on each carcass, probably several dozen kilograms of meat.

While the skin and meat were being removed, a research team was taking measurements and tissue samples from the carcasses. The sex of each animal was determined, and age was estimated. The precise age of each

Figure 5.6. Ecologist Aaron Bhiza cutting fat away from the kidney of an adult cow elephant, prior to weighing it to determine the animal's condition.

individual was determined later by examination of tooth eruption and wear. One kidney from each animal was removed and weighed, as was the surrounding fat, to get an index of the animal's condition (Figure 5.6). From females, ovaries were removed and dissected; corpora lutea and nigra were described, and placental scars were counted. Fetuses were weighed and measured. I assisted in these studies, and I also measured shoulder and pelvic heights, long-bone lengths, and variable bone characteristics such as shapes of mandibles. I dissected out femora and humeri from several carcasses in each cull group after skins and meat had been stripped.

5.4. Meat removal

During meat removal by the experienced crews, no bones were ever cut, deliberately or accidentally. There were instances when other butchering gangs, which had been assigned to skin and strip meat from elephants that had been shot for rations or fence control (to prevent animals from leaving the park and raiding crops outside), used axes to chop off meat masses. Those gangs often were in a hurry and clearly were less experienced than the culling teams. They left carcass bones deeply chopped and cut-marked.

By contrast, the culling crews chopped off only the lower jaw (to be brought back to the base camp for age determination) and the front of the head (to be brought back to base camp for ivory removal). No bones were broken, chopped, or cut for meat removal.

After the skin was removed on the upward side of each carcass, the meat cutters moved in to fillet the skeletons. Usually the cutters filleted the hindquarters of each animal as they became exposed, moving serially through the herd; then they began again as forelimbs were exposed.

Meat cutters separated the hindlimb meat from bone with a few powerful but fast cuts above the knee, up the muscle masses, and around the femoral head. Waiters pulled the meat tightly with steel hooks, stretching it away from the bone so that it could be cut and separated without having to cut against the bone. The meat was fully removed in a few seconds.

After butchering over forty elephants, I have found that one person alone can easily detach a limb bone in a short time after the meat has been removed. The joint attachments in limbs are simple, and little fine-scale maneuvering of the knife is required to separate bones. Driving in any kind of a wedge to separate conarticular bones (as suggested to explain marks on some fossil mastodont bones, see Chapter 6) seems unnecessary.

The front leg (from scapula through to the foot) usually ended up detached from the rest of the body, because meat on the anterior part of the thoracic cage was removed from both sides, around the scapulae and upper humeri, thus separating these bones from the thorax. This meat was taken off the rib cage and along the spine, in rather thin sheets. When one side of the carcass had been stripped, the animal was turned over, and the process was repeated.

The bones that conceivably could have been cut by meat strippers were the ribs, vertebrae, femoral shafts, scapular blades, and iliac portions of innominates. The meat of scapulae and femora was cut away by making long slices just above where the hooked and stretched meat attached to the periosteal bone. The strokes were directed along the long axes of both the femur and the scapula. Cut marking was unlikely, and in fact I never saw it on any elements examined at abandoned cull sites. Even my own disarticulation of the femora and humeri did not cut-mark bones, although

I did cut against epiphyses at times to force the knife edge through connective tissue. Articular cartilage is thick on elephant limbs, and when it had decayed on the elements that I disarticulated, even knife marks that had appeared to be deep were no longer preserved. The cortical bone surface itself has never preserved cut marks. I have also examined hundreds of ribs, vertebrae, and innominates at cull sites where experienced crews butchered carcasses, and none were cut-marked.

5.5.   Cull sites after abandonment

I visited several cull sites three days after abandonment, and later I revisited them after ten days, one year, three years (Figure 5.1), and up to six years. At some fresh sites I had deliberately tried to chop and cut bones so that I could determine how much marking would survive weathering and scavenging by carnivores. At other sites I had stamped numbers into bone shafts so that I could reassemble individual skeletons after several years of bone dispersal by scavengers or trampling animals. All sites were sketched, photographed, and described in notebooks at the time of the cull and during each visit. I documented as much butchering behavior as possible and recorded all details of carcass treatment. I tried at each of my visits to match up skeletons with flesh-and-blood carcasses that I had seen being butchered, and most of the time succeeded. I was thus able to trace the process of carcass and skeletal changes over time at sites of mass deaths.

None of the sites is located adjacent to permanent water sources. Virtually all, in fact, are in thick bush, thickets, or woodland, where elephants are to be found feeding or traveling in the morning hours. In the very warm part of the dry season, elephants may go to water at midday and spend about thirty minutes to an hour drinking, playing, or cooling off. Larger herds spend longer times at water because of competition for access. More often, elephants drink at night or in the morning hours before sunrise. Thus, when the culling team encountered a herd, the animals had already moved several kilometers from water holes.

In the season of plant growth (following the first rains, throughout the rainy season, and up to the middle of the winter dry season), elephants may not go to water every day. Although adults require large amounts of water daily to digest food and maintain body temperature, during lush times they may be able to get enough moisture from their forage. Thus, they may space their water-hole visits: a visit every two or three days, if convenient. Under those circumstances, elephants may spend more time at water holes than they do when they visit water holes every day.

The culling team never shot herds near water, because the disturbance would have been too great for other animals (which probably would have

Figure 5.7. (Left). An elephant trail in Kalahari sand leading to Shabi Shabi. Such trails are traditionally reused for generations. (Right). Another trail leading to Nehimba, passing through thicker grass and shrubs. The trail is covered with a layer of twigs, stems, leaves, and shredded bark.

abandoned that water source permanently) and undoubtedly would have frightened off other elephants for good.

All cull sites lay on major animal trails that had been created and used mainly by elephants. An elephant trail is easy to recognize: It is about 60 cm wide, trampled quite flat and bare of growing vegetation, and often littered with dung boluses or a carpeting of plant remains from trampled elephant droppings (Figure 5.7). Some cull sites have a complex network of crisscrossing trails, especially the larger ones where more than thirty animals were shot in tight groupings. The converging trails probably resulted in part from the existence of the bones and probably in part accounted for the bones being where they are. That is, elephants will investigate unusual features such as cull sites, thus creating new trails to and from them; however, many trails already existed at the time the elephants were killed, and the culled animals had in fact been traveling along those routes when encountered and shot.

5.6. Scavengers

Vultures congregated at cull sites after abandonment by butchering crews. In fact, many vultures learned to follow the spotter's airplane (a Maule or a Piper Supercub) from the base camp to the field locales each morning, so as to find out where the freshest meat would be available each afternoon.

Three days after a typical cull, the site generally contained drying carcasses that had been minimally disturbed by large or small scavengers. Vulture droppings covered some carcasses and ground surfaces, and hyena or jackal spoor could be found. Some carcass feet may have been nibbled, as well as viscera, but by and large the carcasses showed nearly no alteration since the moment of abandonment. Scavengers may have had to travel miles to find fresher cull sites, but apparently preferred them while they were available. When the culling finally ceased each year (usually in September or October, after daytime temperatures became too high for proper treatment of meat and skins), scavengers returned to the older sites to recover edible tissue. Most vulture scavenging occurred within one week to ten days after death; most hyena scavenging occurred several weeks to several months after death.

Elephant carcasses, once stripped of skin and most of their meat, dried quickly in Hwange's bright, warm sunshine. Connective tissue over the thoracic cage dried to a parchmentlike covering, and finally to a mummy-like and very tough enclosing wrap. Foot bottoms dropped off, perhaps because of autolysis of the fat layer and decay of the thick skin attachment to the toenails and legs. The skeletons remained in their original postures, and few bones were moved out of order. Skulls, of course, were already damaged by tusk removal, and the forelimbs may already have been detached, although they lay close to the rest of the skeleton.

Maggots and arthropods were rare on fresh carcasses at cull sites, even when viscera had been exposed. That was because culling occurred in the winter when insect activity was at a low level. Beetles ate soft tissue from the inside of the carcass out, preferring the cooler environment out of direct sunshine. There were no large meat masses left for flyblowing, and what soft tissue there was became dried and very hard within a few days in the sun. However, after the rains began (September or October through March), much of the soft tissue that had dried out again softened up, and insects tried to feed on it. Hyenas, jackals, and lions also ate the soft parts.

A year after death, most bones were exposed and falling apart, because cartilage, tendon, and ligament had been eaten, decayed, or broken. Bones became naturally defatted after several months to a year, although many retained interior oil for years, especially those elements that were shaded by thick vegetation or by other bones.

Hyenas encountered bone sites after the passage of several months or

Figure 5.8. On the left is a large bone flake (showing a scar from an earlier flake removal) refitted to part of the shaft of a small elephant's ulna. The flaking was done by spotted hyena, three years after the elephant's death in a large cull.

several years and fed on bones where previously they had not bothered. Because many elephants in mixed herds had been under thirty-five to forty years of age (when most long-bone growth ceases in females), their epiphyses had not fused to element diaphyses. As a result, their epiphyses separated from shafts when soft tissue and cartilage decayed, and hyenas could therefore gain relatively easy access to long-bone interiors that retained oil. In their attempts to reach ever deeper into these bones, hyenas levered off large flakes of long-bone walls. The largest flakes I measured were over 30 cm long by about 10 cm wide. Each flake had one thick end where a hyena had grabbed the exposed end of a shaft, and one thin or feathered termination where the flake had separated from the shaft (Figure 5.8). Recognizable tooth marking was relatively rare on the bone flakes or fragmented elements. Some fragments undoubtedly had been chewed and eaten, but many denser shaft fragments had been abandoned before they had been gnaw-marked.

5.7. Comparisons with die-off sites

Kill sites and die-off sites have both similarities and differences. The mass kill sites considered in this study are never located near water, that is, within 1–2 km of permanent water sources. They are situated in upland areas where vegetative cover is thick and where game trails are numerous. Some trails were created by curious elephants visiting the bone sites, but many others have existed for generations as traditional travel routes.

The upland site locations provide a relatively rapid overgrowth of grass whenever the rains come; bones are also buried in leaf-fall and are shaded by shrubs and trees, but their potential for burial into the earth is low, because little sediment buildup occurs in uplands. However, during dry periods, when ground cover is reduced, windblown sediments can lead to deep burial of skeletons. Sedimentation rates and burial potential may be higher at die-off sites located adjacent to or within water sources in present-day subtropical Africa.

The kill sites contained a nonselective sample of entire herd groups, occasionally missing one or two unweaned calves that were captured alive. All kill sites contained skeletons of one or more very small calves, and often several fetuses in different stages of development. Small numbers of adult males also were shot along with some mixed herds because at the time of the culling they had been consorting with estrous females or had been unfortunate enough to be temporarily feeding or traveling in company with targeted mixed herds. The die-off sites described in Chapter 4 did not contain adult males (save for one that had been shot illegally elsewhere and then had walked to a seep and died there), but, as mentioned earlier, during the later stages of intense or long-term environmental stress, even adult males will begin dying, and therefore the eventual presence of their bones in this type of site is possible.

The kill sites contain densely clustered bones, with much higher concentrations of elements per unit of surface area than in the die-off sites. However, the kill-site bones constitute a single bedding of large particles, whereas the die-off sites tend to contain vertically dispersed or stratified clusters.

The kill-site skeletons contain few articulated units, with the commonest articulation being short segments of the vertebral column. The kill-site bones show a range of different weathering stages, but most fall within one or two stages of each other (Behrensmeyer 1978), progressing very slowly through those stages as shade and vegetation slow down the weathering process.

There is little sign of carcass use by carnivores at the kill sites, which is similar to the situation at die-off sites, although some kill sites do contain fragmented limb elements broken by large carnivores treating the sites as resource quarries to fall back on when the hunting of live prey fails.

The bone representation at kill sites is similar to that at die-off sites. Skulls and mandibles are rarely removed, nor are innominates, and yet elephants seem to be fascinated by these elements, often picking them up and carrying them short distances, and then exchanging one for another. An innominate and skull may be put into each other's place one day, then replaced the next day, only later to be switched once again. Limb bones may also be moved, as well as removed, but ribs and vertebrae are dispersed or disappear far more often. Caudal vertebrae and foot bones are very difficult for taphonomists to find at the sites, but hyoids tend to remain near skulls. In these ways, skeletons at kill sites are not greatly different from those at die-off sites.

In summary, the main differences between die-off sites and kill sites are (1) the presence of bullet holes and isolated instances of butchering damage on bones at kill sites, whereas such features usually are lacking at die-off sites, (2) the upland locations of kill sites, undergoing eolian burial, in contrast to the stream-channel depressions or spring-fed-pond locations for die-off locations, and (3) the low proportions of adult males at kill sites.

Single-carcass kill sites created by the killing of individual elephants (usually adult males) for meat rations provide another kind of bone site. Such sites may or may not be near water sources, and typically they contain evidence of thorough carcass use by butchers using a heavy hand to chop apart elements and hack into bones to remove meat. Single-carcass sites are discussed further in Chapter 7.

III

THE FOSSIL RECORD

6

Finding meaning in proboscidean sites: the world fossil record

M. Marcel de Serres, a keen observer, calls attention to the great disparity in numbers between the remains occurring in pleistocene beds and those which are now being accumulated.

—H. Howorth, *The Mammoth and the Flood* (1887)

6.1. Introduction

In the preceding three chapters I summarized and explained some of the social behavior of proboscideans, and I described bone sites whose characteristics have a great deal more meaning when viewed in light of an expanded knowledge of elephant ecology. As with all mammals, quirks and eccentric behavior among individual elephants contribute to anomalies in particular death sites. I have highlighted much of the expectable *patterning* in bone accumulations that results from patterning in behavior.

The goal in this chapter is to find patterns in the fossil record, and perhaps to define the anomalies as well. This is done by broadly surveying proboscidean finds and discussing some perceived regularities in them. The sites to which I refer are listed in Tables 6.1–6.5. These tables are not just lists of sites worth knowing about. I am willing to argue the pros and cons of each entry (and of each potential entry that has been left out of the tables), but because of space limitations I refer to only a sampling of the sites to illustrate discussion about a limited number of important topics.

The sites of interest here contain the bones of *wild* elephants; sites selected for this discussion resulted from hunter-gatherer interactions with elephants as prey species. A growing body of literature provides information about sites from the Neolithic, Bronze Age, Iron Age, and historical periods containing elephant bones, teeth, and ivory (e.g., Barnett 1982; Bökönyi 1985, 1986; Hooijer 1978; McClellan 1986; Reese 1982, 1984, 1985, 1988 personal communication; Voigt 1982), but those kinds of sites were occupied by townspeople or, in many cases, "civilized" people. In most of those later sites the elephant bones represent ivory workshop debris more often than food refuse; I do not discuss them here.

195

Table 6.1. *Proboscidean "kills" and butchering sites*

Site name or location	Taxon	Approximate age	References
1. Djibouti (Africa)	*Elephas* (cf. *recki*)	1,300,000 B.P.	1, 2
2. FLK North, Tanzania (Africa) (two loci)	*Elephas*	Early Pleistocene	3–6
	Deinotherium	Early Pleistocene	3, 4
3. Mwanganda's Village, Malawi (Africa)	? *Elephas*	Middle Pleistocene	7, 8
4. Olorgesailie Basin, Kenya (Africa)	*Elephas recki*	Middle Pleistocene	9, 10
5. Torralba and Ambrona (Spain)	*Palaeoloxodon antiquus*	Middle Pleistocene	11–14
6. Aridos 1 and 2 (Spain)	*Palaeoloxodon antiquus*	Middle Pleistocene	15
7. Anza-Borrego, California, (USA)	*Mammuthus imperator*	Middle Pleistocene	16, 17
8. Saxony (W. Germany)	*Palaeoloxodon antiquus*	Riss/Würm Interglacial	18
9. Gröbern (E. Germany)	*Palaeoloxodon antiquus*	Eemian Interglacial	19
10. Northeastern China	*Mammuthus* sp.	40,000–70,000 B.P.	20
11. Skaratki (Poland)	*Mammuthus primigenius*	> 37,000 B.P.	21
12. Grundel, Missouri (USA)	*Mammut americanum*	ca. 25,000 B.P.	22
13. Krakow–Nowa Huta (Poland)	*Mammuthus primigenius*	16,000–30,000 B.P.	23, 24
14. Cooperton, Oklahoma (USA)	*Mammuthus columbi*	17,000–20,000 B.P.	25–28
15. Saskatoon, Saskatchewan (Canada)	*Mammuthus ?primigenius*	ca. 20,000 B.P.	29
16. Taima-taima (Venezuela)	*Gomphotheres*	Late Wisconsin	30–34
Muaco (Venezuela)	*Gomphotheres*	Late Wisconsin	33, 35
El Bosque (Venezuela)	*Gomphotheres*	Late Wisconsin	36–38
17. Tomsk (USSR)	*Mammuthus primigenius*	12,000–20,000 B.P.	39–42
18. Clovis sites (USA):		Terminal Pleistocene	
Angus, Nebraska	*Mammuthus* sp. (hoax?)		43–45
Kimmswick, Missouri	*Mammut americanum*		46–48

196

Site	Species (MNI)	Date	Ref.
Miami, Texas	*Mammuthus columbi* (MNI = 5)		49, 50
Lehner, Arizona	*Mammuthus columbi* (MNI = 13)		51–53
Domebo, Oklahoma	*Mammuthus columbi* (MNI = 1)		54–57
Blackwater Draw, New Mexico	*Mammuthus columbi* (MNI = 8 +)		58–61
Dent, Colorado	*Mammuthus columbi* (MNI = 12)		52, 53, 62–65
Escapule, Arizona	*Mammuthus columbi* (MNI = 1)		56, 66
Murray Springs, Arizona	*Mammuthus columbi* (MNI = 1?)		67–71
Naco, Arizona	*Mammuthus columbi* (MNI = 1)		72, 73
Colby, Wyoming	*Mammuthus columbi* (MNI = 7)		74–78
19. Lange-Ferguson, South Dakota (USA)	*Mammuthus columbi* (MNI = 2)	ca. 11,000 B.P.	79–82
20. Sites lacking Clovis points, of Clovis age (USA)		ca. 11,000 B.P.	
(a) U.P. (Rawlins), Wyoming	*Mammuthus columbi*	11,280 ± 350 (I – 449)	83, 84
(b) Duewall-Newberry, Texas	*Mammuthus columbi*	12,000–10,000 B.P.	85
(c) Pleasant Lake, Michigan	*Mammut americanum*	10,395 ± 100 (Beta – 1388)[a]	86–89
		12,845 ± 165 (Beta – 1389)[a]	86–89
(d) Rappuhn, Michigan	*Mammut americanum*	Atop glacial till	90, 91
(e) Manis, Washington	*Mammut americanum*	12,000 ± 310 (WSU – 1866)[a]	92
(f) South Egremont, Massachusetts	*Mammut americanum*	11,850 ± 60 (USGS – 591)[a]	92
		11,630 ± 470 (GX – 9529)[b]	93
(g) Lake Willard, Ohio	*Mammut americanum*	11,440 ± 655 (GX – 9024 – b)[c]	93
(h) Dansville, Michigan	*Mammut americanum*[d]	9250 ± ? (lab no.?)[c]	94
(i) Selby, Colorado	*Mammuthus columbi*	ca. 11,000 B.P.	95
(j) Huntington Canyon, Utah	*Mammuthus columbi*	Several contradictory dates, spanning from about 9,500 to 11,500 B.P.	96, 97
			98

Table 6.1. *(cont.)*

Site name or location	Taxon	Approximate age	References
21. Valley of Mexico (two sites)	*Mammuthus columbi*	$16,000 \pm ?(C - 204)$[a] $11,003 \pm 500 \ (C - 205)$[a] $9,670 \pm 400 \ (?)$[a]	99, 100 99, 100 99, 100
22. Libyan Desert (Africa)	*?Loxodonta*	Mesolithic–Recent	101
23. Zoo Park, Namibia (Africa)	*?Elephas*	$5,200 \pm 140$ (lab no.?)	7, 102

[a] Not on bone.
[b] Spruce cones.
[c] Bone.
[d] Associated with wooden spear.

Sources: 1, Chavaillon et al. (1986); 2, Chavaillon et al. (1987); 3, Leakey (1971); 4, Isaac and Crader (1981); 5, Bunn (1986); 6 Shipman (1986); 7, Clark and Haynes (1970); 8, Kaufulu (1990); 9, Isaac (1977); 10, Bower (1987); 11, Howell (1966); 12, Klein (1987); 13, Binford (1987); 14. Shipman and Rose (1983); 15, Santonja and Villa (1990); 16, Minshall (1988); 17, G. Miller (1988 personal communication); 18, Movius (1950); 19, Weber 1988); 20, Sun et al. (1981); 21, Chmielewski and Kubiak (1962);22, Mehl (1967);23, Kozlowski (1986); 24, Kozlowski et al. (1970); 25, Bonfield (1975); 26, Mehl (1975); 27, Anderson (1962); 28, Anderson (1975); 29, Pohorecky and Wilson (1968); 30, Bryan (1973); 31, Bryan et al. (1978); 32, Cruxent (1970); 33, Rouse and Cruxent (1963); 34, Tamers (1966), 35, Cruxent (1961); 36, Page (1978); 37, Bryan (1978); 38, D. Stanford (1981 personal communication); 39, Beregovaya (1960); 40, Kashchenko (1896); 41, Kashchenko (1901); 42, Kuznetsof (1896); 43, Figgins (1931); 44, Strong (1932); 45, Wormington (1957); 46, R. Adams (1953); 47, Fowke (1928); 48, Graham et al. (1981); 49, Sellards (1938); 50, Sellards (1952); 51, Haury, Sayles, and Wasley (1959); 52, Saunders (1977a); 53, Saunders (1980); 54, Leonhardy (1966); 55, Leonhardy and Anderson (1966); 56, Stafford et al. (1987); 57, Mehl (1966); 58, Agogino (1968); 59, C. Haynes and Agogino (1966); 60, Hester (1972); 61, Warnica (1966); 62, Figgins (1933); 63, C. Haynes (1966); 64, C. Haynes (1967); 65, Cassells (1983); 66; Hemmings and Haynes (1969); 67, C. Haynes (1976); 68, C. Haynes (1980); 69. C. Haynes (1981); 70, C. Haynes and Hemmings (1968); 71, Hemmings (1970); 72, Haury (1953); 73, C. Haynes (1964); 74, Frison (1976); 75, Frison (1978); 76, Frison (1986); 77, Frison and Todd (1986); 78, Graham (1986b); 79, J. Martin (1987); 80; Hannus (1983); 81, Hannus (1984); 82, Hannus (1986); 83, C. Irwin, Irwin, and Agogino (1962), 84; H. Irwin (1970); 85, Steele and Carlson (1989); 86, D. Fisher (1984a); 87, D. Fisher (1984b); 88, Shipman, Fisher, and Rose (1984); 89, D. Fisher and Koch (1983); 90, Wittry (1965); 91, Kaye (1962); 92, Gustafson, Gilbow, and Daugherty (1979); 93, R. Moeller (1983 personal communication); 94, Falquet and Haneberg (1978); 95, J. Holman (1986); 96, Stanfcrd (1979); 97, D. Stanford (1982 personal communication); 98, D. Gillette (1989 personal communication); 99, Aveleyra and Maldonado-Koerdell (1953); 100, Aveleyra (1955); 101, Deraniyagala (1955); 102, MacCalman (1967).

Table 6.2. *A sample of noncultural single-animal sites*

Site or location	Taxon	Approximate age	References
Soviet Union			
Lena River (Adams)	*Mammuthus primigenius*	34,450 ± 2,500 (lab?)	1–5
		35,800 ± 1,200 (lab?)	
Beryozovka	*Mammuthus primigenius*	29,000 ± 44,000 B.P.	4–11
Taimyr	*Mammuthus primigenius*	11,450 ± 3,500 (lab?)	5, 12, 13
Magadan (Dima)	*Mammuthus primigenius*	ca. 44,000 B.P.	14–18
United States			
Hazen, Arkansas	*Mammuthus columbi*	Late Wisconsin	19
Haley, Indiana	*Mammuthus columbi*	13,840 ± 195 (lab?)	20
Upstate New York (Warren)	*Mammut americanum*	Late Wisconsin	21
Marion County, Ohio	*Mammuthus columbi*	10,340 ± 125 (Di-Carb-799)	22
Springdale, Ohio	*Mammut americanum*	?Early postglacial	23
Orleton Farms, Ohio	*Mammut americanum*	?Late Glacial	24, 25
Saltillo, Pennsylvania	*Mammut americanum*	?Late Glacial	26
Stanton County, Kansas	*Mammut americanum*	?Late Glacial	27
Oak Creek, Nebraska	*Mammuthus columbi*	Wisconsinan age	28
Shelton, Michigan	*Mammut americanum*	12,320 ± 110 (B-9084) to	
		10,970 ± 130 (B-10303)	
Inglewood, Maryland	*Mammuthus* cf. *columbi*	20,070 ± 265 (SI-5357)	29
		29,650 ± 750 (SI-5408)	30
			30

Sources: 1, Dubinin and Garutt (1954); 2, M. Adams (1807); 3, Tilesius (1812); Tolmachoff (1929); 5, Heintz and Garutt (1964); 6, Herz (1902); 7, Herz (1904); 8, Pfizenmayer (1939); 9, V. Garutt (1987 personal communication); 10, N. Vereschhagin (1987 personal communication); 11, Zalyenskii (1903); 12, Garutt (1964); 13, Garutt (1965); 14, Shilo et al. (1983); 15, Vereshchagin et al. (1981); 16, M. A. Zaslavskii (1987 personal communication); 17, N.Vereshchagin (1987 personal communication); 18, Goodman et al. (1979); 19, Puckette and Scholtz (1974); 20, Pace (1976); 21, Warren (1855); 22, Hansen et al. (1978); 23, Ettensohn (1976); 24, E. Thomas (1952); 25, A. Wood (1952); 26, Robinson and Krynine (1938); 27, O'Brien (1978); 28, Barbour (1925); 29, J. Shoshani et al. (1989); 30, G. Haynes (unpublished data).

199

Table 6.3. *Mass accumulation sites*

Site or location	Taxon	Approximate age	References
Western Europe			
Italy			
Castel di Guido	*Palaeoloxodon*	200,000–458,000	1
Fontana Ranuccio	*Palaeoloxodon*		2, 3
Rebibbia Casal de'Pazzi	*Palaeoloxodon*		4
Polledrara	*Palaeoloxodon*		5
Condover, England	*Mammuthus primigenius*	12,700 ± 160 (OxA-1021)	6, 7
		12,290 ± 390 (Birm-1273)	6, 7
La Cotte de St. Brelade, Channel Islands	*Mammuthus primigenius*	Middle Pleistocene (150,000 B.P.)	8–12
Central Europe			
East Germany:			
Taubach-Ehringsdorf	*Palaeoloxodon antiquus*	Middle Pleistocene	13–15
Weimar-Ehringsdorf	*Palaeoloxodon/Mammuthus*	Middle Pleistocene	13–17
Süssenborn	*Mammuthus trogontherii*	Pleistocene	13
West Germany:			
Mosbach	*P. antiquus & M. trogontherii*	Pleistocene	13–15
Mauer	*M. trogontherii*	Pleistocene	13–15
Cannstadt	*P. antiquus*	Pleistocene	13–15
Emmendingen	*P. antiquus*	Pleistocene	13–15
Salzgitter-Lebenstedt	*Mammuthus primigenius*	Upper Acheulean	17, 18

Sources: 1, Radmilli (1982); 2, Bidittu et al. (1979); 3, Segre and Ascenzi (1984); 4, Anzidei and Ruffo (1985); 5, M. C. Stiner (1988 personal communication); 6, Coope and Lister (1987); 7, A. Lister (1989 personal communication); 8, Callow and Cornford (1986); 9, Scott (1980); 10, Scott (1984); 11, Scott (1986a); 12, Scott (1986b); 13, Soergel (1912); 14, Weigelt (1927); 15, Müller-Beck (1982); 16, Günther (1975); 17, Grote and Preul (1975); 18, Staesche (1983).

Table 6.4. *Abundances of taxa in major Eurasian collections*[a]

		Ranked abundance[b]				
Region and site	¹⁴C dates or estimated age	Mammuthus	Rangifer	Equus	Bison	Other
Ukraine and Russian Plain						
Berdyzh	23,430 ± 180 (LU-104)	1				
Pushkari	16,775 ± 605 (QC-899)	1				
Yudinovo	15,660 ± 180 (LU-127) 13,830 ± 850 (LU-103)	2				1 (fox)
Gontsi	13,400 ± 185 (QC-898)	1?				
Mezhirich	15,245 ± 1080 (QC-900) 14,320 ± 270 (QC-900)	1				
Yeliseyevichi	Several, ranging from 33,000 ± 400 (GIN-80) to 12,970 ± 140 (LU-102)	2				1 (fox)
Kostyonki (all sites)	Two dates 30–25,000, six dates 25–20,000, two dates 20–9,000 B.P.	(Abundances vary widely with sites and levels within sites; fox, hare, and *Equus* often dominate)				
Mezin	24,200 ± 1,680 (?) 24,920 ± 1,800 (?)	1	3	4	2	
Western Siberia, Northern Russia, Urals						
Sunghir	Several, ranging from 25,500 ± 200 (Gro-5425) to 14,600 ± 600 (GIN-14)	3?	1	2?		
Byzovaya[c]	18,320 ± 280 (TA-121)	1				
Shikayevka	Estimated 13,000 B.P.	1?	2			
Cherno-ozere II	14,500 ± 500 (GIN-622)			3?	2?	
Achinsk	Estimated Pleistocene/Holocene	?		1?		2? (wolf)

Table 6.4. (cont.)

Region and site	14C dates or estimated age	Ranked abundance[b]				
		Mammuthus	Rangifer	Equus	Bison	Other
Volchya Griva	14,200 ± 150 (SOAN-78)	1		2	3	
Tomsk	Late Pleistocene	1 (No other fauna?)				
Srostki	<20,000 B.P.		2			
Bobkovo	<20,000 B.P.	1			1	
Eastern Siberia (includes Yenesei and Lena river valleys)						
Afontova Gora	20,900 ± 30 (GIN-117) 11,335 ± 270 (?)		1			2 (fox)
Buret	<14,700	2?	1			
Druzhinka	Estimated Pleistocene to Holocene	3	1	2		
Krasni Yar	Estimated 30,000 B.P. in lower level, 14,000 B.P. in upper		2	1	3	
Mal'ta	23,000 ± 5,000 (?) 14,750 ± 120 (GIN-97)		1			
Tashtyk I and II	12,180 ± 120 (LE-771)		1	2		
Verkholenskaya Gora	12,570 ± 180 (Mo-441)			2		1 (fish or deer)
Kokorevo (all sites)	Several, ranging from 15,460 ± 320 (LE-540) to 12,690 ± 140 (RU-629)		1	3	2	
Makarova II	Three dates, ranging from 11,950 ± 50 (GIN-481) to 11,400 ± 500 (GIN-480)			1	2	
Novosyolevo staroe VI	11,600 ± 500 (GIN-403)		1			
Novosyolevo staroe VII	15,000 ± 300 (?)		1?			

	Dates				
Aldan River basin					
Ikhine I & II[d]	Several, ranging from 31,200 ± 500 (GIN-1020) to 12,200 ± 170 (LE-953)	3		1 or 2	2 or 1
Ust-Mil	Several, ranging from 35,600 ± 900 (LE-955) to 12,200 ± 170 (LE-953)	3		1	2
Verkhne-Troitskaya	Several, ranging from 18,300 ± 180 (LE-955) to 14,530 ± 160 (LE-864)	3		2	1
Diuktai Cave	Several, ranging from 14,000 ± 100 (GIN-404) to 12,100 ± 120 (LE-907)	1[e]	3 or	3	2 (fox or hare)
Far East, Northeast Asia					
Berelyokh	Several, ranging from 13,700 ± 400 (?) to 10,440 ± 100 (?)	1[f]			1[g]
Bochanut (redeposited finds)	Late Pleistocene	1?		2	1?
(Paleontological expeditions)					
Novosibirsk expedition of 1885–6	Late Pleistocene	1			1
Northern Siberia collections of 1900–8	?24,800 ± 120 (GIN-162)	2		2	1

[a] References are too numerous to cite here, but can be found in G. Haynes (in press).
[b] 1 = most abundant; 4 = least abundant (of taxa listed).
[c] May be a noncultural site.
[d] Very few stone tools found. Fauna is consistent throughout.
[e] See text.
[f] In nonarcheological deposit.
[g] Hare in archeological site.

Table 6.5. *Mass assemblages from the New World* (*Not considered fully cultural*)

Site or location	Taxon and MNI	Approximate age	References
Big Bone Lick, Kentucky	*Mammuthus/Mammut* (dozens)	Late Wisconsin?	1
Jackson County, Ohio	*Mammuthus/Mammut* (MNI =?)	Late Wisconsin?	2
Kimmswick, Missouri	*Mammut americanum* (>100)	Late Wisconsin	3–5
Boney Spring, Missouri	*Mammut americanum* (31)	Late Wisconsin	6
Hot Springs, South Dakota	*Mammuthus columbi* (>43)	ca. 26,000 B.P.	7–10
Lamb Spring, Colorado	*Mammuthus columbi* (>30)	Late Wisconsin	11–15
Friesenhahn Cave, Texas,	*Mammuthus* (>100), few *Mammut*	ca. 20,000 B.P.	16–18
Aguas de Araxa, Brazil	*Haplomastodon* (30–40)	?Late Pleistocene	19
Waco, Texas	*Mammuthus columbi* (14)	Late Wisconsin	20, 21

Sources: 1, Cooper (1831); 2, Briggs (1838); 3, R. Adams (1953); 4, Fowke (1928); 5, Graham et al. (1981); 6, Saunders (1977b); 7, L. Agenbroad (1987 personal communication); 8, L. Agenbroad (1989 personal communication); 9, Agenbroad and Laury (1984); 10, Agenbroad and Mead (1990); 11, Stanford et al. (1981b); 12, Stanford et al. (1981a); 13, Rancier et al. (1982); 14, Elias (1986); 15, Elias and Toolin (1990); 16, Evans (1961); 17, Graham (1976); 18, Graham (1984); 19, Simpson and de Paula Couto (1957); 20, C. Smith (1986 personal communication); 21, G. Haynes (unpublished data).

Also, and unfortunately, I have been unable to evaluate gomphothere sites in South America, and Asian elephant sites in India, Sri Lanka, and southeast Asia. The problem with the Asian sites is that so little is known about the prehistoric ecological relationships between humans and elephants there,[1] and also that the finds of fossil and recent bones have provided only ambiguous information. In India, *Elephas* bones have been discovered in locales that also contain flaked stone artifacts, but the discoveries generally have not been made in stratified or undisturbed contexts. The main problems in understanding South American gomphothere sites are (1) confusion in the taxonomy, (2) scattered and abbreviated reporting of finds, and (3) lack of information about gomphothere ecology and behavior.

Armed with knowledge from the actualistic studies, I cast a critical eye on interpretations of the fossil proboscidean record. My bottom-line impression about the evidence for human hunting and processing of proboscideans is that it appears to be far less substantial than has been claimed. The sites and assemblages of interest have inspired unreconciled and contradictory interpretations, because the excavation techniques and the types of data sought in each case were widely variable, resulting in much variation in the supportability of interpretive statements.

6.2. Meaning and ambiguity in the fossil record

Not all elephant bone sites or assemblages are alike, of course, but neither is each site unique. In North America, some late Pleistocene proboscidean sites have yielded flaked stone artifacts, but others have not; some have yielded broken and scattered bones, whereas in others no bones have been broken, and skeletons have been in anatomical order.

In spite of the differences in sites regarding what is and is not found, observations in themselves do not provide superior or inferior evidence about the past. It is the way the findings are *interpreted* that may be judged good or bad. Interpretations vary from insupportable to strongly supported, from low quality to high. In general, overinterpretation of data from sites is an exercise in making one story plausible at the expense of other stories. In some cases, quite different sets of data have resulted in virtually identical interpretations, whereas in other cases similar data have been interpreted in distinctly different ways. Like Rorschach inkblots, the data can have different meanings, depending on the individualized desires and wishes of the interpreter. Therefore, it may seem virtually impossible to standardize for everyone the meanings to be found in the fossil data.

---

[1] Deraniyagala (1955) remarked on the scarcity of elephant bones in archeological sites in Sri Lanka (then called Ceylon), attributing that to human disinclination to hunt such large prey.

A fundamental aspect of interpretation involves deciding whether or not a proboscidean was killed by humans. I am often surprised to see the widely divergent (even contrasting) findings that have been interpreted as reflecting deliberate kills. For example, consider the terminal Pleistocene mammoth sites in North America. A bit more than a dozen sites in the United States have yielded typologically distinctive Clovis projectile points closely associated with the bones of mammoths (or possibly mastodonts, in two cases) (Table 6.1). The radiometric and geologic ages of the sites show great synchroneity of deposition (11,500–10,500 years B.P.). Many mammoth bones at the sites are in partial articular order, unbroken and unmarked. Some of these sites also contain fire areas (possibly hearths), artifacts other than Clovis points, and bones of species other than proboscideans, thus sharing characteristics with each other in addition to the simple association of Clovis points and mammoth bones. We must decide whether or not these sites are examples of the strongest possible evidence that Clovis people killed, butchered, cooked, and ate mammoths.

The Duewall-Newberry site in Texas (Steele and Carlson 1984, 1989) contained buried bones of one mammoth in a spatially limited cluster that lay bedded in an alluvial deposit. The geochronological age of the bones was estimated at 10,000–12,000 years B.P. Prior to discovery, at least one large limb element had been broken by impact along the shaft, and no stone tools or fire features were uncovered. At this site there were no conventionally recognizable artifacts other than some modified bones, although the modifications could not clearly be proved to have resulted from human behavior alone rather than any other process. Does this site provide acceptable evidence that prehistoric people killed and butchered a mammoth?

Possibly similar cases include a mammoth from Marion County (Ohio), a mammoth from the Cooperton site (Oklahoma), the Rappuhn mastodont (Michigan), and the Grundel mastodont (Missouri) (Table 6.2). The Marion County mammoth (*Mammuthus columbi*) (M. Hansen et al. 1978) was recovered from a silty sand that also contained wood, twigs, shells, and spruce cones. The assemblage included broken long bones, bone fragments, and vertebral fragments, some of which had "nonrecent broken edges," due to preburial breakage (M. Hansen et al. 1978:104). Wood preserved at the site was radiocarbon-dated $10,340 \pm 125$ years B.P. (Di-Carb-799). The skeleton originally lay in a shallow near-shore area of an early postglacial lake that was at least 500 m in diameter.

The Cooperton mammoth site in Oklahoma contained some bones from a young animal concentrated in an area 2 by 3 m (A. Anderson 1962, 1975). The radiocarbon dates for bone apatite were $20,400 \pm 450$ years B.P. (GX-1216), $19,100 \pm 800$ years B.P. (GX-1214), and $17,575 \pm 550$ years B.P. (GX-1215). Large rocks and boulders were closely associated with the bones, some of which were fragmented. Bonfield (1975) thought that the bones had been fresh when broken; Mehl (1975) thought that some bones had been "hand-placed" or stacked up unnaturally.

The Rappuhn mastodont (*Mammut americanum*) site in Michigan contained bones from an old male animal lying atop glacial till, covered by peat in a swamp formed in a kettle hole (Wittry 1965). The bones were scattered, although the rib cage was nearly in anatomical order as if it had simply collapsed in place. The bones of the right rear leg were also in order, but not articulated, whereas the bones of the left rear leg were. Several bones were not found, including the pelvis, both scapulae, some vertebrae and ribs, and several leg bones. Foot bones were scattered widely and "charred in a manner which indicated that flesh and cartilage [had been] present" (Wittry 1965:17). Other bones also appeared partly charred. Scratch marks on the skull and some vertebrae were interpreted as knife cuts (Wittry 1965:18). A layer of wood poles, each 7–8 cm in diameter, lay between the skull and the postcranial bones, a distance of roughly 6 m. Some beaver-cut sticks were lying over the pole layer, which was thought to be a platform used to support the butchers, although in fact it may have been a beaver lodge or dam. Kaye (1962) described early postglacial beaver dams or lodges in New England deposits that seemed similar. Kaye also added that "some logs in the beaver structure are partly charred, undoubtedly as a result of forest fires" (Kaye 1962:906).

The Grundel mastodont site in Missouri contained bones from one old adult animal lying in a loess deposit, but no stone tools were in association. One tusk had suffered a "green-bone fracture," and several bone fragments may have been produced by "percussion" (Mehl 1967:21). The ^{14}C date on charcoal found at the bone level was 25,100 ± 2,200 years B.P. (I-559). All large long bones were missing, as were both scapulae, the foot bones, and most of the skull.

Some scholars have no intellectual problem in accepting the Clovis evidence as a clear indication of human behavior. On comparison, should the finds at Duewall-Newberry, Marion County, Cooperton, Rappuhn, and Grundel be admissible as evidence of human behavior? It has been argued that if we accept as valid only those prehistoric sites that provide a full range of well-preserved evidence in ideal contexts, then we run the risk of overlooking many less well defined traces of past human behavior (Morlan 1980; Morlan and Cinq-Mars 1982).

In an earlier study (G. Haynes and Stanford 1984) I identified three different kinds of potential links between human behavior and preserved animal remains; in evaluating proboscidean sites, I have tried to interpret the evidence from each site using those three categories:

1 Contemporaneity, referring to geochronological coexistence within time periods defined by the available geologic, radiometric,

and stratigraphic data, with no direct archeological evidence in site collections or deposits to demonstrate that humans and animals encountered each other face to face.

2   Association, in which archeological evidence indicates that human behavior affected the animal remains; for example, bones may have been used in dwelling construction or as raw material for tool manufacture, and so forth. However, in these cases there is no strong evidence that humans actually killed or butchered the animals; the evidence indicates only that bones were used by people.

3   Killing and utilization, referring to human involvement in butchering animals after having slain them. The evidence for killing may be the existence of tools, such as projectile points found bedded within the scatter of bone remains; the evidence for butchering may be the existence of cut marks or impact fractures on bones. Utilization may also refer to human scavenging of carcasses, as distinguished from the actual killing of animals. Recent literature seems to display a fixation with making the interpretive distinction between hunting and scavenging, although foraging people themselves may not have seen much difference in terms of effort or expected returns.

The last category may seem to be based on idealized, arbitrary, and unnecessarily conservative standards, but these standards have been derived from examination of actual ethnoarcheological situations in which documented cultural activities have left material traces that seem to mimic prehistoric traces (e.g., Binford 1978, 1981; Crader 1983; O'Connell 1987; Yellen 1977). Often these standards may be scarcely attainable, but that is not sufficient reason to lower them to accommodate inadequate sources of data.

6.3.   An archeological story: mammoth hunters of northern Eurasia

Archeological stories have simple standards, none stronger than believability or data accommodation. Few people in the world, and probably no archeologists, are unaware of the interesting stories told to accommodate the large features of mammoth bones in late Pleistocene sites in central and eastern Europe. Abundant artifactual remains are associated with these features, and human behavior undoubtedly is somehow reflected in the assemblages. But the question is this: How much human behavior is revealed, and how much is devised by our imaginations?

Most of the sites created by Upper Paleolithic Eurasian cultures that are thought to have specialized in mammoth hunting are not interpreted as actual death sites of proboscideans, but as human encampments or processing stations. The large dwelling sites in central and eastern Europe, such as Dolní Vestonice, Czechoslovakia (Klíma 1954, 1963), Předmostí, Czechoslovakia (Absolon and Klíma 1977), Krakow-Spadzista Street, Poland (Kozlowski et al. 1974), and Mezhirich, USSR (Pidoplichko 1969), contained stacked tusks and bones from dozens of mammoths, hundreds of stone artifacts, bone and ivory items that were drilled, cut, carved, polished, and painted, as well as burned bone, baked clay figurines, multiple hearths, human bones, and the remains of many steppe-dwelling or taiga-dwelling mammalian taxa, such as arctic fox, wolf, reindeer, wolverine, horse, willow grouse, lion, and bear. The stone inventories usually were of the classic Upper Paleolithic types, including microblades, burins, microburins, raclettes, drills, bifaces, Levallois flakes, unifacial points, end scrapers, microcores and macrocores, backed blades, notches, and so forth.

These huge, rich sites, interpreted as encampments of mammoth hunters, often were located at a confluence of rivers or near springheads. Many radiocarbon dates for these bones cluster around 15,000–25,000 years B.P. (e.g., Kubiak and Zakrzewska 1974). It is possible that the occupation of such sites spanned many centuries, perhaps even millennia (Klíma 1954; Shovkoplyas 1965). Some sites have been geologically dated to later times, and artifact inventories also place some components to later times. Great piles of animal bones often surrounded the hearth areas and dwelling loci in these sites, with most elements disarticulated and found in heaps. Many bones showed signs of carnivore gnaw damage (Shovkoplyas 1965:101, 1976:27) or different degrees of weathering (Klíma 1954; Shovkoplyas 1965, 1976). The archeological consensus seems to be that the remains derived from animals that had been hunted, butchered, and transported back to a homesite. Following the opinion of Vereshchagin (1967:379–80; Klein 1973; Soffer 1985), I propose that the bones were naturally abundant nearby before humans became involved with them, because the heavier elements, such as skulls with tusks, would seem to have been excessively troublesome to transport any distance, although a shortage of brush and wood for building shelters possibly could explain long-distance transport.

Bone accumulations in the western part of the Soviet Union may have been hoarded by humans from redeposited bone heaps found along the larger rivers, perhaps similar to the situation recorded in the Volga River valley of the USSR or the Old Crow River valley in the Yukon today. Vereshchagin (1967) suggested that mammoth remains accumulated in ravines after spring floods had transported fresh carcasses downriver; additionally, enormous catastrophic death episodes may have resulted from heavy snowfalls or vast seasonal floods.

Because the large accumulations of mammoth bones in eastern Europe and northern Asia, such as those at Berelyokh and Volchya Griva, have not been well described in the English-language literature, it will be worthwhile to reexamine our interpretations of these kinds of sites. There are over 1,000 Upper Paleolithic sites or collecting localities in the Russian plain, which is the southern part of the European USSR (Beregovaya 1960, 1984). Yet fewer than fifteen Upper Paleolithic sites in the region are mammoth-dominated and contain huge piles of bones from dwelling structures. Nevertheless, these mammoth-centered sites sometimes are considered typical of Upper Paleolithic sites in the USSR and adjoining Europe, to judge by the disproportionate coverage of them in the available literature. These sites clearly are unusual and in fact are numerically rare. The largest of them or the largest collections from multilayered sites are also the most recent, dating to the final four to seven millennia of the Pleistocene. Faunal collections from most Upper Paleolithic sites contain far fewer mammoths than other taxa, but more information has been published on the spectacular mammoth-centered assemblages, making their attributes pre-eminent in the awareness of prehistorians. For example, the famous Kostyonki (usually spelled Kostenki) series contains twenty-five sites, of which ten are multilayered and only five have mammoth-centered levels (Table 6.5) (Praslov and Rogachev 1982).

The pattern is similar in the next major geographic region to the east: western Siberia, which lies between the Ural Mountains and the Yenesei River (Figure 6.1). Mammoths are not the most abundant taxon in the vast number of sites from the Late Sartan time interval (approximately coeval with the Late Wisconsin in North America). Horse, deer, bison, and sheep outnumber mammoths at most bone sites, with a few notable exceptions:

1   Volchya Griva, a ridge top where test excavations yielded 1,380 bones representing eight mammoths, of which only one was adult. There were also three *Bison* bones, five horse (*Equus*) bones, and five bones of wolf (*Canis lupus*) in the area tested. Many limb bones were broken. Only two stone flakes were found at the site. Tseitlin (1979) suggested that the mammoth bones had been deposited thousands of years before humans broke them for tool manufacture, in which case perhaps only the bones of the smaller animals are evidence of subsistence.

2   Shikayevka, where one whole mammoth skeleton was found in anatomical order, along with bones from another mammoth and thirty-five stone flakes or blades lying within the bone bed, as well as bones of reindeer, antelope, hare, fish, and wolf. Shikayevka may be a mammoth kill site, but the faunal list sounds more like a homesite, perhaps a camp established briefly around the slain (or scavenged) mammoths. Alternatively, the site may be a time-averaged or palimpsest deposit, mixing cultural and noncultural bones.

3   Tomsk, where the bones of one still-growing mammoth (an animal

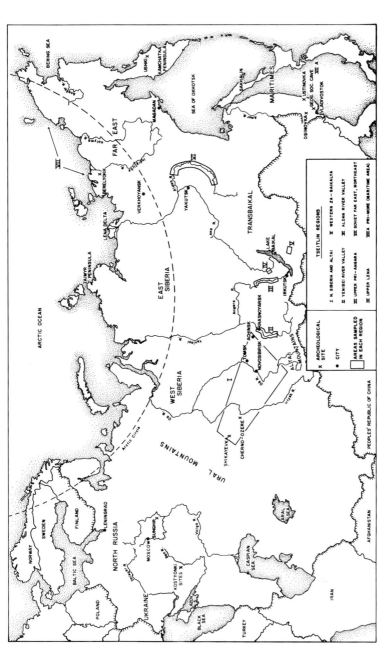

Figure 6.1. Map of the Soviet Union showing Tseitlin's (1979) designation of northern Asia's main archeological regions (enclosed within irregular boxes) and sites. Note how relatively small the sampled regions are. Drawn by E. Paige.

less than thirty-five years old) were found associated with a hearth area (a "bonfire" layer) and about 150 cores and flakes.

The dates for these three sites are estimated to have been between 14,000 and 11,000 years B.P., virtually the end of the Pleistocene. No earlier archeological sites in western Siberia dominated by mammoth have been reported.

Farther east, in the valleys of the Yenesei and Lena rivers, the known Paleolithic sites containing fauna and unequivocal artifacts have been dated between 25,000 and 11,000 years B.P. (Abramova 1979a,b; Oklad nikov 1953). These faunal collections are emphatically dominated by reindeer (*Rangifer*). Reindeer predominate at sites dated early (21,000 years B.P. at Afontova Gora) and at late sites (11,000 years B.P. at Novosyolevo-staroe). Usually reindeer bones far outnumber the bones of the next most abundant taxon; for example, one layer at a certain site contained sixty-seven *Rangifer* bones and only two *Bison* bones, although the MNIs had not been so different. This may have been an effect caused by low total numbers of bones. The sites sampled may have been base camps, where lighter bones were carried, so that the actual kill sites for the larger animals are poorly known. Another potential explanation for the bone proportions is that reindeer were staples, and mammoth, bison, and horse were not heavily exploited.

South of the Lena River, one site was described in such a way that casual readers of its descriptions translated into English would consider it mammoth-dominated. This site, called Mal'ta, contained mammoth bones that may have formed the bases of dwelling structures (Okladnikov 1953). Also, many art objects were made of ivory. However, there were many times more reindeer than mammoths in the faunal assemblage (more than 400 reindeer, versus three mammoths).

In the Aldan River region, at the earlier sites [dated by Mochanov (1977) to 35,000–23,000 years B.P., although some researchers, such as Abramova (1979c), doubt the early ages] the bones of bison and horse far outnumber those of mammoth by ratios as great as 16:1 per level. The total numbers of stone tools at the sites are quite low. Also present are reindeer, wolf, deer, moose (*Alces*), and woolly rhino (*Coelodonta*). Not intending to second-guess the excavators unfairly, I would caution that large bone deposits in open-air and cave settings should be considered as resulting from time averaging or mixing during sediment deposition or excavation, especially if artifact inventories are notably low. Recent ethnoarcheological observations suggest this possibility (Binford 1981; Yellen 1977), and recent taphonomic observations of noncultural deposits indicate it as well (Behrensmeyer 1982a, b; G. Haynes unpublished data).

In the Aldan region, the only stratified archeological site with abundant mammoth remains is Diuktai Cave, dating to about 13,000 years B.P.

(Mochanov 1977; Mochanov, Fedoseyeva, and Alekseyev 1983). In nine cultural levels, some of which are subdivided in the published descriptions, there are 1,180 identified mammoth bone fragments, compared with one to four bones per level for bison, horse, moose, sheep, or wolf. *However, all but two pieces of the mammoth material is comminuted ivory, fragments of tusk that are not worked artifacts.* Admittedly, some unidentified bone fragments in the assemblage may be from mammoth. Diuktai Cave should not be interpreted as a mammoth-centered assemblage, because the ivory break- age may be due to postdepositional destruction, not to tusk processing.

Farther east, few archeological sites containing faunal assemblages have been reported. One of these, the Geographical Society Cave, may also be an example of mixed deposits (Okladnikov 1953; Vereshchagin 1981a). A mammoth tooth and an antler from a deer of unidentified taxon were collected from the surface of the cave-fill, and within the deposits were bones from a horse and a deer, lying over bones from mammoth, bear, hyena, wolf, lion, woolly rhinoceros, horse, deer, and bison. A few pebble cores and large flakes were found in the deposits. Tseitlin (1979) estimated the date at 18,000–19,000 years B.P.

Another Far East site is Berelyokh, a massive accumulation of over 8,500 mammoth bones eroding out of a riverbank north of the Arctic Circle (Vereshchagin 1977, 1981a). More than 100 mammoth individuals were represented in the original bone assemblage, only a fraction of which is now curated at the USSR Zoological Institute in Leningrad. Female mammoths outnumbered males nearly 2:1 (Vereshchagin 1977). Table 6.6A and B presents data from my studies of the collections in Lenin- grad, comparing age profiles determined from teeth and limb bones. Figure 6.2 is an age profile based on teeth in the collections. Figure 6.3 is a far less detailed age profile based on limb bones. The site also contained fifteen bones of other taxa, including wolverine, horse, reindeer, woolly rhino, bison, and cave lion. The deposit probably was not of cultural origin (Vereshchagin 1977). Vereshchagin suggested that the bone assemblage resulted from recurring deaths of individuals and herds over several thousand years. This is the only megafaunal assemblage in the Far East that is dominated by mammoth, but the heart of it contains no artifacts. In an archeological site 100–200 m away there were stone blades, flakes, possibly unutilized wedge-shaped micro-blade cores, and bifaces associated with the bones of hare (the predominant taxon), wolf, fox, goose, and mammoth. There were also bone and ivory items interpreted as percussion-shaped knives, points, and scrapers (Mochanov 1977). The mammoth bones in this archeological site probably had been gathered from the nearby large natural accumulation.

Apart from the archeological assemblages, most noncultural collections in northern Asia (Table 6.4) are dominated by taxa other than proboscideans.

Table 6.6A. *Age profile from Berelyokh sample curated in the Zoological Institute, Leningrad*

Age class (years)	Upper teeth (MNI and %)	Lower teeth (MNI and %)
0–12	8 (42%)	6 (32%)
13–24	8 42%)	7 (37%)
25–36	1 (5%)	1 (5%)
37–48	1 (5%)	4 (21%)
49–60	1 (5%)	1 (5%)
Total	19	19

Table 6.6B. *Berelyokh age profile based on curated elements: number of individuals represented (and % of total)*

Element	Age group 0–24 years	Age group > 24 years
Humerus (MNI = 21)	15 (71%)	6 (29%)
Ulna (MNI = 13)	7 (54%)	6 (46%)
Femur (MNI = 20)	11 (55%)	9 (45%)
Tibia (MNI = 38)	23 (61%)	15 (39%)
Upper teeth (MNI = 19)	16 (84%)	3 (15 – 16%)
Lower teeth (MNI = 19)	13 (70%)	6 (30%)

I suggest that the Paleolithic cultures of northern Eurasia were subsisting on and "specializing" in the numerically more abundant animals in their home ranges and were only opportunistic utilizers of mammoth.

The dramatic image of mammoth-hunting specialists living during the Upper Paleolithic in the cold steppes of northern Eurasia has become firmly established in world archeological thought. It is an image based on a very selective sample. The current folklore about mammoth hunters is an imaginative interpretation that lacks a balanced perspective on the paleoecology and archeology of the whole of a poorly sampled world region. One would expect to find some kill sites or butchering stations left by mammoth killers, not just their dwellings and storage pits (Soffer 1985, 1989), if that is what the bone piles actually are, in our understanding of the words "dwelling" and "storage pits" as they pertain to Stone Age foraging people. As a result of efforts to fit each mammoth-bone-containing site into a preconceived script about mammoth hunters, our current archeological understanding of these and contemporaneous sites in Eurasia neglects to explain much about the paleoecology of the animals represented at the sites. Few meaningful questions about them are being asked anymore, the most important of which may be why there are so many of these sites. We have no real knowledge about how all these animals died, nor about how

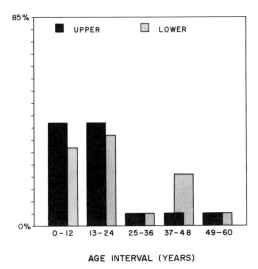

AGE INTERVAL (YEARS)

Figure 6.2. Comparison of different age profiles based on upper and lower teeth from the Berelyokh site, USSR, curated at the Zoological Institute in Leningrad.

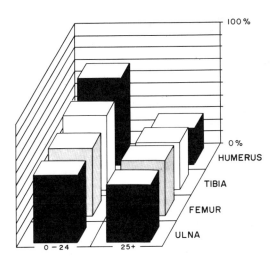

AGE INTERVAL (YEARS)

Figure 6.3. Different simplified age profiles for the Berelyokh site, based on limb bones curated at the Zoological Institute in Leningrad.

many lived at any one time. The dozen or so large sites may be natural die-off locales utilized by nomadic people, or they may be central settlements created by mobile mammoth killers who had developed social and economic inequality (Soffer 1985, 1989). The romance of hunters who lived inside the bones of their prey is exciting and apparently accepted by archeologists and novelists with equal conviction, but it is a case of architectural anthropology: The bone features themselves have fully defined a prehistoric way of life described in the literature. Our ideas about the late Pleistocene people of northern Eurasia are anchored to their houses and larders. But their houses used to be live animals, and it cannot be spontaneously assumed that Eurasian foragers in frozen steppes speared them all so that they could shelter behind skulls and scapulae.

6.4.   Finding meaningful patterns in the fossil proboscidean record

In spite of strong doubts and skepticism, as expressed earlier, I have not completely rejected all earlier interpretations of the sites in the tables. If cornered, I must confess to thinking that many of the stories about the sites probably are correct; I do object to the casual way the data often have been transformed into "evidence," as with the Eurasian mammoth-hunters legend. Some sites will be question marks forever, but others still harbor the additional data needed to solve their puzzles. In the following sections I discuss several potentially meaningful features of fossil sites, including age profiles of mass assemblages, characteristics of element breakage and marking, and body-part representation.

By no means is the discussion to follow intended to be a well-rounded survey of everything to seek before attempting to interpret a proboscidean site. For instance, I do not try here to describe the potential strengths and weaknesses of interpretations of sedimentary data when dealing with proboscidean assemblages.

6.4.1.   Types of age profiles and their meanings

Following the lead of Saunders (1977a, b, 1980, 1984), I have assigned proboscideans (of all geologic time periods) ages in African elephant years (AEY), which is the life-age based on comparison with dental criteria in modern African elephants; note that, following Roth and Shoshani (1988), it is also possible to assign fossil proboscideans ages in Asian elephant years.

I have been able to examine only a part of the fossil sample. Unfortunately the life-age of specimens in many important assemblages was never determined or reported in the literature sources, so few assemblages provide enough data for an age structure to be developed.

In earlier studies I assigned jaws and teeth to two-year intervals

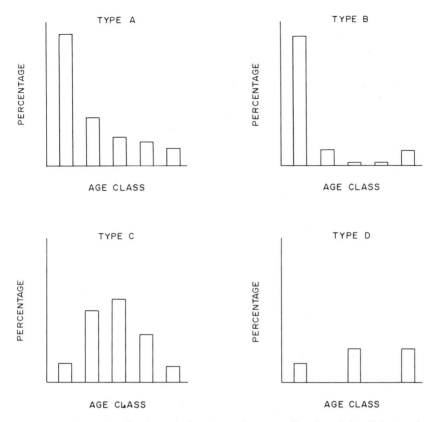

Figure 6.4. The four main types of age profiles found for living and extinct proboscidean bone assemblages. Type A results from nonselective mortality in stable populations; type B results from selective mortality affecting mixed herds; type C results from selective mortality affecting males only, or from nonselective mortality affecting declining populations; type D is patternless, and the causes may be varied.

(Conybeare and Haynes 1984; G. Haynes 1985b); but because of disagreements over ages assignable [using different criteria, such as those of Laws (1966), Roth and Shoshani (1988), or Sikes (1971)], I now refer individuals to twelve-year intervals. The biological significance of the intervals is explained in Chapter 3.

Elsewhere (G. Haynes 1987c) I have described four different types of profiles seen in fossil assemblages (Figure 6.4). In one type (called type A), subadults predominate, and the other age classes are represented in decreasing proportions, as is to be expected in a stable or expanding population.

In type B, subadults are again predominant, greatly outnumbering the mature animals that could have borne them. Therefore they must have been selectively killed.

In type C, subadults are conspicuously rare, and prime-age adults predominate. This type represents selective mortality over an extended period of time. In theory, it may also result from less lengthy nonselective mortality affecting a declining population.

Type D would be seen with small assemblages containing the bones of few individuals, too small a sample or too patternless for an age profile to be clearly interpretable.

Recent field observations have shown that mass deaths of large herbivores are more frequent in nature than had commonly been thought (Berger 1983; Child 1968, 1972; Coe 1978; Coues 1897; G. Haynes 1981, 1982, 1987a, c; Sinclair 1977; Weigelt 1927), although some researchers (e.g., Western 1980) have proposed that such assemblages are in fact rare and that the year-to-year "normal" addition of elements from individual deaths contributes a relatively greater proportion of bones to fossil deposits.

Table 6.7 shows types of age profiles that I have been able to find for each proboscidean taxon in Pleistocene and Holocene sites. Note that there are several known mass sites of fossil *Palaeoloxodon*, such as Torralba, Ambrona, and Ehringsdorf, but they probably are cumulative sites, judging from geologic and taphonomic evidence.

I should point out that age profiles in bone assemblages may or may not reflect age profiles in live populations. And even when two such age profiles are congruent, the potential meaning of the congruence may be contrary to common sense. For example, "youthful" age profiles may not be healthy for proboscideans, although in theory they are thought to be indicative of vigorously growing populations. Moss (1988) studied a somewhat isolated and occasionally stressed population of *Loxodonta* in Amboseli National Park, Kenya, and concluded that if human hunters or ivory poachers removed older adult females (i.e., animals that had borne calves and learned how to rear them successfully), the result would be a disastrous population crash, because inexperienced young elephant mothers tended to lose their calves to death quickly because of inattention or bad decision making. Therefore, although an elephant population with a high proportion of youthful adults (those animals barely over twelve years old) and a very low proportion of females over fifteen to twenty years old fits the model of a "young," reproductively active, vigorously growing population, in reality such a population may be dramatically doomed to die out, unable to recruit new members, because first-time elephant mothers are relatively inept and need older females for guidance and protection. If the maturing females in such a youthful population finally do succeed in rearing a few calves to maturity, those individuals may not be well educated in all the ways of survival, and their calves will be at an added disadvantage.

Table 6.7. *Age-profile types and examples*

Taxon	Holocene sites (Types)	Pleistocene sites (Types)
Loxodonta and *Palaeoloxodon*	A. Cull sites B. See Chapter 4 C. Never found D. Pygmy-killed elephants	A. Never B. Ehringsdorf (?); may be time-averaged C. Torralba/Ambrona; Mosbach (?); possibly time-averaged D. Zoo Park; FLK North
Mammuthus	Extinct	A. Time-averaged: Fairbanks Muck; Berelyokh; La Cotte de St. Brelade; Kostyonki (?); Mezin (?); Mezhirich (?); perhaps Předmostí B. Single event, or short-term: some Clovis sites; Friesenhahn Cave; Mauer C. Short-term: Lamb Spring; Hot Springs; Süssenborn (?); Mosbach (?) D. Most common: some Clovis, many paleontological sites (e.g., Cooperton, Adams mammoth, etc.)
Mammut	Extinct	A. Never (?) B. Never (?) C. Many mass occurrences (?); Boney Spring; Aguas de Araxa D. Most common (Rappuhn, Pleasant Lake, etc.); mysteries: Big-Bone Lick, Kimmswick

Another factor that would depress elephant recruitment is the cessation of ovulation during stress, such as drought (Moss 1988). There may be no estrus cycle and no breeding going on during prolonged periods of stress, although elephants make a quick recovery and show unusually high pregnancy and lactation rates when environmental conditions return to normal.

6.4.2 Age profiles in mass assemblages of Elephas, Palaeoloxodon, and Loxodonta

Mass death sites of middle or early Pleistocene proboscideans are difficult to evaluate simply because the bones usually are discovered during quarrying or other unsystematic processes that destroy stratigraphic contexts. Most often, fragmentary or unassociated teeth form the bulk of the material available for analysis, and only taxonomic study is feasible. Fortunately, a number of European assemblages have provided valuable information about Pleistocene elephant age profiles.

Torralba and Ambrona in Spain are sites with similar sequences of sediment deposition, containing evidence of human occupation dating from the early or middle Acheulean (400,000–300,000 years B.P.) to the late Acheulean (Howell 1966; Santonja and Villa 1990). Both sites contain a series of fluviolacustrine sediments partially redistributed by colluvial action, and a relatively low density of stone artifacts. Freshwater springs may have been intermittently active during the times of human occupation, providing a major attractant for large mammals and people. Fauna from the sites include at least thirty straight-tusked elephants [*Palaeoloxodon* (= *Elephas*) *antiquus*], twenty-six horses, twenty-five red deer, ten aurochs, ten steppe rhinos, and four carnivores (Howell 1966).

Howell's 1966 report was preliminary and has been supplemented by numerous other published papers, but I refer to it here as a main source because it described the assemblages well. Later analyses by Howell and associates will provide some refined interpretations and corrections (F. Howell 1987 personal communication).

At Torralba, semiarticulated and unbroken remains of a large elephant's left side were found in one locus. The pelvis and skull were missing. In association were four flake tools. Other parts of the elephant and the bones of other species were found nearby, along with charcoal, pieces of wood, bifacial stone cleavers, and stone flakes. Two other areas were found containing "extensively broken up elephant bones, representing different individuals, and a few flake tools, especially side scrapers, and a single "biface" cleaver (Howell 1966:123).

Ambrona's lower levels yielded remains of at least thirty *Palaeoloxodon* individuals, some of whose limb bones were missing. It has been proposed

(Howell 1966) that the elephant bones represented numerous different individual kill sites, as well as occasional mass kills. There were clusters of bones, skulls, and teeth; in addition, there were associations of broken and complete elephant bones and the bones of bovids, cervids, and equids. Some areas contained many large limestone rocks, but other areas were clear of rocks. Artifact densities were low, and most artifacts were found near broken bones. Pieces of wood were found that showed marks interpreted as cut scars, use polish, beveling, and possibly burning.

Binford (1981, 1987) has argued that the Torralba/Ambrona assemblages are palimpsests representing bone input from millennia of noncultural processes and occasional cultural activity. A wide range of activities besides butchering seems to be suggested by the diversity of tool forms, and several different episodes of occupation have been suggested to explain the artifact assemblage (Santonja and Villa 1990). I do not see reason to disagree with these new rounds of interpretation. The age profile for the Ambrona elephants (Klein 1987) is dominated by prime-age adults (type C), typical of selective mortality over a drawn-out period of time. Shipman and Rose (1983) and Klein (1987) found data to support the idea of a complex depositional history. Detailed and fine-grained taphonomic examinations of the assemblages may be forthcoming, in an attempt to distinguish the causes of mortality events through time.

Svoboda (1989), in reviewing the literature, suggested that elephants and rhinos were important prey for Lower and Middle Paleolithic hunters in central Europe (*Homo erectus* and archaic *H. sapiens*), but the nature of the sites and their assemblages, as reviewed by Svoboda (1989), makes this kind of statement weakly supportable; many bone collections were made unsystematically, and the behavioral contemporaneity or association of human tools with the bones has not been clearly demonstrated in all cases. Postdepositional processes, such as fluvial disturbance, have affected many of the middle Pleistocene collecting localities.

Famous (and productive) collecting localities in Germany are Ehrings-dorf (Taubach-Ehringsdorf and Weimar-Ehringsdorf, in East Germany), Süssenborn, also in East Germany, and Mosbach, Mauer, Cannstadt, and Emmendingen (all in West Germany). Soergel (1912), using age-determination methods that perhaps would be considered inaccurate today, determined that only 16 percent of the *Palaeoloxodon* individuals (originally designated *Elephas*) at Taubach-Ehringsdorf were "adults," that is, animals having third molars (M6 teeth) present in the mouth at the time of death. Likewise, the Mauer collections were thought to contain a high proportion of "calves" and subadult *Mammuthus trogontherii* (originally designated *Elephas*). However, the Süssenborn and Mosbach assemblages contained 61–78 percent "adults," and the Mosbach *Mammuthus* as-semblage contained over 58 percent "adults." Soergel blamed human hunting for the high proportion of younger animals in the Mauer and Taubach-Ehringsdorf collections. In contrast, according to Soergel, most of the single-animal proboscidean sites in central European loess contained adults and hence were more likely to reflect noncultural deaths. Note that the geochronological ages of these assemblages vary widely; some are Interglacial, whereas others are Glacial.

Günther (1975) examined the quarried *Palaeoloxodon* and *Mammuthus* teeth from Weimar-Ehringsdorf and noted differences between age profiles for the two taxa. At least 100 animals were represented in these samples. The *Mammuthus* sample contained a much lower percentage of individuals less than twenty years old, whereas the *Palaeoloxodon* sample contained a rather evenly distributed spread of animals in most age classes. Müller-Beck (1982) questioned that the differences reflected human procurement preferences and abilities, and he proposed instead that the behaviors of the two taxa probably accounted for the dissimilarity.

Cannstadt and Emmendingen were termed die-off areas (or "dying grounds") by Soergel (1912), because young animals were not overrepresented, as at Mauer and Taubach-Ehringsdorf. However, as I have shown in Chapter 4, often it is die-off locales that provide disproportionate numbers of young proboscideans, as compared with adults.

Sites that contain redeposited or time-averaged assemblages of bones must be carefully analyzed to separate out potential contributory agencies. Unfortunately, the quarried proboscidean bones in these older German collections seldom were sampled with modern scientific standards in mind, and in hindsight their completeness and the distinctions between different chronologically meaningful layers cannot be determined. It seems possible that the materials were derived from paleontological (nonarcheological) palimpsests that also incorporated redeposited or occasionally in situ archeological specimens. The ecological or behavioral meaning of the collections is not evident, and the published interpretations are as yet unsupported hypotheses.

6.4.3.   Age profiles in mass assemblages of Mammuthus and Mammut

The Mosbach (West Germany) *Mammuthus* sample was dominated by "adult" animals. The Weimar-Ehringsdorf *Mammuthus* sample also contained a low proportion of "subadults." Other German sites and assemblages, such as Salzgitter-Lebenstedt, have been briefly described by several authors in German (e.g., Grote and Preul 1978; Staesche 1983; Tode 1953), and even more briefly in English by Müller-Beck (1982). At this particular site, at least sixteen individual *Mammuthus primigenius* were represented, along with eighty reindeer, six or seven bison, four to six horses, two woolly rhinos, and several other individuals of various taxa, including birds, fish, cave lion, deer, and jerboa. About 75 percent of the identified bones and fragments, most of which are broken into small pieces, are from reindeer (Staesche 1983). The geologic age of the assemblage (which is probably time-averaged, although it is considered a hominid midden that has been redeposited downslope) may be over 80,000 years. Neanderthal-type hominids (or pre-Neanderthals) (Staesche 1983) were credited with hunting and butchering all these animals.

Soergel (1912) proposed that the hundreds of Upper Paleolithic Předmostí (Czechoslovakia) mammoths (*Mammuthus primigenius*) were killed by an epidemic, because, according to him, a whole herd was represented, and calves were not abundant, as they would have been in an assemblage created by human predation.

La Cotte de St. Brelade, in Jersey (in the English Channel), a coastal cave located at the bottom of a sheer cliff, contained bones from numerous mammoths (*Mammuthus primigenius*) and woolly rhinos (*Coelodonta*) in several excavated levels (Callow and Cornford 1986). A layer of loess separated two bone subassemblages thought to be short-term accumulations resulting from human activity, such as the butchering of animals driven over the cliff. Hundreds of Middle Paleolithic artifacts were found, and certain characteristics of the bones, such as stacking, smashing, and cutlike incisions, were interpreted as indications of human involvement. Foot bones were missing, and vertebrae were rare. The site has been interpreted as a fall where mammoths in herds were chased over a cliff (Callow and Cornford 1986), but one should not too quickly reject the possibility that the mammoths died on the flats in front of the cave, and their carcasses later floated into the cave or were carried there by people.

Most of the animals were subadults or young adults (Scott 1980, 1984, 1986a, b). Table 6.8 presents data on the age structures from levels A, B, C, 6, and 3 at the site. Figure 6.5 shows age classes by layer and Figure 6.6 treats them all as one sample (i.e., time-averages them). Notice how time averaging changes the various age profiles into shapes typical of stable or expanding populations. According to the site report, the different levels were stratigraphically separated and clearly represented discrete episodes of death and deposition. Older adults are entirely missing from the sample, possibly indicating that a selective mortality process sampled these animals. It is also a possibility that nonselective death events sampled a very young population that lacked older adults, because of earlier selective mortality processes.

Earlier, Table 6.6 presented data from my studies of the Berelyokh site collections curated in Leningrad, comparing age profiles that were determined using teeth and limb bones. Figure 6.2 showed an age profile based on teeth in the collections. Figure 6.3 showed a far less detailed age profile

Table 6.8.   *Age profile from La Cotte de St. Brelade*

| Age class (years) | *n* in layer | | | | | % in layer | | | | |
	6	3	A	B	C	6	3	A	B	C
0–12	2	2	3	3	1	50	25	75	75	50
13–24	2	4	0	1	1	50	50	—[a]	25	50
25–36	0	2	1	0	0	—	25	25	—	—
37–48	0	0	0	0	0	—	—	—	—	—
49–60	0	0	0	0	0	—	—	—	—	—
Total	4	8	4	4	2	100	100	100	100	100

[a] Zero.
Source: Scott (1986a, b).

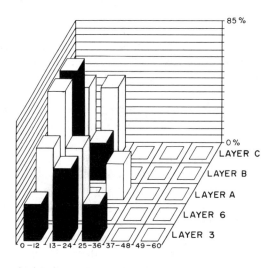

AGE INTERVAL (YEARS)

Figure 6.5. Age profiles from five different strata in La Cotte de St. Brelade.

based on limb bones. The age profile is similar to what I would expect to be produced by a healthy proboscidean population that was nonselectively sampled, such as by mass drowning. Time averaging of isolated deaths might also create such an age profile (G. Haynes 1987c).

The Hot Springs (South Dakota) site contained the bones of at least forty-three mammoths (*Mammuthus columbi*), based on tusk counts, and

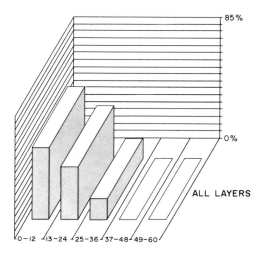

AGE INTERVAL (YEARS)

Figure 6.6. The result of adding all the samples in the different La Cotte strata together and refiguring a time-averaged profile.

perhaps far more (L. Agenbroad 1987 and 1989 personal communications; Agenbroad and Laury 1984; Agenbroad 1990), as well as a few possible *M. primigenius* teeth (L. Agenbroad 1990 personal communication). The bones were in sediments deposited in the warm waters of a spring-fed pool occupying a sinkhole. Some elements in a few associated skeletons were articulated. A few teeth or bones of other taxa, such as wolf (*Canis* cf. *lupus*), coyote (*Canis* cf. *latrans*), American camel (*Camelops*), a peccary, and a raptor, probably washed into the deposit from adjoining land surfaces and may not represent animals that died in the pond, although several bones of an individual short-faced bear (*Arctodus*), were recovered in one part of the deposit (Agenbroad 1990). A very small sample of microfauna was recovered after great effort, but the paleoenvironmental meaning of the subassemblage is difficult to specify. Radiocarbon dates for the apatite in mammoth bones were $26,075 \pm 975$ years B.P. (GX-5895-A) and $21,000 + 700, - 640$ years B.P. (GX-5356-A). Because apatite was used in the dating, both age determinations may be subject to doubt.

Most mammoths from the site were young adults (probably male, judging by their large sizes) that apparently had become stranded or stuck in a pond with steep sides and slick walls (Agenbroad and Laury 1984). Figure 6.7 shows the age profile for the sample uncovered to date, as determined by my methods (see the Appendix), not necessarily identical with the profile determined by Agenbroad (Agenbroad 1990).

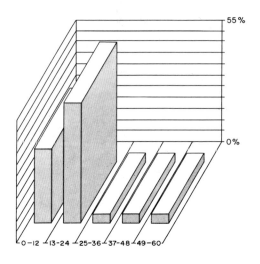

AGE INTERVAL (YEARS)

Figure 6.7. Age profile for the *Mammuthus columbi* sample uncovered to date in the Hot Springs mammoth site. Data from L. Agenbroad.

The Lamb Spring (Colorado) site contained spring-laid and fluvial sediments overlain by windblown and colluvial deposits. The basal layer was an aquifer, through which a spring-fed stream channel was cut in late Pleistocene times (Stanford, Wedel, and Scott 1981b). Within the channel lay a large quantity of mammoth (*Mammuthus columbi*) bones, as well as fewer elements from camel (*Camelops*), horse (*Equus*), canids, cervids, birds, and other small vertebrates. The radiocarbon dates were 13,140 ± 1,000 (M-1464), 12,750 ± 150 (SI-6487) (dating plant fragments in a level below the bones), and 11,735 ± 95 years B.P. (SI-4850). Accelerator radiocarbon dates of 17,850 ± 550 (AA-3574), 16,790 ± 310 (AA-2823), and 14,500 ± 500 years B.P. (AA-3575) for fossil insects indicate that dry and cold steppe or grassland was followed by wetter and cold tundra. Subtle reworking of organics by the spring-fed flow complicated sedimentary analyses of the site (Elias 1986; Elias and Toolin 1990). Some large bones were broken and flaked, and it has been suggested that human actions accounted for the modifications (Rancier, Haynes, and Stanford 1982; Standford, Bonnichsen, and Morlan 1981a). The mammoth teeth and jaws from the site generally were small, but a few of the same life-age were large, indicating that some males died at the site. A few pathological elements were present, including two co-ossified cervical vertebrae, a healed broken rib, and several maloccluded teeth. The frequency of such pathological findings was low in the collection, and none of the abnormalities seems to have been serious or

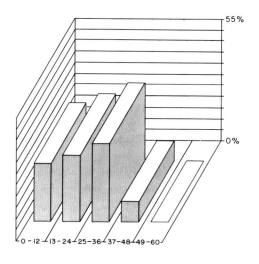

AGE INTERVAL (YEARS)

Figure 6.8. Age profile for the *Mammuthus columbi* sample excavated from the Lamb Spring site.

debilitating. Figure 6.8 shows the age profile that I have determined from analysis of dentition.

Friesenhahn Cave in Bexar County, Texas (Evans 1961), contained the bones of numerous taxa, but the assemblage was dominated by hundreds of teeth from subadult *Mammuthus* and bones from *Homotherium* (*Dinobastis*) *serum* (the scimitar cat). Graham (1976, 1984) determined the age structure for the hundreds of mammoths represented at the site (approximately of Glacial Maximum age) and found a nearly exclusive preponderance of individuals under five years of age. In contrast, the scimitar cat population showed a bimodal age distribution containing both juveniles and old adults. Graham interpreted the assemblage as the result of *Homotherium* predation on young proboscideans and transport of body parts back to the cave, which was being used as a den by the cats. Tooth marks appear on some bones, possibly inflicted by *Homotherium* incisors (Marean and Ehrhardt 1990).

The Waco (Texas) site was discovered in 1979, and excavations continued through 1988. Details have not yet been published. At least fourteen individual mammoths (*Mammuthus columbi*) are represented by bones at the site, possibly lying in a buried arroyo or stream channel. Many animals may be female. Efforts to obtain radiometric dates have produced different results of 17,700 ± 400 and about 28,000 years B.P. (C. Smith 1987 personal communication). I determined the ages of ten individuals, and the

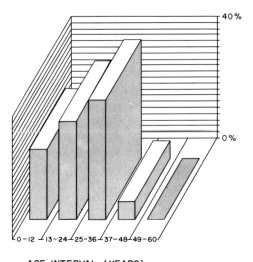

AGE INTERVAL (YEARS)

Figure 6.9. Age profile from the Boney Spring *Mammut americanum* assemblage.

age profile for the assemblage is weighted toward younger animals, most of which probably were subadults.

Saunders (1977b) reported on excavations at Boney Spring (Missouri) in which the remains of Pleistocene fish, ostracods, amphibians, reptiles, many rodents, horse, tapir, deer, and thirty-one mastodonts (*Mammut americanum*), were found along with wood and chert fragments in a gray clay bed 4 m below the present ground surface. The spring probably was active intermittently; most of the bone bed was formed between 16,500 and 13,500 years ago (Saunders 1977b:110). Saunders postulated a ponded environment of bone accumulation and preservation, the waters later becoming unconfined in a shallow depression subject to evaporation and irregular recharge by the flooding Pomme de Terre River. Mastodont trampling may have embedded some bones in older sediments.

The age structure for the mastodont sample (Figure 6.9), if it resulted from nonselective mortality, is characteristic of a declining population. Saunders interpreted the assemblage as resulting from a short-term interval of mortality due to drought and decline in habitat quality (King and Saunders 1984) during the last deglaciation of the Pleistocene.

An undated (but probably late Pleistocene) site in Minas Geraes state, Brazil, yielded bones of thirty to forty mastodonts (specifically gomphotheres), *Haplomastodon waringi* (Simpson and de Paula Couto 1957). I refer to the site here only for comparison with mammutid sites such as Boney

Spring, because a comparative survey of South American proboscidean assemblages is not yet possible. The sample was considered to be from one population of animals that had died in a mineral spring forming a pothole in a streambed. No skulls were found, and few complete bones. About 2 percent of the bone sample recovered was not mastodont. There was no bimodality of tooth sizes, which perhaps indicates a unisexual sample, because the known proboscideans are sexually dimorphic. Most mastodonts from the site were over fourteen years old but younger than fifty years, based on my rough extrapolation from the data of Simpson and de Paula Couto (1957:139).

6.4.4. *General observations on the profiles*

As can be seen from this survey of the better-known sites of mass occurrences dating to the last part of the late Pleistocene, the extinct taxa are more often found in abruptly formed mass assemblages than are the extant taxa *Elephas* and *Loxodonta*. It should be remembered that mammoth skeletons and teeth probably have been found more frequently in single or type-D sites than in mass sites. Some apparently "catastrophic" (type-A) mammoth assemblages are time-averaged accumulations, although others could be short-term or abruptly created assemblages.

Proboscidean mortality quite often is density-independent; that is, the processes killing elephants are not brought on by a high density of elephants in local ranges. Drought or environmental stress will kill particular animals whether there are few or many of them. Elephants die when they run out of food, as documented by Corfield (1973), Hanks (1979), Conybeare and Haynes (1984), G. Haynes (1985b, 1987a, c), and Moss (1988), but in many recent cases the elephants have not been to blame for the loss of resources – environmental causes have been behind the die-offs. Of course, this does not mean that all prehistoric mass mortalities were density-independent.

Most modern elephant mortality results from selective processes. For example, human hunting for meat and ivory usually results in the death of adult males (see Chapter 7). In some modern cases the age profiles for elephant kills carried out by hunting people over the course of several years have been biased toward adult males, simply because ivory tusks are valued commodities, and male tusks are much larger than female tusks.

The information available from studies of recent elephant hunters does not provide an especially good base for modeling age profiles resulting from ancient proboscidean hunting, but it needs to be considered. Elephants are not the preferred prey of the Luangwa Valley Bisa people of Zambia, but when elephants raiding their crops are shot, those elephants are utilized (Marks 1976:205). In the past, an elephant hunter was a high-status individual, but that status seems to have derived as much from the procurement of ivory and tail hairs (which were political or commercial by-products) as from procurement of the meat, which was used not only for subsistence but also for fulfilling social obligations. The age profiles for elephants killed by Bisa during the past century probably would be dominated by adult males. A detailed age profile for recent crop raiders shot by game guards, plus the rare elephant hunted by the Bisa, is not available for comparison, but the shot sample does appear to be biased toward adult males (Crader 1983:Table 1; Marks 1976:Table 39, Table 41). Likewise, it is to be expected that recent Pygmy hunters in Zaire have also tended to kill elephants with relatively large tusks, because ivory may be as important a harvested product as is meat. In fact, some ethnographers have speculated that Pygmy hunting of elephants is a very recent activity, motivated by trade agreements with neighboring blacks. Big game of any sort may have been rare in the diet of Mbuti people before they acquired iron for spear points. Schebesta (1932) doubted that the Mbuti ever hunted large game until neighboring blacks traded them iron in exchange for ivory, meat, and other goods, with the understanding that the iron was to be returned if these valued commodities were not provided on a continuing basis. Bahuchet and Guillaume (1982) also discussed the changes in Pygmy hunting strategies that resulted from contact with Iron Age blacks and colonial whites. J. Fisher (1988 personal communication) noted that in the cases of elephant killing that he witnessed in Zaire, sedentary, village-living horticulturalists got more meat from elephants killed by Efe Pygmies than did the Pygmies themselves.

Because the information available on the preferences of elephant hunters is inadequate, it is impossible to determine if African hunters more often deliberately selected older male elephants for their ivory, but I believe that to be a reasonable proposition.

In addition to the problem of holes in the factual record, the literature on African hunting people has been transformed from straightforward knowledge into dramatic, imaginative "ethnographies" written by visitors with little to report but lore. Many descriptions of elephant hunters have reconstructed their hunting abilities and their relationships with game on the basis of sources that have never been specified. For example, D. Holman (1967) lyrically described an elephant-hunting people in east Africa, deeming their economy essential to the ecological health of elephant populations because their hunting acted as the only check on the growth of elephant populations. According to that portrayal (D. Holman 1967:31), the Liangula of eastern Kenya literally lived off the elephant – "hunting it was once their entire life." They allegedly hunted solely for food, never for ivory, and they "would have gone preferably for the younger, more tender animals," leaving the big males "to develop into tremendous tuskers."

Vrba (1980), Brain (1981), and Klein (1982) have argued that predators (such as human hunters) will preferentially kill juvenile big-game animals, as well as some senior or old-age adults. Other predators, such as the gray wolf (*Canis lupus*), select these same age classes (R. O. Peterson 1986). Animals in these classes tend to be less experienced, smaller, or weaker. Elephant-hunting people seeking only subsistence from their prey, not maximized profit from commercial products, if such a fabled people existed, logically would have selected mostly subadults or senescent adult elephants as prey, and possibly more females than males. Age profiles would have been dominated by animals under twelve to fifteen years of age, the typical age of sexual maturity in modern elephants (Eltringham 1982:Table 4).The age of sexual maturity would have varied from habitat to habitat, and might have been somewhat different for *Mammuthus* and *Mammut*, but the distinctive age profile resulting from meat hunting should be clearly recognizable.

Abrupt mass-death processes kill off mainly subadults and older female adults. Single old adults die from accidents or disease; single subadults tend to die from predation. However, if these selective mortality processes create events that are time-averaged within single localities, the resulting age profiles probably will be of type A, very similar in form to the profile produced by an abrupt, nonselective type of mortality event, such as a mass kill. Finely detailed analyses of sedimentary evidence would be necessary to distinguish whether bone assemblages were formed abruptly or over a lengthy time span and would be the necessary factors distinguishing two quite different kinds of mortality processes that create two similar age profiles.

Abrupt die-offs create type-B age profiles. Time-averaged selective deaths create type-C age profiles. I suggested earlier that type-C profiles could also result from nonselective die-offs affecting a population that lacked young animals, in other words, a declining population destined to die out eventually.

Time-averaged (but nonselective) deaths create type-A age profiles, and abrupt, nonselective kills also create type-A age profiles. A series of repeated single deaths (selective, type-D) could be time-averaged into type A.

In North America, the largest mass assemblages of mammoths (such as the Clovis sites) are most often of type B *when chipped stone artifacts are clearly associated with the bones*, or type C *when artifacts are not present* (such as at Lamb Spring or Hot Springs). These are interesting correlations, although they may not hold up as new sites are discovered and old sites are reanalyzed. It is deceptively easy to find a cause – effect relationship between an unmistakable human presence and an overrepresentation of young animals in site assemblages, but I caution that both of those characteristics may themselves be the effects of causes that are as yet unknown.

In Eurasia, mass mammoth sites usually are of type A or B, although so little is known about the age profiles of large assemblages that possibly more type-C sites exist. I am especially interested in determining the age structures for the hundreds of mammoths at Dolní Vestonice, Předmostí, Mezin, and Mezhirich.

As for mastodonts, few mass sites are of type A or B; most mass occurrences seem to be of type C. Two big mysteries concern the age structures for Big-Bone Lick and Kimmswick, two of the largest (but insufficiently reported) mass occurrences in North America. The age structure for the Brazilian gomphothere mass death site Aguas de Araxa should also be studied in depth.

Elephas, Palaeoloxodon, and *Loxodonta* left no Pleistocene mass assemblages that were abruptly formed, or none that have been reported yet. The known instances of mass occurrences apparently involved extended time spans in the creation of the assemblages. Other prehistoric elephant assemblages are of type D. Modern elephant bone sites resulting from human kills are of type D, or perhaps type C when ivory hunting is heavy. Environmental die-offs produce type-B sites. Type-A sites result from recent culling. Early ivory hunters rarely shot whole herds or family groups together, so the existence of type-A bone sites is most likely a phenomenon of only the past thirty years or so, when culling (as opposed to selective cropping) began as a technique for managing elephant populations.

Is there any general meaning in this array of age profiles? It should be kept in mind that the existence of many type-D and type-C sites may be a result of poor bone preservation; smaller bones of younger animals could have been filtered out of the assemblages by diagenetic processes after deposition, thus altering type-A or type-B age profiles into type-C profiles. There are so many potentially complicating factors and so little information available about the fossil record that an attempt here to define large trends in age profiles would be premature and of limited reliability.

I do not accept the theory that *Elephas* and *Loxodonta* were almost never hunted in prehistory, which could account for the relative scarcity of fossil "living floors" containing their bones. As support, it would not necessarily be wise to refer here to painted, pecked, or incised rock art depictions of elephant hunts – which to many archeologists indicate that elephants were sought by human hunters (Figure 6.10). As Garlake (1987) has discussed, the large mammals depicted in some African rock art need not have been actually attacked: The art may have been symbolic, mythical, or otherwise not a direct depiction of hunting events; see the breakthrough studies of Lewis-Williams (1984).

The basis for my belief that African and Asian elephants were indeed

0   4   8 cm

Figure 6.10. Rock painting of a prehistoric elephant hunt in Zimbabwe. Redrawn from Peter Garlake (1987). *The Painted Caves*, Figure 52, published by Modus Publications (Harare, Zimbabwe).

killed from time to time in prehistory is mainly the plausibility of the idea; admittedly it is an article of faith, because fossil data are rare. *Elephas* or *Loxodonta* individuals most likely were not often killed at water holes, which are the types of sites where bones of mammoths and mastodonts are commonly found. Water-hole sediments bury and preserve bones well and are prime locales for bone discovery. Upland, single-carcass sites preserve bones only poorly, because sediments rarely build up quickly enough to protect bones from subaerial weathering, an inexorably destructive process. There are few upland (non-water-hole) mammoth bone sites to compare with ethnographically documented upland *Loxodonta* sites (which form the overwhelming majority of modern and late prehistoric *Loxodonta* bone sites). Likewise, there are few *Loxodonta* water-hole bone sites to compare with the large numbers of *Mammuthus* bone sites (which were most often formed in water holes or adjacent streams). We have many different samples, with quite different characteristics, formed under different conditions – and the human behaviors supposedly behind the creation of these sites may have been variable as well. The fact that we know of so many mammoth sites containing the bones of numerous individuals does not mean that mammoths were more heavily hunted than *Loxodonta* or *Elephas*; one cannot simply count the identified bones or MNIs and translate those numbers into some set of general human behaviors, such as specialization in mammoths, when the taphonomic biases affecting the samples of the different taxa are so different. These different biases must be explored in future taphonomic analyses before any generalizations about human evolution or hunting behavior can be formulated.

The extinction of mammoths and mastodons was preceded by the creation of die-off bone assemblages within water holes; the assemblages were dominated by subadults when mixed herds were affected (type-B age profiles) or by adults if male groupings were affected (type C) (Table 6.7). We know of virtually no *Loxodonta* or *Elephas* mass death sites dating from prehistory, although in recent times some have been created by human killing in upland (non-water-hole) sites or by noncultural die-offs at water holes. These recent sites resemble the last sites involving mammoths and mastodons before they became extinct. At this point we might begin to fear that the signs are indicating that extinction is near for *Loxodonta*. Perhaps the future success or failure of a proboscidean taxon (i.e., its destiny to survive or die out) leaves its signature in the bone record by the dominant type of preserved age profiles associated with it.

6.5   Bone modifications: distinguishing cultural from noncultural

When proboscidean bones are found fragmented, scratched, or cut-marked in fossil sites, the question arises whether the breakage was done by people or by another agency such as postdepositional trampling. The fragmentation patterns under contention in the literature are mainly spiral (actually helical) fractures, a kind of breakage that may occur in dry bones, but which is found more often in bones that are "fresh." An undefined degree of flexibility must remain in compact bone tissue if it is to fracture in the spiral manner. Fossilized bone breaks in linear patterns if it is brittle, as is often the case, whereas fresh or nearly fresh bone tissue breaks in more helical or curving patterns, because of the preservation of collagen. Yet bones that are thousands of years old may retain enough flexibility to break as if fresh. I shall mention three examples of fossil bones retaining this capacity: the Warren mastodon (New York), the Oak Creek mammoth (Nebraska), and the Inglewood mammoth (Maryland). Discussions of the mechanics and causal factors behind different types of breakage can be found in the work of Bonnichsen (1979), E. Johnson (1985, 1989), Morlan (1980), and G. Haynes (1981), as well as other sources.

Warren (1855) described the discovery of a nearly complete mastodon (*Mammut americanum*) skeleton lying in a marl buried under peaty deposits in Orange County, New York. The bones lay virtually in anatomical order, with the forelegs "extended under and in front of the head, as if the animal had stretched out its arms in a forward direction to extricate itself from a morass, into which it had sunk" (Warren 1855:5–6). The bones of two or three other animals, supposedly recently mired nonproboscideans, were also recovered from the bog. According to Warren (1855:150–1), the mastodont bones retained sufficient organic matter to make them nearly as elastic as recent bones.

Barbour (1925:108) described the Oak Creek (Nebraska) mammoth (*Mammuthus columbi*) site as containing a "probably complete skeleton" enclosed in a "tenacious clay." The skull was broken into several large and many small pieces. Also found were the scapulae, the humeri, and other elements that were not identified because they had been removed earlier for use as fertilizer. The bones were "black, perfect and tough as modern bones, and splinters of them [were] elastic" (Barbour 1925:108). Barbour remarked on the fact that the relatively thin scapulae were preserved without breaks, whereas other elements, such as one humerus, had suffered damage.

The Inglewood (Maryland) mammoth (*M. columbi*) was recovered below the water table in a very heavy black clay rich in Cretaceous fossils and Pleistocene plant parts (G. Haynes unpublished data). About half of the skeleton, lacking all of the limb bones except one, was found in anatomical order. Most of the lower dentition was well preserved; the animal was about twelve to fifteen years old, which is near the age at which modern elephant males leave herds to go off on their own. The one long bone and the skull were fragmented and flaked by heavy equipment being used to dig a drainage ditch.

Other bones may have been fragmented while still embedded in the matrix under the heavy weight of bulldozers and a dragline, which caused waterlogged sediments to deform and bend. Figure 6.11 shows the site's bedded strata, where a small stream drained the ditch excavated by the heavy equipment; this is the main area of the bone bed. Figure 6.11 also shows the site's highly disturbed strata located 5 m away. If only the top part of this figure were published in a site report, the most likely agency for bone breakage would *not* seem to be sediment disturbance, and in fact a hypothesized cultural origin probably would be favored.

One complete rib was radiocarbon-dated at $20,070 \pm 265$ years B.P. (SI-5357). The bones were associated with bits of preserved vegetation, some of it of probable Cretaceous age, as were invertebrate fossils found in the sediments. Twigs, roots, and leaves in the sediments may have been masticated by the mammoth. A radiocarbon date for the organics placed them at $29,650 \pm 750$ years B.P. (SI-5408), perhaps reflecting contamination by Cretaceous floral materials mixed with the 20,000-year-old mammoth bones and mammoth-deposited plant parts.

These three examples of Pleistocene bones still able to break as if fresh should be sufficient to reintroduce caution into any interpretation of

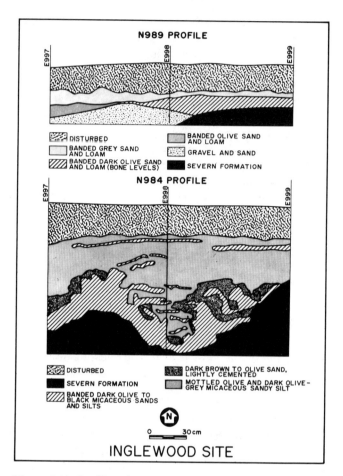

Figure 6.11. Profiles of two vertical strata cuts at the Inglewood mammoth (*Mammuthus columbi*) site in Maryland. At the top (N989) is an exposure adjacent to the loci where the bones were recovered; at the bottom (N984) is an exposure 5 m away, showing extremely disturbed and contorted deposits.

spirally fragmented specimens. This caution is needed to prevent overactive imaginations from creating scenarios involving human behavior as a nearly exclusive explanation for patterned bone breakage and flaking.

Examples of bones interpreted as flaked tools have become common in the literature of the past twenty years. Some bones from the middle Pleistocene sites Torralba and Ambrona showed striations, polish, and "wear," as well as "trimming" and persistent patterns of breakage. Bones of elephant, horse, and aurochs were flaked into a "variety of recognizable forms" (Howell 1966:140), similar in shape or edge treatment to types of stone artifacts. Other problematical middle Pleistocene specimens have been found at central European sites, as discussed by Svoboda (1989). Lebenstedt, in Germany (Tode et al. 1953), contained pointed ribs of *Mammuthus*, among other fragmented bones from other taxa; Bilzing-sleben, in Germany, produced a variety of middle Pleistocene broken bones, including specimens interpreted as scrapers, push planes, knives, flakes, scaled pieces, and anvils (Svoboda 1989).

There are at least four Lower Paleolithic sites in west central Italy that have yielded fragments of elephant bone flaked into hand-ax shapes or "retouched" along the edges: Castel di Guido (Radmilli 1982), Fontana Ranuccio (Biddittu et al. 1979; Segre and Ascenzi 1984), Rebibbia Casal de'Pazzi (Anzidei and Ruffo 1985), and Polledrara (M. Stiner 1988 personal communication). The flaked bone fragments (from *Palaeoloxodon* or *Elephas antiquus*, among a few other taxa) were excavated from Pleistocene deposits dated to 200,000–458,000 years ago by the potassium-argon method. The sites that have been reported in some detail contain what might be secondary deposits. Molar teeth and tusks were proportionately overrepresented in Rebibbia and found mainly in coarser sediments deposited by moderately strong river flow (Anzidei and Ruffo 1985:71–5). The bones at Fontana Ranuccio were very fragmentary, as well as scratched – characteristics quite common in lag or secondary deposits left by fluvial/colluvial processes (Badgeley and Behrensmeyer 1980; Behrensmeyer 1975; Voorhies 1969), though they also can result from differential weathering destruction. The flaked bones at Ranuccio were found just below a layer that was soliflucted and possibly cryoturbated, and the bones were scattered (Biddittu et al. 1979:53; Segre and Ascenzi 1984:231), suggesting that the spatial distribution and specimen morphology were not necessarily direct reflections of human actions alone.

The few illustrated specimens that are purported to be deliberately flaked or utilized bone tools are quite similar in morphology to the flaked bones illustrated from Olduvai Gorge, Tanzania (Leakey 1971), and do seem to look like flaked stone tools. For example, one specimen looks like an Acheulean hand ax. To critically evaluate these bone specimens, which are called examples of a flaked-bone industry, it will be necessary to examine the complete assemblage and compare specimens from the entire range of bone modifications, from those least likely to be artifacts to those most likely to be artifacts. Evidence or arguments that would strongly support the idea that no random processes affected the bones have yet to be presented.

In the decades of the 1970s and 1980s, W. Irving and R. Morlan and their associates recovered scores of broken fossil bones from the Old Crow Basin in Canada's Yukon Territory and described them in numerous publications (e.g., W. Irving and Harington 1973; W. Irving, Jopling, and Kritsch-Armstrong 1989; Morlan 1980, 1986). Included in the sample, and interpreted as evidence of a mid-Wisconsin and earlier human presence, were broken megafaunal bones and several unmistakable artifacts, such as a caribou tibia whose proximal end had been removed and whose shaft had been modified into sawteeth, similar in form to fleshing tools used by Recent northern native cultures. An assay on the tibia's apatite fraction yielded a radiocarbon date of about 27,000 years B.P. (W. Irving and Harington 1973); a later AMS date on this specimen's collagen is about 1,300 years (Nelson et al. 1986). The earlier date is congruent with other dates on fragmented mammoth bone found in the same collecting localities. W. Irving et al. (1989) categorized perceived modifications to the proboscidean bones into several classes, including carving, cutting, polishing, and flake removal. Proboscidean bone specimens classified as flakes and cores, well illustrated by R. Morlan (1980) and W. Irving et al. (1989), led Irving to propose the presence of a "paralithic technology," in which large bones were used as raw material for toolmaking. This "industry" appeared perhaps as early as 300,000 years ago. Stone tools are extremely rare in the Old Crow collections. Although the scarcity of stone artifacts does not fit our typical archeological expectations, even for sites in the middle Pleistocene, there is no theoretical reason to expect that middle Pleistocene people always made stone tools everywhere they lived. However, the broken bones from the Old Crow Basin may not be the technology of humans, and may have been nonculturally modified, a possibility that Irving and Morlan considered unlikely.

Other sites with flaked and fragmented proboscidean bones in North America include Owl Cave, one of three rockshelters in the Wasden Site of southern Idaho. S. Miller (1989; S. Miller and Dort 1978) described this lava tube's deposits, which contained in its earliest archeological levels three fluted points, several small flakes and flake tools, and associated bones of birds, herptiles, small mammals, and large mammals, including proboscideans (probably mammoth). Mammoth bone fragments illustrated by Miller (S. Miller 1989) certainly appear to have been broken when fresh, as well as incised, but some incisions are similar to trample markings and rodent gnawings, and the agencies that caused the fragmentation may not have been solely human. Obsidian-hydration "dates" on artifacts (Green 1983, cited in S. Miller 1989) are earlier than most of the radiocarbon dates associated with Clovis mammoths in other American sites. The dates, including obsidian-hydration determinations on Folsom artifacts, as well as radiocarbon assays on mammoth and bison bones, seem to place the assemblage between about 12,800 and 9,700 years B.P. Because the wide span of dates and the faunal list are not easily reconcilable with Folsom data from elsewhere in the American West, this assemblage may be in need of analytical methods to provide sharper resolution in separating elements mixed together in a late Pleistocene palimpsest. The mammoth bones may have been broken by the people who left the fluted points, or by bison-killing people who came after them, or perhaps by noncultural Late Glacial agencies, such as rockfall.

Provisionally calling all flaked bones artifacts may not be wise, because we simply do not know enough about bone breakage in nature. For example, we do not know how fresh a proboscidean bone must be to fracture in a spiral pattern, nor do we know what natural processes can produce such fractures. Bonnichsen (1973, 1979), Morlan (1980), and Stanford (1979) have argued strongly that humans broke the modified bones found in some fossil assemblages. Biberson and Aguirre (1965) carried out replicative "experiments" on modern elephant bones and demonstrated that human actions could fragment proboscidean bones into shapes similar to those seen in the middle Pleistocene Torralba and Ambrona assemblages. Stanford et al. (1981a) performed similar experiments and reached similar conclusions about deliberate human flaking of proboscidean bones. However, nonhuman causes of breakage have been suggested by other scholars to explain the same specimens. For example, Binford (1981) suggested that thrashing about by dying proboscideans (a behavior documented in Africa) (Coe 1978) could have produced fractured leg bones. Supporting evidence for either side of such arguments will depend on analogies from replicative experiments or studies of living elephants.

Although the arguments and data presented by Biberson and Aguirre, Bonnichsen, Morlan, and Stanford, among others, have not succeeded in proving conclusively that humans broke proboscidean bones in certain fossil assemblages, they have expanded the range of questions we ask of the fossil record. Unfortunately, the viewpoint that only human actions could have modified certain fossil proboscidean bones seems to have become canonical wisdom for some prehistorians, who no longer address the possibility that other agencies broke the bones. The plausibility of human agency was somehow transformed into probability, which was then treated as fact (e.g., Bonnichsen 1981; E. Johnson 1985). Their hunches about some data now actually produce or "confirm" other data, rather than further their interpretations (Hempel 1966).

Publications written by the faction advocating human agency to explain bone breakage emphasize that very specific features of humanly modified modern proboscidean bones are also found on fossil specimens (Bonnichsen 1979, 1981; Stanford et al. 1981a), but quantified and explicit descriptions of these features on fossil specimens have not been published. In my experience of examining the fossil materials, I have found that some features are indeed similar to the features a human can create on modern bones, but not all features are always identical, and quite often some are entirely missing from fossil specimens considered close matches with modern "replicated" specimens.

For example, a bone that a human has deliberately flaked off the thick cortical wall of a modern elephant limb element may have a heavily hammered proximal end, where the striking platform was impacted and deliberately shaped by percussion. According to scholars experienced in replication, the creation of such a platform is an essential step (Bonnichsen 1979; Stanford et al. 1981a) in flaking bone. The tissue is compressed and deformed, the result of platform preparation or of the heavy blow necessary to detach a flake (elephant cortical bone is very strong, because its structure is laminar). Yet specimens diagnosed as flakes and cores from locales such as Lamb Spring (Colorado), Wasden (Idaho), or Old Crow (Yukon) very infrequently possess deformed (compressed) striking platforms. If such platforms are in fact present, they have not been adequately illustrated and described. If originally there had been high-impact, hard-percussor deformation on these fossil fragments, the telltale morphology would not have disappeared after it was produced. And while not all specimens such as cores will necessarily retain platforms, it seems that a relatively high proportion should. Lacking such modifications, these cores and flakes could well have been produced by an agency other than direct percussion.

In addition, certain features of broken bones that have been proclaimed diagnostic of percussion flaking (dynamic impact), such as the presence of hackle lines on the ventral surfaces of flakes, supposedly converging where impact occurred, can in fact be created on bones by processes other than dynamic impaction. For example, the tusks of live elephants will break if bent too far, and features identical with hackle lines sometimes appear on such fracture surfaces, and yet no direct impact was involved in the fracturing (Figure 6.12).

Certainly it has been amply demonstrated that humans could have broken bones and manufactured tools out of them. My reaction to replicated specimens is that such tools would have made poor cutting implements in comparison with flaked stone tools. In the first place, the edges become dull relatively quickly. Second, and more important, bones cannot be broken to make sharp-edged tools unless some kind of very heavy and hard object is on hand to throw or strike against them. Lugging a cobble to a butchering site seems more difficult than carrying a few flakes, but in at least a few cases cobbles have been found near bones (e.g., at Lamb Spring). Most important, sharp-edged stone tools would have had to be already available to cut through skin and muscle in order to expose the bones to hard impactors, unless old bones were being picked up at kill sites and such bones had been cleaned naturally by decay and carnivore feeding. If old bones were being picked up and used for making tools, the fracturing would not always have been spiral fracturing typical of fresh bones.

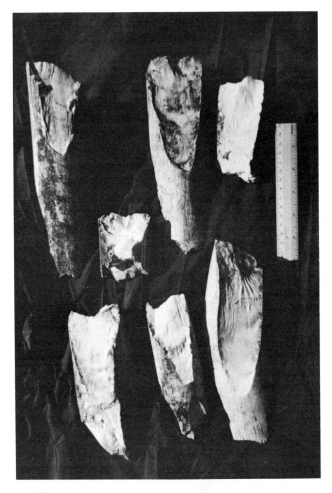

Figure 6.12. A sample of broken tusk midsections showing features identical with hackle lines and concentric ripple marks.

It has also been amply demonstrated that modern noncultural agencies, such as trampling or carnivore feeding, can break long bones spirally. Now that we know that humans could have broken bones, and nonhuman processes also could have broken them, it is appropriate to ask, So what? We certainly shall not prove any favored hypothesis merely by adding more plausibility to it, which is the function that current replicative (and actualistic) studies seem to fulfill. In this controversy, one side seeks to establish the existence of generic processes or agencies that invariably produce easily distinguishable bone modifications; the other side devotes its energy to replying to the challenge by finding exceptions or by finding several different agencies that can produce identical ranges of bone modifications. Both types of studies – the replication of bone tools, and the observations of noncultural bone fragmentation – show us that different agencies plausibly could have affected bones in the past. Lacking in our literature is a well-planned program to prove or demonstrate that one or the other agency in fact did affect specific bones.

We are not yet able to define the "symptomatic" component of the phenomenon in question, which is bone breakage under all the different noncultural conditions that existed in the past. We are therefore not able to determine when we have an eccentric variation, which would be the special cases in which people broke bones, interrupting the bones' trajectory through the process of natural destruction or preservation.

In the symptomatic instance, we may think of the bones' innate responses to certain conditions such as trampling, carnivore feeding, violent transport in water, or even natural weathering in protected microenvironments – the response is to break, oftentimes spirally. There is an inevitability to the breakage that is built into the bone because of the characteristics of structure and the response of the structure to patterned agencies such as trampling. In the second or eccentric instance, the bones' responses depend on special conditions – the bones may or may not break when the special conditions affect the predictable attributes of the bones. For example, a trampled bone that should break spirally may in fact not break spirally, after all, if it has been burned or heated – it may break transversely or longitudinally. Or a bone that is not trampled or fed upon by carnivores may not crack and weather into its "programmed," predictable shapes if the process is interrupted by the actions of an additional agency, such as, for example, humans making bone tools.

We do not know what kind of agency is predictable, and what kind is eccentric, resulting from special cases. For elephant bones, we simply do not yet have the ability to distinguish typical and atypical responses.

In North America some of the important fossil proboscidean sites share similar factors of deposition and bone preservation, as well as bone modification, and these similarities may perhaps provide crucial keys to understanding which bone modifications are eccentric and which are predictable. One important reason that animals were attracted to many sites (which would later give up their bones as fossils) and that the bones were preserved was the presence of water – standing on the surface, flowing through the sites, or found underground in saturated sediments. Wet areas are precisely the locales where *ex vivo* bones are preserved longest in a nearly fresh state, and precisely the areas where relatively large numbers of herbivores traditionally congregate. The weathering deterioration of proboscidean bones is slowed considerably in shaded, moist environments. Bones in these microenvironments during the Pleistocene undoubtedly would have suffered much less stress from temperature and moisture fluctuations than did the elephant bones I examined in the modern drought locales. Modern studies have focused on unusually heavy local die-offs due to drought and forage overuse by elephants (Coe 1978; Corfield 1973), situations in which carcasses are plentiful on dry ground whose plant cover has been seriously depleted.

The bones that are contained within extant or dried water holes and channel deposits probably will look considerably different than do bones lying on dry, bare ground surfaces. Aside from the field studies reported in Chapter 3, no other research seems to be dedicated to finding significant accumulations of modern elephant bones recovered from within existing African water holes or other preserving microenvironments. Yet these are the microenvironments most closely analogous to the major fossil-bearing deposits of late Pleistocene North America. The vegetation surrounding areas where bones are highly visible in substantial numbers usually has been overbrowsed and overgrazed in times of drought, so that most cover has been destroyed, and most of the bare surface and near-surface sediments around the water holes, as well as the exposed water-hole bottom sediments, have been heavily trampled and disturbed by animals seeking the dwindling water. Weakened animals occasionally become mired in the mud and muck of water-hole bottoms or banks (Matthiesson and Porter 1974; Murie 1934; Shipman 1975; Sinclair 1977), and such may have been the fate of Pleistocene mammals. Many of their bones would have become incorporated into the muds and sediments and would have been well preserved, in contrast to the situation for bones left lying atop ground surface during times of drought.

The next step in taphonomic research should be to begin expending much more energy investigating the surficial deposits of the earth in regions where recent disturbance has been minimal. Future studies will look at extant water holes that contain bones contributed by wild animals for which ecological data are available. If data concerning climate and plant communities are available for the past 50–100 years, then we must look for 50–100-year-old bone assemblages in deposits that would have preserved organic material, so that we can see how environmental change or stability affects mortality among living mammals and how the bones themselves have been affected by processes that can be studied in detail. It will be this kind of research that eventually will reveal probable causes of bone modification in nature.

6.6. Worked bone fragments

Shaped items made from mammoth bone are uncommon in North and South America, in contrast to the abundance of implements made from mammoth bone and ivory in Eurasian Paleolithic sites. Shaped implements made from elephant bone are also uncommon in African and south Asian archeological assemblages, although carved ivory specimens are not rare in mid-Holocene and later sites. The use that prehistoric people made of mammoth bones far surpassed their use of elephant bones, judging by the numbers of reported implements. Semenov[2] (1976) examined hundreds of modified specimens made from mammoth bone and ivory from Upper Paleolithic sites (as well as older and younger contexts). Although his studies of wear traces and marks left by certain tool types were valuable contributions, some of his interpretations were presented as proven facts concerning the manufacturing of bone implements, without demonstration that agencies other than deliberate human actions could not have produced the same marks on the specimens. For example, a mammoth cuneiform bone and a carpal bone that appear to have been carnivore-gnawed (Semenov 1976: 176) were called anvils used in the percussion retouching of stone tools. In general, human actions proposed to account for bone or ivory modifications are plausible and quite possible, but in many cases [such as a mammoth humerus "trepanned" by Paleolithic humans (Semenov 1976: 144) or fragments of mammoth innominates used "as vessels" (Semenov 1976: 169)] the evidence in favor of the interpretations is too briefly and unconvincingly presented.

Semenov described and illustrated ivory specimens that had been chopped, notched, chipped, cut, and carved (Semenov 1976: 144F.). He also discussed specimens of mammoth bones that apparently had undergone similar modifying processes. For the most part, even with a mind full of skepticism or edified by extensive exposure to actualistic studies, it is impossible to find alternative agencies other than human actions that would have left the marks and traces described on the specimens. For example, longitudinally split mammoth tusks bearing lengthwise grooves probably were deeply incised with burin-edged stone implements; I know of no noncultural agency that produces such damage. Many illustrated cut marks (Semenov 1976: 171) do not appear to be noncultural created. However, the inferred functions of some sliced, flaked, and broken specimens are very much open to question.

C. Haynes and Hemmings (1968) reported a shaped bone implement (presumably mammoth bone, judging from its size and density) from the Murray Springs Clovis site in Arizona and inferred its use as a shaft straightener (Figure 6.13), based on the plausibility of the idea as well as on interpretive possibilities offered in the literature to explain the function of similarly shaped items from European Paleolithic sites (Figure 6.14).

A whittled mammoth tusk tip from Blackwater Draw (Saunders 1985; Saunders et al. 1990) is reminiscent of ivory sectioned in European

---

[2]More correctly spelled Semyonov in English.

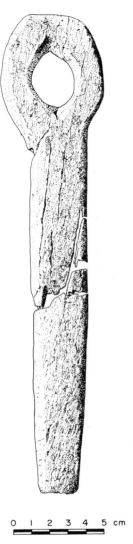

0   1   2   3   4   5 cm

Figure 6.13. (left). A bone tool found in the Murray Springs Clovis site, Arizona. Note the finished shape and the similarity to European bone tools shown in Figure 6.14. Drawn by T. Duggan.

Paleolithic sites and perhaps indicates that ivory working was more common in Clovis times than the relatively poorly preserved American record allows us to know.

"Foreshafts" or "points" of shaped ivory and bone are known from a few North American Paleo-Indian sites or locales, such as Sheaman, Wyoming

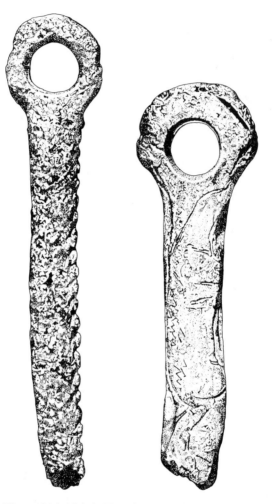

Figure 6.14 (right). Two bone tools from the Pekárna Cave Upper Paleolithic site in Moravia, Czechoslovakia. Redrawn from plate 39 of Augusta and Burian (1960), *Prehistoric Man*, published by Paul Hamlyn (London). There was no scale in the original. Note that these specimens do not have the beveled edge inside the circular hole that was seen on the Murray Springs tool shown in Figure 6.13. Drawn by T. Duggan.

(Frison and Stanford 1982), Blackwater Draw, New Mexico (Hester 1972), Anzick, Montana (Lahren and Bonnichsen 1974), and the Fairbanks, Alaska, mining district (Rainey 1940). Many authors interpret such specimens as foreshafts, which functioned to connect stone points to wooden spear shafts. Some specimens have beveled "bases" and conical "tips"; roughened surfaces on the bases could have been bound against other roughened and beveled surfaces to let friction improve the grip and prevent slippage of the separate parts of a compound spear.

6.7. Miscellaneous other questionable bone artifacts

References to roughly modified proboscidean bones have become more frequent in recent reports from North and South America. Holen and Blasing (1990) and May (1990) reported green-fractured mammoth bones and possible tools made from mammoth bones associated with two small stone flakes, found in loess and dated (on bone collagen) to about 12,000 years B.P. Bullen, Webb, and Waller (1970) described one Florida find of a mammoth vertebral fragment with grooving and "chop" damage considered to be butchering marks. Cook (1931:102) described "broken fossil mammoth bones showing cutting and abrasion marks made when the bones were 'green,' as well as fossil bison bones bearing artificial abrasion marks and cuts along their faces," found in Colorado blowouts where Paleo-Indian projectile points were also found. Dillehay (1986, 1989; Dillehay and Pino 1986; Dillehay et al. 1983) mentioned gomphothere bone fragments considered to have been worked into implements in the Monte Verde (Chile) site, but the specimens have not been fully reported; the importance of this site is its early date of 12,000–13,000 years B.P. Leechman (1950:158) described a bone fragment from Manitoba exhibiting "tool marks" that appear as whittling-chatter ridges and grooves, but this specimen has now been correctly identified as antler, not proboscidean bone (Buchner and Roberts 1990). Rogers and Martin (1989) reported an unusually shaped and modified proboscidean bone fragment recovered from a sandbar in the Kansas River, Kansas. I have already referred to the Duewall-Newberry site in Texas, where no stone artifacts or cultural features were found associated with fragmented long bones of a single *Mammuthus;* the excavators postulated that cultural activity accounted for the breakage (Steele and Carlson 1984, 1989). I have also mentioned the Grundel mastodont from Missouri; one of its tusks had suffered a "green-bone fracture," and several bone fragments may have been produced by "percussion" (Mehl 1967:21).

Large rocks and boulders were closely associated with the bones of the Cooperton (Oklahoma) mammoth, and some elements were fragmented. Bonfield (1975) thought the bones had been fresh when broken.

The Lange-Ferguson site in South Dakota contained the bones of two *Mammuthus columbi* and one closely associated small stone flake. The left front and rear foot bones of one large mammoth were excavated in anatomical order, oriented as if the feet had been buried before soft-tissue decay began. The bone-bearing layer was up to 6 m thick, and the rear feet were about 2 m above the level of the skull and mandible (J. Martin 1987); bones of a smaller mammoth were mixed with the partly scattered larger individual's bones. Some bones had been fragmented and flaked. Three Clovis-type projectile points were found about 15 m from the bones, although enclosed in a similar sedimentary unit. The excavator assigned a cultural cause to the bone modifications (Hannus 1983, 1984, 1986, 1990), based on the presence of the one flake and the nearby Clovis points. Hannus proposed that the broken bones had been butchering tools, although it is equally possible that the bones had been broken to make "blanks" for the manufacture of implements such as foreshafts or shaft wrenches. It is not impossible that the bones were broken by noncultural agencies such as traumatic injury or postmortem trampling.

Sediments overlying the bone layer were dated $10,670 \pm 300$ (I-11,710) and $10,730 \pm 530$ years B.P. (I-13,104). The animal may have been mired with its front end deeper than its rear, although it is also possible that the silty sediments were redeposited into the steep-walled paleovalley at some time after the death of the large mammoth, when the carcass was still partially articulated.

Other materials from around the world are not as well known to English-language readers, especially in North America, but few of the published reports have been sufficiently thorough to be of much use here. For example, a Paleolithic site in northeastern China contained stone cores and flakes, as well as fragmented, scratched, and flaked mammalian bones and teeth, some interpreted as scrapers, points, and a "digging" implement (Sun, Yuzhue, and Peng 1981:291). The extreme fragmentation of the bones perhaps lends them the appearance of numerous different useful objects, but the surface scratching and other signs of redeposition may indicate that noncultural processes modified them.

References to "cut-marked" proboscidean bones have become nearly as commonplace as references to flaked bones. For example, Shipman (1986:703) believed that several early Pleistocene *Elephas* bones from the FLK North site in Olduvai Gorge had been cut-marked. No limb elements were marked, according to Shipman, who speculated that marks on metacarpals and rib shafts were cuts made during attempts to butcher out kidney fat and the fatpad of the forefoot.

Some animal bones from the middle Pleistocene sites Torralba and Ambrona in Spain were scratched, cut, and striated, in addition to the fragmentation referred to earlier.

Four possibly cut-marked bones from a middle Pleistocene (Irvingtonian) mammoth (*Mammuthus imperator*) were discovered at a southern California spring site that did not yield stone tools (Minshall 1988). Photographs and detailed descriptions of the marked bones have not yet been published, but the scholars responsible for the site's discovery are on record as being certain that the marks are not the result of carnivore gnawing or trampling by animals. Minshall (1988) did not shy away from acknowledging that a posssible implication of the site (if it is indeed cultural) is that hominids (presumably *Homo erectus*) were present in western North America between 300,000 and 460,000 years ago. I have not seen the specimens and cannot offer an opinion on the marks, though I am skeptical.

Bones of a single mammoth (*Mammuthus primigenius*) were recovered at Skaratki, Poland, in a sandy silt that also included peat and gravel below the groundwater table. The recovered bones were concentrated in two clusters, one of which contained mostly vertebrae and the bones of one foot in anatomical order, whereas the other contained some bones from the other front foot and fragments of other elements. One bone fragment discovered prior to the excavation had "traces of cutting" (Chmielewski and Kubiak 1962: Plate 8, Figure 1); the modifications on this element appear similar to carnivore gnaw damage, specifically rotational scarring, as described by Kitching (1963). One small flint flake and some small crumbs of charcoal were recovered in the same thick sedimentary level containing the bones, but at least 1–2 m separated from the bones. The bone layer may have contained an ice-wedge cast. Cryoturbation sediment structures also were found in test pits into overlying peat layers. A radiocarbon date on the peat (which may have formed in an interglacial period) was reported as greater than 37,000 years B.P., potentially establishing a minimum age for the bones. The unrecovered bones of the mammoth may have been removed by decomposition, scavenger activity, or human activity (Chmielewski and Kubiak 1962) after the animal became stuck in wet ground. According to this interpretation, the bones that remained in the ground were those that would have been too difficult to retrieve from the peat, or those body parts that human butchers stood on while dismembering the carcass. In my opinion, the possibility that the mammoth was killed and butchered is, unfortunately, vanishingly small, and the cut marks were very likely noncovery created.

Krakow–Nowa Huta, Poland, has been interpreted as a kill site of a *Mammuthus primigenius*, thought to have been slain and butchered in Middle Würm (II) times [Upper Paleolithic (= Gravettian), 30,000–16,000 years B.P.], and then naturally buried in a loess. The animal died not in a bog or mire but on a well-drained terrace surface. The bone inventory includes the left tusk, four upper molars, cranial fragments, four vertebrae (some lacking spinous processes), the sacrum, and the pelvis (lacking parts of the pubis and ischium, and with damage to sacral and coxal tubers). The bones of a collared lemming were also found in the loess, in association with the mammoth. Twelve flint artifacts were recovered from the site, which was discovered during construction within the city of Cracow. The artifacts include nine blades struck from two single-platform cores (Kozlowski 1986). Kozlowski, Welc, and Kubiak (1970) assumed that the butchering humans took away the hide, meat, long bones, mandibles, ribs and right tusk. Three bone fragments have possible traces of cutting and planing; one is a wedge-shaped object made of a scapula fragment, and two other zygomatic arch fragments are thought to have been grooved by stone implements, after which insects burrowed into the bone surfaces. This is another example of the slimmest of cases for the killing and butchering of mammoths by humans, resting on the spatial nearness of articulated bones and artifacts and the probable misinterpretation of noncultural marks as butchering damage.

The disarticulated and broken bones of the Saskatoon mammoth from Saskatchewan were recovered from an industrial sand pit, in addition to pieces of wood, hundreds of bones of other taxa, very small chipped chert flakes, and stones interpreted as artifacts. Several bones had been broken and scratched, but many of the modifications obviously were fresh (Pohorecky and Wilson 1968:39). Some chert flakes were worn smooth along fracture edges. The sand had been deposited by retreating glaciers and was overlain and underlain by till. The bones and chert may not have been in primary depositional contexts, and the evidence for human involvement may actually be the end effects of noncultural processes. A radiocarbon date for one bone was 20,000 ± 500 years B.P. (S-483), which is thought to be too young, because wood in the sand pit was dated at more than 34,000 years B.P. (S-426).

Some bones were missing from the skeleton of a gomphothere excavated at Taima-taima, Venezuela (Bryan et al. 1978), and some bones were "cut" at tendon attachment points (which are the same places that any carnivore would leave tooth marks when attempting to detach soft tissue). The skeleton was lying in a buried fine clayey sand over a cobble-and-pebble pavement, laid down in a late Pleistocene water hole. Bits of wood that may have been chewed by the mastodont were radiocarbon-dated at 14,000–12,500 B.P. Bryan 1973). The pelvis had been "splayed by collapse," and the lower part of the spinal column had been "forcibly turned to the left" (Bryan et al. 1978:1,275). The right-hind-limb bones were in correct anatomical order, but the left-hind-limb bones were disarticulated. All caudal vertebrae and most foot bones were missing. No bones had been fragmented by impaction. The midsection of a quartz El Jobo projectile point was found within a cavity in the right pubis; a utilized jasper flake was found 3 cm from the left ulna. "Rough" stones found around the skeleton were interpreted by Bryan et al. (1978) as pounders, pryers, or other butchering implements. Bones of other mastodonts were found, as well as elements of glyptodont, horse, felid, and bear. Water-worn fragments of mastodont bones were embedded in the underlying stone pavement. Femora of adult mastodonts in the of pavement showed "cuts and striations that were produced by use of the bones as chopping blocks" (Bryan et al. 1978:1,277), damage that perhaps was similar to that on bones illustrated by Rouse and Cruxent (1963: Plate 4b, c) from the same locale. The water wear on some specimens is an important attribute that perhaps should suggest that the assemblage is not pristine, not an instantaneous slice of time and a clean sample of an animal community. Because the projectile point was found next to the animal's pubic bone, it was interpreted as having been thrust into the live animal's body. Using that line of reasoning, it would be equally "logical" to propose that the pebbles and cobbles around the bones had also been thrust into the gomphothere. That is clearly absurd, but it does point out that the practice of picking and choosing materials to be declared behaviorally associated out of a possible palimpsest deposit seems to be no more than sleight of hand.

E. Johnson (1989) compared the bone and artifact assemblage from six Paleo-Indian localities in the southern plains of the United States and concluded that her interpretations of certain characteristics of the bones served to discern the presence of humans and to distinguish bone fractures made by humans from those made by other processes, such as feeding by large carnivores. Of the important localities that she examined, two contained associated Clovis points and mammoth bones (Miami, Texas, and Blackwater Draw Locality Number 1, New Mexico), and another yielded a Clovis point that was not found directly associated with mammoth bones (Lubbock Lake, Texas). Johnson interpreted scratches and incisions on mammoth ribs from Miami as cut, but saw no fragmented elements; Blackwater Locality Number 1 contained incised (possibly cut) and fractured limb elements of *Camelops, Equus,* and *Mammuthus columbi,* but the illustrated "cut lines" (E. Johnson 1989:446) on mammoth elements look similar to trample marks, and therefore the possibility remains that the fragmentation of limb bones may have occurred when they were trampled after slight weathering. Some of Lubbock Lake's mammoth bones were also fractured, but the breakage and scratches on bone surfaces appear similar to the modifications that I have recorded on African elephant bones seen in field studies of noncultural assemblages.

The Pleasant Lake mastodont (*Mammut*) find in Michigan yielded the partial skeleton of one animal, excavated from peat. A radiocarbon date for wood taken from tusk cavities was $10,395 \pm 100$ years B.P. (Beta-1388), and a date for wood lying below the bone level was $12,845 \pm 165$ years B.P. (Beta-1389). Some bones appeared cut, burned, or "worn," and some were fragmented; it was proposed that they had been broken by humans in order to make tools (D. Fisher 1984a, b; Shipman, Fisher, and Rose 1984). D. Fisher and Koch (1983), D. Fisher (1984a, b, 1987), and Shipman et al. (1984) argued that this mastodont (and several others in the Great Lakes area) had been killed by people during preferred seasons of the year, specifically autumn, as opposed to late winter or spring (when natural deaths are expected).

The evidence for human involvement at Pleasant Lake is equivocal. The bones were discovered during dragline excavation, and it is possible that heavy equipment contributed to the bone modifications by compacting the overburden, by causing the embedding matrix to deform and bend, or by directly impacting elements. It appears that some "cut marks" on ribs are in fact not as darkly stained as are the unmodified bone surfaces, perhaps indicating incision after the bones had been degreased and stained by postmortem processes (D. Fisher 1984b: Figure 4c, g). The fracture surfaces of some specimens do show features characteristic of fresh-bone breakage, but bones buried thousands of years in certain substrates can retain this ability. The bones apparently were embedded in an anaerobic, waterlogged environment for thousands of years without interruption.

The elaborate suite of butchering actions proposed to accommodate various modifications to the Pleasant Lake mastodont's bones and body parts seems, when viewed from the perspective of my own experiences with elephant butchering, hardly necessary or logical. For example, a butchering crew that already had sufficient tools to extract a mastodont's stylohyoid, a bone enclosed deep within the soft tissue of the neck, and located below and behind the skull, clearly did not need that particular bone (or any other) to further their butchering of the carcass. Yet, because of the presence of some rounded edges, the one stylohyoid recovered at Pleasant Lake has been interpreted as a butchering tool, used to probe or cut soft tissue. The amounts of time and energy needed to get a stylohyoid out of the body would be excessive and certainly would require the use of sharp-edged tools, whose presence would make the stylohyoid itself redundant as a butchering tool. Furthermore, I can see no economic logic in the postulated combination of actions, such as wedging apart some conarticular bones, but not others, burning disarticulated body parts with soft tissue covering them, but not burning others, cutting and scraping already disarticulated elements, dynamically fracturing some cleaned bones, and leaving some body segments in anatomical order while violently disrupting others. The Pleasant Lake mastodont, if in fact it was butchered by humans, was the focus of far more attention and energy than were needed just to get food or bone tools out of the carcass. Ultimately, the argument in favor of human actions affecting these bones fails, for me, not so much because of utter impossibility but because of its extravagance.

6.8.   Bone representation

For comparison with the bone representation at the Shabi Shabi die-off site in Zimbabwe (see Chapter 4) I have graphed the bone representation at

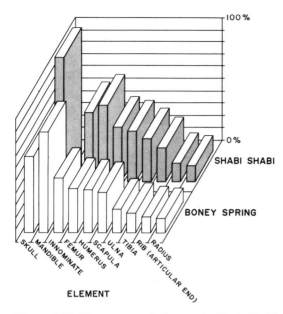

Figure 6.15. Bone representations at the Shabi Shabi die-off site's central area and at the late Pleistocene Boney Spring mastodont site in Missouri.

Boney Spring, a late Pleistocene site in Missouri, where at least thirty-one mastodonts died around a spring during the final deglaciation of the Pleistocene (Saunders 1977b). The representations for the fossil and modern sites are similar, in spite of differences in depositional environments and proboscidean taxa (Figure 6.15).

The bone representation at Shabi Shabi was affected by several factors: Carnivores, which ate the elements (although the process was much reduced compared with what is seen in single-carcass sites) or carried them off for caching; trampling, which destroyed elements; burial, which removed elements from my count; and weathering. The same factors may not have affected the Boney Spring collection in the same ways, but the features of bone robusticity and resistance to mechanical or chemical destruction were similar in the two assemblages, resulting in similar end products of subtractive processes (Behrensmeyer 1988).

I have also graphed the bone representation in a third collection: the Berelyokh mammoth "cemetery" in northeastern Asia (Vereshchagin 1977) (Figure 6.16). Over 8,000 bones and fragments were uncovered at the site, but only about 1,000 were collected and curated in the Zoological Institute of the USSR Academy of Sciences in Leningrad. About one-third of the limb bones and jaws were brought back to Leningrad (A. Tikhonov 1987 personal communication). Figures 6.2 and 6.3 compared age profiles from my analyses of the long bones and teeth; clearly, the jaws and teeth

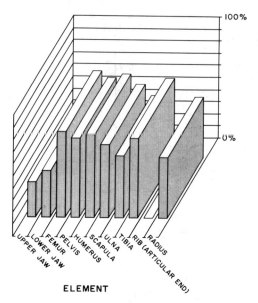

Figure 6.16. Bone representation at the Berelyokh site in the USSR.

represent a larger proportion of the youngest animals than do the curated limb bones.

I suspect that the smaller limb bones of very young animals had been broken in the site more often than had the larger bones of adults and maturing subadults and therefore were not collected in representative proportions. According to A. Tikhonov (1987 personal communication), the curated collection contains the most complete or measurable specimens from the site. The bone representation in the Berelyokh collection is biased by collector preferences. However, Vereshchagin (1977:18–19) reported the total numbers of all elements found at the site, and these can be graphed as bone representation. The numbers were presented in two different tables, with some minor disagreements between totals. I have relied on Table 4 (Vereshchagin 1977:18). Vereshchagin estimated the minimum number of individual mammoths at the site to be 140, although no bones listed in his tables give an MNI higher than 100.

Figure 6.16 shows the bone representation for the Berelyokh assemblage. Although I cannot fully justify doing so, I have taken "upper jaw" and "lower jaw" in Vereshchagin's Table 4 to be analytically equivalent to skull and mandible, respectively.

Notice that the bone representation for this frozen northern Asian site is unlike the representations for Boney Spring and Shabi Shabi. Skulls and

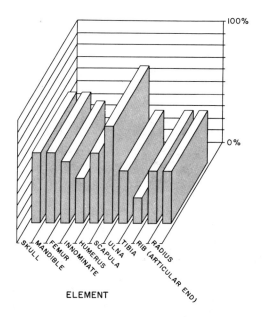

ELEMENT

Figure 6.17. Bone representation at the Colby Clovis site in Wyoming.
Bone elements are graphed in the same order as in Figure 6.16.

mandibles are underrepresented, and tibiae and radii are overrepresented.
The reasons for the differences probably are complex; for example, the
Berelyokh bones may have been affected by ice and flowing water, as well as
by carnivorous animals, weathering, and trampling (the factors that
affected the bones at Shabi Shabi and Boney Spring). The mammoths at
Berelyokh are believed to have died by drowning after breaking through ice
(Vereshchagin 1977). Whether the assemblage represents a single group or
many groups that died over a long period of time is an unsettled issue
(Baryshnikov et al. 1977; Vereshchagin 1977). Perhaps freezing in a
permafrost zone has dramatically altered bone survival.

Because an archeological site is located within 200 m of the mammoth
bone accumulation (Mochanov 1977; Mochanov et al. 1983; Vereshchagin
1974, 1977), it is possible that some bones may have been removed from the
site by prehistoric people, for uses ranging from toolmaking to building
shelters. Skulls with tusks may have been a collector's preference.

Figure 6.17 shows the bone representation at the Colby mammoth site in
Wyoming, a partly redeposited assemblage of disarticulated and articulated body parts (Frison 1976, 1986; Todd and Frison 1986). At this site
the remains of seven mammoths were found in an arroyo, much of which
had been eroded away. Scattered bones and bone fragments were found

throughout the arroyo bottom, and two stacked piles of mammoth bones were found 100 feet apart, both with skulls on top. In one pile of bones, artifacts were scarce, but they included two flakes, one flake tool, one hammerstone, and a small number of flakes produced by use or resharpening. Both piles may have been meat caches, stacked after butchering was complete. The artifacts found included four unusual Clovis-like points, a mammalian humerus shaft with a lengthwise slot missing from it, a *Camelops* radius "chopper," and a fragment of a possible bone foreshaft. Also found were a possible granite chopper, a possible sandstone abrader, another possible bone tool, and a grinding stone with pecking and use striations. The articulated carcass parts may have been caches protected by piled bones.

Note that scapulae were the best-represented large bones in the assemblage, possibly accounted for by their selection as cache guards. Innominates were poorly represented; humeri and femora were better represented than were lower leg bones, following the pattern of Shabi Shabi.

The bone representations at modern African elephant butchering sites documented by Crader (1983, 1984) and J. Fisher (1988) are interestingly similar but also different in several ways. Fisher noted that humeri and femora were less well represented than were tibiae and radii at Efe Pygmy kill sites because these upper leg bones were transported to camps for extracting oils. Crader (1983:128) saw no examples of limb-bone removal by the Bisa people of Zambia, who took only feet back to their village. Crader recorded better representation of humeri and femora than of the distal limb elements, perhaps reflecting no need for limb-bone fat when feet were available. Crader noted that forelimb elements were about twice as common as hindlimb elements, but she accounted for the disproportion by blaming scavengers that followed the Bisa butchering. According to Crader (1983:127), no limb bones were intentionally broken to extract marrow fat.

The mammoth bone representation at La Cotte de St. Brelade, a middle Pleistocene stratified-rock shelter cave on the coast of a Channel Island between England and France, was similar in some respects to the bone representation at Shabi Shabi; for example, skulls were well represented in most excavated levels (Scott 1986a, b). Innominates were either well or poorly represented in different levels. In other respects, such as the relatively higher proportion of scapulae, the La Cotte collection was similar to the Colby Clovis-site assemblage.

The two "dwellings" constructed of mammoth bones at the Mezin site in the USSR (Sergin 1987:50–1) were built with different quantities of bones, but in both of them skulls and scapulae were well-represented elements. Likewise, in the large bone dwelling at Mezhirich in the USSR (Pidoplichko 1969:124), skulls and scapulae were the best-represented elements, appar-

Table 6.9. *Bone representations at three features in the*
Paleolithic sites Mezin and Mezhirich

Element	Number of bones or fragments		
	Mezin dwelling 1	Mezin dwelling 2	Mezhirich
Skull	18	8–9, plus fragments	42
Mandible	21	3	—[a]
Tusk	14	2	—
Scapula	58	6	30
Humerus	See ULB	See ULB	11
Radius	See ULB	See ULB	1
Ulna	see ULB	See ULB	1
Vertebra	—	—	47
Pelvic bone	41	4	11
Femur	See ULB	See ULB	25
Tibia	See ULB	See ULB	3
Bone fragments	37	—	—
ULB[b]	83	24–28	—

[a] No data available.
[b] Unidentified limb bone.

ently followed by femora. The pelvic bones were poorly represented at both Mezin and Mezhirich, as were distal limb elements (Table 6.9).

Figure 6.18 shows the bone representations for two modern (noncultural) bone assemblages that do not contain elephant parts. Included in this graph are data from a bone assemblage of free-roaming Canadian *Bison* that drowned in a flood in 1974 (G. Haynes 1981, 1982, 1988c) and data from an assemblage of large African bovids (weighing over 150 kg) killed by lions and spotted hyenas at a water source in Hwange National Park, Zimbabwe. The bones are not graphed in the same order as on the proboscidean graph with "skull" on the left and "radius" on the right. Notice how the representations are quite different for some elements. For example, tibiae and radii are rather strongly represented in both the African and Canadian assemblages, as they are in the Berelyokh assemblages, but not in the other two proboscidean assemblages. The differences may have to do with bone size and architecture, or with the relative difficulties of joint separation in elephants and large bovids. The Ngamo collection was subjected to more intense carnivore action than was the Lake Claire bison assemblage. The Ngamo/Lake Claire bone representation is most similar to that for Berelyokh, supporting the suggestion that Berelyokh was more

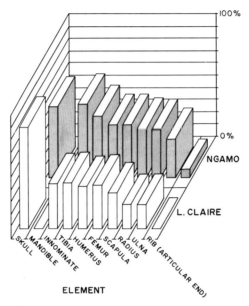

Figure 6.18. Bone representations at two sites containing no proboscidean bones, one in Africa (Ngamo Pan complex in Hwange National
Park, Zimbabwe) and the other in Canada (Lake Claire in Wood Buffalo
National Park). The elements are arranged from most representative
(skull) to least (rib articular end); note how the order is different from the
order of elephant bones in Figures 6.15, 6.16 and 6.17.

heavily affected by carnivores than were the other proboscidean assemblages under consideration here; see the photographs of possibly
gnawed bones published by Vereshchagin (1977).

The bone representations differ from the elephant die-off figures for
numerous fossil assemblages, such as, for example, a noncultural mass death
site of titanotheres dated to the middle or late Eocene epoch (35–45 million
years ago) and located in Wyoming (Turnbull and Martill 1988). The
assemblage may represent a mass drowning. Scapulae are relatively rare in
this assemblage, as are humeri and innominates. However, radii and tibiae
are much better represented than they are in the proboscidean assemblages
at Shabi Shabi and Boney Spring. In this respect, the bone representation in
the titanothere accumulation is similar to the representation at Berelyokh,
although innominates, humeri, and scapulae are better represented at
Berelyokh. Berelyokh also has poor representation of skulls and mandibles.

Differences in body-part representations among various bone assemblages may result from variations in bone sizes, besides the factors of
scavenger activity and weathering destruction rates. A key influence on

body-part representation may be the degree of carcass decay and natural disarticulation that was reached before fluvial disturbance to the remains. For example, some of the mammoth carcasses at the Colby Clovis site may have been partly unbutchered before burial in the ancient arroyo (Frison 1978; Todd and Frison 1986). If the scenario proposed by Frison is true, then the mammoth remains were partially articulated by ligament, tendon, muscle meat, and skin when fluvial disturbance redeposited or buried them. Hence, the taphonomic experiments (Todd and Frison 1986) to measure the flotation potential of elephant bone were not fully valid for determining the degree to which the fossils were transported or redistributed by stream action. The experiments measured the transportability of defleshed, disarticulated bones, not the transportability of partial carcasses.

6.9.   Conclusion: What is evidence, and what is knowledge?

> *There is much information in a little book by J. P. MacLean,*
> Mastodon, Mammoth and Man, *but the reader must not accept all its statements unhesitatingly.*

> —F. A. Lucas, *Animals of the Past* (1929)

Because there are no eyewitness accounts of the distant past, all interpretations of ancient events must rely on the surviving bits of evidence that we gather from the fossil record. Not all those bits of evidence will have equal appeal to different interpreters. Some evidence may be overlooked, and some may be deliberately ignored or dismissed, depending on personal preferences.

The evidence provided by fossil data may be strong, weak, or contradictory, but *it is not knowledge about the past*. The word "evidence," as used in modern legal systems, refers to a medium of support for arguments offered concerning an event: When the observations and details have been submitted to a competent jury, the body of evidence is then evaluated and weighed in an attempt to determine the truth. In other words, the evidence is evaluated in people's minds; it does not in and of itself automatically reveal the truth.

Direct evidence is the result of specifiable agencies that are the only possible causes for certain facts about the fossil record. Circumstantial evidence may suggest that certain facts about the record were created by agencies that (according to empirical generalization) usually are accompanied by certain other circumstances, thus affording a basis for reasonable inference that those other circumstances did indeed exist. Circumstantial evidence does not provide proof, but heightens the probability that a specific agency was in operation.

A chronic lack of codified interpretive standards often allows very poor evidence to result in unshakable interpretations. These unproven interpretations may then be transformed into canonical wisdom that becomes equated with knowledge about the past and continues to be referred to as "proof" of other interpretations. And once these unproven assumptions are mentally transformed into truths, it becomes quite difficult to demote them.

In an earlier study (G. Haynes and Stanford 1984) I found reason to doubt nearly all earlier claims that twenty-one North American sites with *Camelops* bones contained evidence of human predation on (or "utilization of") *Camelops*. My method of reasoning was not necessarily stronger than the methods employed in earlier studies, but it was more cautious. The scaled sets of described standards had been intended to suggest increasing chances that the bones had been added to the record by human actions, based on impressionistic criteria that were not necessarily more testable than earlier intuitive standards, but which at least were more explicitly stated. To make my caution plausible, I offered descriptions of noncultural bone modifications observed in taphonomic fieldwork. The same types of bone modifications in the fossil assemblages had been described in the original site reports as culturally produced. Nonetheless, I found enough reason to doubt that humans had been proven to have had something to do with the deaths of the *Camelops*.

The proboscidean sites reported in the scientific literature (Tables 6.1–6.5) have also been evaluated against a single set of standards, and as concluded in the *Camelops* study I have found many of the published interpretations of the proboscidean sites to be insupportable. They are good stories, but they are not "knowledge" about the past.

Most of the *Camelops* sites that I evaluated lacked closely associated and unquestioned artifacts such as chipped stone tools. Many proboscidean sites do contain stone artifacts in close association with bones. It is axiomatic to many archeologists that the presence of stone tools spatially close to clustered bones justifies an interpretation of behavioral association, especially when the geomorphological or radiometric evidence shows no obvious sign of time averaging within the sedimentary matrix. However, "time averaging" is a geologist's term, and the length of time required for sediments to show evidence of it may be several decades or longer. Many noncultural animal deaths may occur during those years, as well as several cultural activities. Human kills of animal groups usually are interpreted as *events* that last for hours, or days at most; evidence of time averaging in sediments results from a *process* that must last much longer. If a number of human artifacts and the bones of butchered animals are added to a sedimentary surface that is relatively unstable, and if other (noncultural) animal bones are also added to the surface within a few months or years, there will be no verifiable way to determine whether or not all bones are behaviorally associated with the artifacts. The land-surface sediment, if buried and preserved, is a single unit of deposition.

Informed archeologists accept this uncertainty about bone deposits, especially because there now are plentiful published examples of modern bone inputs at localities such as lakeshores and stream channels (Behrensmeyer 1988; Conybeare and Haynes 1984; G. Haynes 1985a; G. Haynes and Stanford 1984) that attract nonhuman as well as human predators, and where bone inventories mix the remains from cultural and noncultural events. Binford (1981) called such mixed deposits palimpsests, a word derived from Greek, referring to a surface overwritten several times without thoroughly erasing earlier writing.

Yet whenever archeological reports interpret clustered bone sites as the remains from single-episode cultural events, rarely do they present explicit arguments to support such an interpretation. Many of the large bison kill sites in the North American High Plains are reported as if they were self-evidently cultural *events*. This assumption seems to be based on belief in the unlikelihood that certain site attributes would have been created by time averaging – for example, stone projectile points of a single defined type may occur within the bone bed. Some such sites contain fire features. A single taxon of mammal may dominate the bone assemblage, and the age profiles may be suggestive of a single herd killed together; an example is the Casper site in Wyoming (Frison 1974). However, even at the Casper site, bones of several animals were found that probably were not the victims of human hunting; some of these animals almost certainly were nonfood taxa (M. Wilson 1974). Perhaps some of the food taxa (such as bison) also had not been slain by humans at this site, but died nonculturally, as had the other animals.

Some bones in these kinds of interpreted kill sites may be broken or cut-marked, as if processed by humans, lending weight to the assumption that the animals had been deliberately slain together. If features that are in fact rather unlikely in a kill or butchering site *do* occur in association with other attributes such as thick hearths or ash deposits, postmolds, and artwork, the clustered bones may be considered *not* the result of kills but rather of house construction using animal bones, as, for example, in the Mezhirich site in the Soviet Union (Pidoplichko 1969) (Table 6.4). Yet still the bones may be thought to represent single episodes or single events of human activity.

There are few lines of evidence possible in archeological data that are capable of indicating that a single event created an assemblage, as opposed to a succession of events over extended time periods, more than, say, several days or weeks. Radiocarbon dating has a margin of error that often is more than a human lifetime wide; recognizably distinct artifact types such as projectile points may enjoy popularity for decades or longer. In the future, detailed sedimentary studies may be able to factor out very brief depositional episodes, but at present the technology necessary for such studies is not widely available or familiar to fieldworkers. And many of the sites that were sources for the data lying behind our current thinking about human behavior, prehistoric culture history, and site-formation activities no longer exist. Many were destroyed in the digging, and in others the contextual data concerning critical finds were not recorded with the necessary detail for new methods to be applied.

Because often it cannot be clearly proved that archeological sites were created by time averaging rather than by one-time events, it would appear that archeological interpretations will continue to depend on articles of faith that are rarely challenged. One such article of faith is the belief that single events create single archeological components or single sites and assemblages.

The evidence from the proboscidean sites listed in the tables herein spans a range of many different degrees of strength, from the sound to the laughable. I am realistic enough to know that however bluntly I criticize the arguments for a particular site interpretation, those arguments will continue to be accepted by archeologists who believe in the canonical wisdom. Believers in orthodoxy or conventional wisdom feel true horror when contemplating the possible ways in which chronic doubting can affect our state of knowledge – they realize that if the articles of faith are wrong, then many (most? all?) Stone Age archeological sites are practically useless as sources of meaningful information. The sites may be smeared, time-averaged palimpsests, and the evidence they contain is confused or contradictory. Must we therefore ignore them and seek others, discarding any potentially tainted evidence? In a court of law, there is a strong preference for not eating fruit from a tree known to be poisoned.

The answer, of course, is not to discard, but to be critical when reexamining evidence, with an eye to *testing* the articles of faith and interpretations founded on them, not to giving *testimonials* to interpretations. Interpretations are too often good stories, but bad science proclaimed as fact, practically graven in stone, and eventually becoming folk wisdom, or myths about the past (Binford 1981; Eiseley 1945, 1946). We must continue to develop new methods of analysis that will rely on statistical probabilities, although even the improbable can be shown to be inevitable (Hsü 1989). I recommend that far more (and more creative) actualistic studies be carried out. Most of all, we must be skeptical, and never believe an interpretation so strongly that we are blind to its weaknesses.

The fossil record has been able to fertilize imaginations for a long time over the course of human history, and whatever has been harvested here out of my own thoughts may or may not be satisfying to everyone reading this.

Many more aspects of the fossil record could be reviewed than those selected in this chapter. The descriptions and discussions here may inspire archeologists or paleoecologists to rethink the potential meaning of proboscidean assemblages, although some scholars will entrench more deeply and fire back. I encourage students to ask difficult questions and keep searching for better methods to find the meaning in bone assemblages. As for drawing all the threads together and passing final judgment on the sites discussed earlier, I offer only a few pronouncements, none of which can be the last word. In Chapter 7 I discuss North American mammoth kill sites; here I shall say that those kill sites do not provide the best evidence in the world, insofar as most fail to yield unequivocal proof that humans slayed proboscideans before processing the carcasses.

It is considered philosophically parsimonious to attach cultural explanations to sites containing associated artifacts and animal bones; in other words, the presence of implements in a bone bed surely indicates that killing occurred. Likewise, butchering tools must mean that processing took place.

The geochronologically earliest associations of elephant bones and artifacts – the Djibouti *Elephas* (Chavaillon et al. 1986, 1987) and the FLK North *Elephas* in Olduvai Gorge (Bunn 1986; Isaac and Crader 1981; Leakey 1971) – are dated to the early Pleistocene, and both contain broken and scattered bones, incomplete skeletons, "background" bones from other animals, stone flakes, and possible hammerstones, as well as other artifacts. Isaac and Crader (1981) favored the view that the FLK North elephant had been scavenged by hominids, not hunted or killed. A *Deinotherium* in the same site was considered by some only fortuitously associated with stone tools, but Leakey (1971) judged it to be the butchering site for a mired animal.

Take your pick of favored views. If one studies the reports of all sites containing proboscidean bones, those with and those without associated lithics, one finds few sites indeed that can confidently be called kill sites. Some were noncultural, and some possibly were scavenged by humans. Some may have contained escaped animals that took embedded spear points with them to their deaths.

What one can say confidently about fossil proboscidean sites remains descriptive rather than explanatory. I have provided some suggested lines of evidence to use in making the explanatory cases, and in the next chapter I apply some of these to argue a slightly different interpretation of New World mammoth kill sites.

7

Extinction in North America at the end of the Pleistocene

Nature herself has not provided the most graceful end for her creatures.... They must perish miserably.

— H. D. Thoreau, *A Week on the Concord and Merrimack Rivers* (1868)

7.1. Part One: Climate the culprit?

At the end of the Pleistocene, most of the proboscidean taxa then living disappeared forever from the biotic record. The preceding several million years of range contraction, range expansion, and global redistribution of species present us many different types of paleoecological problems, such as finding a satisfactory explanation for the disappearance of *Elephas* from Africa but not from Asia, but it has been the phenomenon of extinction without replacement and without evolutionary descendant (which defined the Pleistocene/Holocene boundary) that has excited the most interest among Quaternary paleoecologists. Extinction seems to be our most tempting mystery in need of a solution.

Because proboscidean extinctions at the Pleistocene/Holocene boundary often have been blamed on human hunting [e.g., see the various essays in P. Martin and Wright (1967) and P. Martin and Klein (1984)], it is appropriate to reexamine the topic in relation to a critical survey of the proboscidean fossil record.

According to radiocarbon dates, a proboscidean crisis occurred 10,000–12,000 years ago, the interval when mammoths and mastodonts disappeared from the world.[1] Clearly, there was an event or series of events that we call environmental change, triggered by climatic transformation, and that event or process was global in scale, although different regions of the world underwent their own unique variations of the change (COHMAP Members 1988). In this chapter I discuss these environmental changes, centering my attention on the New World, specifically North America, which lost about two-thirds of its mammalian genera at the end of the

---

[1] Some would say they disappeared later.

264

Pleistocene. The main task here is to determine whether or not the fossil record preserves unambiguous evidence that climatic changes can account for proboscidean die-outs and extinction at the end of the Pleistocene. One complicating factor affecting the way we perceive the fossil evidence is the presence of Upper Paleolithic human groups in the same regions that supported mammoths and mastodons, during the same time periods that extinction occurred. It has not been clearly established that humans preferentially preyed on proboscideans, nor that they increased their proboscidean hunting near the end of the Pleistocene, just before the extinction of mammoths and mastodonts, but these possibilities have been raised by numerous scholars.

To answer a simple question (Who or what killed off the mammoths and mastodonts?) we first have to marshal some facts about proboscidean abilities to survive different mortality processes. We have to know whether proboscidean mortality rates at the end of the Pleistocene (when extinction without replacement occurred) were higher or lower than they had been during earlier time periods (when no extinction occurred, although speciation did take place). We also have to know if reproductive rates just before extinction were relatively high or low. And we have to know whether or not human hunting increased during the time period preceding extinction. Our simple question has become quite complex.

The evidence needed to address the parts of this complex question may be ambiguous because of the biased way in which we have historically sampled the fossil record. Proboscidean bone assemblages from geologic time intervals earlier and later than the major extinction period probably have not been well sampled or described in the literature familiar to archeologists. Also, edaphic and environmental factors that affect preservation and fossilization of proboscidean bones may have been quite different at other times, resulting in a poor bone record from earlier (and later) periods. Thus, *even if mortality rates were higher* in these different times, fewer bones may have been preserved. The Late Glacial period may have been a time of optimal bone preservation because of relatively rapid burial of large death assemblages, as compared with recent or earlier times, and thus the absolute numbers and sizes of Late Glacial assemblages need not necessarily reflect increased numbers and intensities of mortality events.

Were it not for this possibility, we would validly be able to compare overall numbers of mass assemblages, patterns of age profiles, relative sizes of death assemblages, and other characteristics represented from different time intervals to gain a general impression of changes in mortality rates in the past. Because we cannot assume that fossil proboscidean samples from different time periods are comparable, we must be extremely cautious when examining Late Glacial rates.

I argue here that Late Glacial environmental changes stressed the proboscidean populations, and I make my case by first discussing Late Glacial climates and environments, based on published interpretations of fossil data, and then inferring the effects that such changes would have had on proboscidean behavior and biology. These inferences are based on speculations as well as deduced knowledge about fossil proboscideans and modern elephants. The inferences may or may not prove supportable in light of future research, but at the moment I think they are warranted.

7.1.1.   Were they killed off, or did they die off?

All mammoths and all mastodonts died by the end of the Pleistocene, of course; the causes that have been suggested to account for their deaths are innumerable. To simplify the array of choices, and also to show the muddle these "contradictory" arguments have created, I have subsumed the possible causes of death (and ultimately extinction) into two alternative scenarios: In the first, mammoths and mastodonts were actively killed by various means: New disease, increased predation, sudden and inescapable dietary deficiency, physiological dysfunctions, or severe weather. In the second scenario, they all eventually died, but without suffering attack from new or violent means. In the latter instance, during what is usually envisioned as a drawn-out process, recruitment and reproductive rates failed to keep pace with mortality rates, so that even though the animals were in fact dispatched by various agencies, such as disease or old age or accidental mishap, they were not, in effect, killed off. Instead, they died off. In both scenarios, deaths had to outnumber births, and philosophically it is difficult to find important differences between the scenarios.

The major difference is this: In the kill-off scenario, proboscideans and the other taxa that disappeared at the end of the Pleistocene would have survived had not some agency or set of agencies, acting temporarily and deliberately, destroyed them. In the die-off scenario, populations faded away inexorably as they lost members to death faster than new ones were added by births. They had no chance to survive. The most likely cause of die-outs would have been environmental change.

I state these scenarios as separate cases because in my mind they are the simplest forms of single-cause deterministic explanations for extinction. Stating them as different positions highlights the annoying confusion created by the question of what caused extinction. The two scenarios sound alike, in fact. Instead of focusing our efforts on describing the events and processes undergone by Late Glacial mammals, we far more often end up trying to explain or find a cause for the end effects of those processes. We are ready to pin the blame for extinction somewhere, but we have not adequately tracked the members of the doomed taxa through their final time on earth. In the following section I try to make a start in that direction.

7.1.2. Late Glacial climates in North America

Late Pleistocene environmental transformations occurred worldwide, resulting from changes in climatic parameters such as global circulation patterns, annual temperature ranges, and peaks of seasonal precipitation. The details of such changes differ from one world region to another. Could such a variety of changes have stressed large mammals to the same extent everywhere?

The Late Glacial climatic changes are difficult to describe in satisfying detail because the evidence concerning their nature often is circumstantial. And as Vereshchagin and Baryshnikov (1984) emphasized, the immensity of the range in which extinction of the mammoth occurred makes it impossible to isolate only one natural factor that could have caused extinction (Vereshchagin and Baryshnikov 1984:493). For example, in some parts of northern Eurasia, boggy tundra and taiga replaced the well-drained steppelike Pleistocene habitats, whereas in other parts the extent of solid-ground grasslands may have expanded. A few bits and pieces of information from wide areas of the world perhaps indicate some overall trends in extinction. The latest radiocarbon dates for mammoths from Europe are around 12,000–13,000 years B.P. (although two finds have been dated 9,000–10,000 years B.P.), and those for Siberia are around 10,000–11,000 years B.P. (Berglund, Hakansson, and Lagerlund 1976; Coope and Lister 1987; Lister 1990; A. Lister 1989 personal communication; Vereshchagin 1982; Vereshchagin and Baryshnikov 1984). No mammoths younger than 20,000 years B.P. have been found and radiometrically dated in China (Liu and Li 1984:526). Thus, keeping these dates in mind, it appears that there may have been a west-to-east and south-to-north gradient for the time periods of mammoth disappearance, with central and southern Eurasian mammoths dying out first, followed by Siberian and American mammoths. A relatively late radiocarbon date for one north Asian mammoth provides a single instance of post-Pleistocene age, but it does not clearly indicate that northern Asian mammoths were the last surviving on earth; neither do the occasional late American dates for mammoths and mastodonts (Mead and Meltzer 1984).

Numerous analytical and synthesizing studies of the changing environments in North America have been published over the past twenty years. Many have generally emphasized one or another scientific discipline, such as palynology, geology and stratigraphy, paleontology, or archeology. I shall rely here on a few heroic attempts to synthesize information. Such syntheses are also available for other areas of the world where proboscideans became extinct. For example, Soffer (1985) attempted to draw all the different types of studies together in her analyses of the central Russian plain; Abramova (1979a, b) and Mochanov (1977, Mochanov et al.

1983) attempted similar feats in their analyses of the archeological record in very limited parts of the northern Asian region. Grichuk (1984) provided a vegetational history of the last glaciation in the Soviet Union, and Velichko (1984) attempted reconstruction of late Pleistocene climates in the territory of the USSR.

Guthrie (1968, 1982, 1984) and Hopkins (1982) have published valuable synthesizing studies of Beringia. Barry (1983) summarized interpretations of late Pleistocene climates in the United States. C. V. Haynes (1970, 1980, 1982, 1984, 1987) has provided a stratigraphic and geomorphic framework for understanding late Pleistocene environments in interior North America.

The interval from 25,000 to 16,000 B.P. has been termed the Late Pleniglacial (Nilsson 1983:256) or Glacial Maximum; the interval 16,000–10,000 B.P. is here termed the Late Glacial, during which time the deglaciation process was occurring (Porter 1983; Ruddiman and McIntyre 1981). The two time periods together (25,000–10,000 years B.P.) correspond to Stage 2 of the oxygen isotope record from Pacific ocean core V28-238 (Shackleton and Opdyke 1973). The first interval, the Glacial Maximum, was the time of maximum continental glaciation in the Northern Hemisphere; the second interval, the Late Glacial, was a time of rapid disappearance of most of the mass of continental ice sheets. Several warm–cool oscillations occurred during this period.

The climatic oscillations of the Late Glacial would have involved not only warm–cool temperature fluctuations but also changes in effective moisture regimes, air masses and prevailing wind directions, seasons of precipitation, frequency of winter storms, and so forth, because the positions of the continental ice sheets were continually changing, having profound meteorological effects (Barry 1983; Budyko 1975).

Barry (1983), citing numerous primary sources, summarized inferred trends in summer temperature and annual precipitation across the continental United States at about the time of the Glacial Maximum. Much of the interior south of the Laurentide ice sheet apparently was colder and drier than today, with differences in summer temperatures ranging from 5°C to 10°C lower than the modern averages. The southwest and southern plains were relatively wetter, and the central plains and prairies were also wetter, but much of what is now the middle and eastern United States was drier than in recent times. The Pacific northwest was relatively dry, compared with today, and southern Florida was very dry. The Great Basin and the southwest were wetter, because of both more precipitation and reduced evaporation, following from cooler air temperatures and possibly also more cloudiness in warm seasons (Spaulding, Leopold, and Van Devender 1983).

Between 18,000 and 11,000 years ago, climates throughout the United

States markedly changed from glacial conditions to those of the Holocene, but not with homogeneous trends. Different geographic regions of the United States reacted differently to the main climatic shifts: The relatively rapid shrinking of the continental glaciers, the consequent changes in jet-stream location, and abrupt increases in seasonal difference in precipitation and temperatures between 15,000 and 9,000 years ago (COHMAP members 1988). There have been numerous scholarly studies of regional trends in paleoclimates and paleoenvironments during that time interval, but here I shall simplify my impressions of those studies, with apologies to the scientists who labored to test and refine the kinds of inferential statements that I may seem to be advancing as demonstrated fact. Recent published reviews, such as those by Graham (1987), L. Davis (1987), and Semken and Falk (1987), should be consulted for additional discussions of floral and faunal communities in the southwestern, central, and northern plains, respectively.

The interval between about 14,000 and 10,000 years ago is the most critical for an understanding of the actual process of the dying out of the proboscideans. Judging from interpretations of faunal assemblages dated to that interval, many paleoecologists think that summer temperatures between 13,800 and 11,800 years ago were still colder, on average, than today, but climates were generally equable, changing to continental, which change peaked about 9,000 years ago; see, for example, Barnosky (1989), COHMAP members (1988), and Waters (1989). Evaporation was exceeded by precipitation over much of the western and central United States, except for a brief time around 12,500–11,800 years ago. The central plains were increasingly tree-covered, especially by closed spruce forests (Watts 1983), replacing "boreal grasslands" (Rhodes 1984) in the north (note that the northern plains of Late Glacial times are today's central plains, because the Laurentide ice sheet moved as far south as Iowa, effectively eliminating the more northern states and Canadian provinces from consideration as places where plants and animals lived, except in the so-called ice-free corridor). In the southern plains, grasslands were not replaced by forests, but the nature of grasses growing in those open areas changed from "cool-season" types to "warm-season" types. Guthrie (1982), Graham (1981), and Rhodes (1984), among others, have speculated (based on field evidence and theoretical grounds) that warming climates in various parts of North America led to a change in the proportions of annual plants using different photosynthetic pathways, with one possible end result being that the relatively more nutritious cool-season grasses (the species with "C-3" pathways) were replaced by warm-season grasses (species with "C-4" pathways). Many late Pleistocene grasslands remained open environments throughout the Late Glacial period, and yet their carrying capacity in regard to herbivores decreased greatly. Hence, there is reason to question the argument advanced by some scholars to refute Pleistocene–Holocene environmental change as a major cause for megafaunal extinction, which argument often takes the following form: Grasslands did not shrink in extent at the end of the Pleistocene, and may even have increased in size, the main implication being that Holocene grasslands should have been able to support megafauna as well as or better than Pleistocene grasslands. Even if Holocene grasslands were more extensive than earlier grasslands, they did not have the high percentages of C-3 species that characterized Pleistocene grasslands, and therefore their carrying capacities may have been lower. However, it must be noted that modern grazing livestock do well on either C-3 or C-4 grasses, and C-4 grasses support a large proportion of tropical African grazers today. And it seems quite possible that the major replacement of C-3 grasses occurred *after* the interval of extinction.

Evaporation probably exceeded precipitation in the western states between 12,500 and 11,800 years ago, resulting from warmer summer temperatures. There may have been greater seasonality in midcontinental and western climates, as well as more freeze–thaw periods in winter when snows melted or softened and then froze solidly into ice-crusted masses. Winter storms and blizzards would have been more numerous than before, and summers warmer and drier. Shrubs may have invaded some areas, such as spring-outlet overflow channels or floodplains, a process that occurs when previously wet sedgelands, meadows, and grasslands dry out during much of the growing season. This kind of shrub invasion further lowers water tables. Riparian woodlands would have died back with the decrease in effective moisture, further encouraging shrubland expansion.

Between about 11,800 and 11,100 years ago (or 11,500–10,700 years ago, depeding on preferred radiometric dates), midcontinental and western climates may have become equable again, with overall precipitation exceeding evaporation once more. Average annual temperatures probably were cooler than during the preceding time interval, but with winter storms again reduced in severity and frequency. There would have been an increasing arboreal component to mosaic vegetational communities.

C. V. Haynes (1967, 1970, 1975, 1984, 1987), on the basis of his studies of the paleohydrological record from the west, central and southwestern United States, concluded that local water tables dropped dramatically during deglaciation, reflected in net stream degradation. Water sources may have seasonally dried up altogether, especially low-order streams or small spring-fed ponds. But during the Clovis interval (about 11,500–10,700 years ago), those streams were once again aggrading, although incised much lower than before. Organic-rich deposition characterizes some sediments from that interval in parts of the American west.

7.1.3. Vegetation during the Late Glacial

Many North American environments during the Glacial Maximum were complex mosaics of different plant taxa that today do not co-exist. Before about 14,000 B.P., far northern landscapes were open and treeless, lacking snow cover in the winter because of constant winds and low precipitation, as described by Hopkins (1982) and summarized by Guthrie (1982).

In those more diverse habitats, shortgrasses, tallgrasses, and herbs grew near each other, but not in altitudinally restricted zones as they do today in North America. Mixed in with those complex grass and forb communities were shrubs and trees along stream courses, or scattered in the open ground much like heterogeneous savannas (Guthrie 1984). The biotic associations of the Pleistocene, not only in Beringia but also in much of the rest of what is now the United States, were quite different from the modern types, and their rearrangement at the end of the Pleistocene involved the rapid development of greater biotic zonation that favored "simplified faunal assemblages more precisely adapted to a specific habitat" (Guthrie 1984:267). Some plant associations may have expanded greatly – and in fact grasslands and open plains did (Guthrie 1984:289; McDonald 1984) – while others shrank or disappeared.

At the end of the Pleistocene, megamammals in those regions were faced with loss of plant diversity and shortened growth seasons. The loss of dietary diversity meant that many herbivores could not cope with their increased intakes of toxic antiherbivory compounds, which had become concentrated in their diets because of lack of alternative food. Animals also suffered from a decline in the nutritional balance available from less diverse diets (Guthrie 1982, 1984). As Graham and Lundelius (1984) pointed out, the loss of plant diversity would have led to heightened feeding competition among herbivores seeking the highest-quality feed, which may have survived in very low densities in some habitats. In a dissenting opinion, McDonald (1984) argued that North American environments were *improved* as habitats for large mammals after 18,000 B.P., in comparison with the Glacial Maximum interval, which would have been the time of colder air temperatures, greater "simplicity of habitats" (MacDonald 1984:432), limited patchiness, and reduced primary productivity. Mehringer (1967) concurred in that view.

7.1.4. Possible effects of changing climates: range abandonment, dwarfing, mineral deficiencies, winterkill

During the period from about 15,000 to 10,000 years ago, environmental fluctuations probably were choppy and relatively wild. Late Pleistocene environments of the Late Glacial interval were subject to cycles of dearth and plenty. Climatic conditions fluctuated, and in some years the growth seasons may have been unpredictably shortened or delayed. Low productivity may have stimulated exploratory migrations, although wild animals under stress do not automatically embark on long treks in search of food and water, because they have no way of knowing what resources will be found in unfamiliar ranges. However, there is some evidence that normal seasonal migrations carried out by herding African ungulates, such as buffalo (*Syncerus caffer*) and wildebeeste (*Connochaetes taurinus*), in search of water and food may be lengthened during unusually dry years (Child 1965, 1968; Davison 1950; Duckworth 1972; Higgins 1969; B. Williamson 1979a, b). Perhaps Pleistocene proboscideans also expanded their ranges during hard times, at the same time abandoning parts of their traditional ranges.

In the relatively unstable environments of the Pleistocene/Holocene transition, certain species would have become "dominant," especially those that colonized quickly and reproduced and matured rapidly. Mortality for such species often is density-dependent and catastrophic, with the result that the population size does not attain equilibrium. These characteristics are associated with r-selected mammals (Pianka 1970), usually small-statured and fast-developing species such as rodents. Late Pleistocene megamammals may have been adjusting themselves to unpredictable resources and variable climates by maximizing their rates of population increase (or becoming more like the r-selectionists). One way they could have done that was by decreasing their body sizes, an evolutionary strategy to deal with shortened plant growth seasons and poorer quality of vegetation (Guthrie 1984; King and Saunders 1984). Kurtén (1968) has reasoned that dwarfing is an adaptive development that allows the population density to be kept up at an adequate level for reproduction, in the face of limitations on food and habitat. Dwarfing may result from heavy predation, as well. Mammals may respond to the removal of population members by maturing sooner, allowing reproduction to occur and new members to be recruited into the population. Thus, a reduction in body size may occur either in stressful environments or in richer environments where there is hunting pressure. Note that recent studies of modern African elephants indicate that they respond to resource scarcity or drought stress by reducing their reproductive rate (females may cease to ovulate) and by increasing juvenile mortality (see Chapter 4). Clearly, large individuals seem better able to survive such periods of environmental stress. Hence, *Loxodonta* reacts to unreliable or scarce resources *in the short term* by allowing younger members of the population to die, or by ceasing to produce young altogether. Small adult body sizes may result from hunger or illness in the growing years, but under very severe conditions of environmental stress the animals that are still growing probably will die, resulting in the survival of full-size (not dwarfed) adults that will be able to reproduce following the end of the environmental-stress intervals. It is not clear how these kinds of population responses to stress could eventually result in evolutionary dwarfing – it would require long periods of stress, involving centuries.

Another feature of Late Glacial climates was their possible increase in continentality, and loss of equability. An increase in seasonality would have led to significant time periods within the year when grasses and forbs were dry and not growing, providing poor-quality food sources. Mammoths would have been able to subsist on available browse in dry seasons or winters, although if vegetational communities had changed from mosaic to zonal, monospecific stands of plants would have provided an inferior selection of feed for large herbivores such as mammoths (Guthrie 1984). Modern elephants feed on the move, or sample a wide range of different plants throughout the day, thereby avoiding overuse of plant stands and also avoiding excessive intake of plant compounds that may be toxic or may reduce digestibility (Freeland and Janzen 1974; Janzen 1975). The loss of variety in the diet could have overloaded the digestive system with too much of some toxins and may have provided inadequate nutrition (Guthrie 1984; Olivier 1982). Seasonally excessive intake of some minerals, such as potassium or calcium, may have forced mammoths and mastodonts to seek out mineral licks to restore dietary balance (Jones and Hanson 1985). Imbalances would also have depressed reproductive rates (Koen 1988; Koen et al. 1988).

More emphatic extremes of seasonal temperatures, precipitation, cloud cover, and storminess would affect large mammals, whose seasons of breeding and parturition are times of stress for large proportions of the population. These sensitive times of the annual round no longer would be synchronized with nutritional cycles (such as vegetational plenty in late summer) or with climatic mildness (such as early spring thaw) (Kiltie 1984). The taxa with lengthy gestation periods and long calving intervals would not quickly recover from disastrous spring blizzards that would kill much of the year's calf crop. Autumn ice storms could kill pregnant or lactating females in the herd.

McLean (1986) proposed that the abrupt increase in air temperature during Late Glacial summers would have resulted in reproductive dysfunc-tion among Pleistocene fauna. Hyperthermia would have caused blood flow to the uterus to be reduced, leading to embryo death, dwarfing, or skeletal abnormalities. If during the deglaciation interval hot summers resulted from high levels of solar radiation in midyear, much greater than modern levels (COHMAP members 1988), then many mammals would have been too well adapted to cooler glacial conditions to escape dysfunction. Large, hairy mammoths or mastodonts would have been heat-stressed, especially in environments where tree cover was scattered or sparse.

If some or all of these stresses occurred, there would have been localized die-offs during the last millennia of the Pleistocene, resulting from drought conditions and starvation consequent to growing seasons shortened by the onset of winter weather or the icing over of winter feeding ranges, among other potential causes. Many mammoth and mastodont ranges would have been abandoned because of a series of die-offs, declining numbers of water sources, or a series of poor growing seasons. However, after about 11,800 years ago, some of those abandoned ranges may have been recolonized, as climatic, hydrological, and floral conditions improved somewhat. Yet at some time between 11,200 and 10,800 years ago, with the resumption of "deglacial" climates and vegetational changes, more severe localized die-offs would have occurred. Catastrophes such as occasional drought usually are stochastic, insofar as their effects on animal populations are not regularly or predictably repeated. In the case of small or isolated populations, stochastic events have severe outcomes (and in fact provide deterministic explanations for die-outs).

There is unmistakable evidence of *Mammuthus* and *Mammut* die-offs or mass mortality events toward the end of the Pleistocene – for example, at Boney Spring, Lamb Spring, and several Clovis-associated sites such as Lehner and Dent (see Chapter 6). However, the theory of single agencies, such as drought or winter icing, accounting for the killing in all such sites is not firmly established.

Specialized climatic terms such as "drought" are relative. The Australian Bureau of Meteorology defines a drought period as three months or more when rainfall is measured within the lowest 10 percent of similar periods ever recorded; a severe drought is acknowledged when measured rainfall falls into the lowest 5 percent of recorded totals for similar periods in the past (Voice 1983). The vegetation and animal communities in some parts of the world are in many ways already adapted to arid or subarid conditions, so great moisture deficiencies in those areas from year to year may have little significant impact on biota. Even in normal years, such regions are much too dry to support agriculture, and any decrease in available moisture from year to year may be an order of magnitude less than the kinds of decreases that would cause disastrous droughts in more humid farming regions of the world, such as temperate North America. What I wish to emphasize here is that a drought at the end of the Pleistocene may have been rather mild compared with droughts today in some parts of the world. And in fact much of the mortality resulting from climatic stress would have been density-independent. For example, ice-crusted snow or late winter blizzards would have killed animals regardless of their abundance or scarcity, although such agencies may have selected certain age or size classes more often than others. Similarly, dry heat in summer and autumn would have decreased the availability of palatable food, affecting animals without regard for population density.

On the other hand, loss of water sources because of hot, dry climatic conditions would have led to density-dependent deaths as a result of crowding around the remaining water sources. Migration routes would have been badly affected by overcrowding and overgrazing. Many overused feeding areas and shrinking water sources may have provided salt-heavy diets to stressed animals, and in the absence of renal or other physiological adaptations for coping with excessive sodium, animals would have been further stressed.

Certain processes leading to death would have had both density-dependent and density-independent features. If the mosaic habitats of the Late Glacial interval were effectively altered into discrete, separated zones of plant communities that previously had grown interfingered with each other, many large mammals would have lost the use of large areas that earlier they could have migrated through to reach areas that were less heavily grazed or less dried out. King and Saunders (1984) described this phenomenon as increased "insularity." Sustained long-distance travel through vegetationally poor areas is possible for big game, but such moves are ecologically taxing, and, as mentioned earlier, there is no guarantee of better forage ahead.

C. V. Haynes (1984) summarized the end of the Pleistocene in the unglaciated western United States as follows: Within a time interval of less than 1,000 years, stream floodplains became inactive, and streams began incising their channels. Over-bank soil deposition ended, because rivers were no longer overflowing their banks. Water tables were lowered, and many lower-order streams became intermittent or ephemeral. Spring flows were reduced or eliminated. When over-bank deposition began again at about 10,900 years B.P., streams were much smaller than they had been before. Ephemeral streams dissected areas where ponds and marshes once existed.

We should not expect the conditions of water-table lowering, increased warmth and aridity, and loss of big game to have been synchronous or identical throughout North America. The local features of air masses, interference with or encouragement of precipitation (rain shadows and the proximity of large bodies of water), and aridity are the factors that control vegetation, and hence the animals that will be found locally. The climatic, floral, and faunal changes of the Pleistocene/Holocene transition were not identical in different parts of the world, or in different parts of any continent, because of local variations in substrates, topography, latitude, distance from water bodies, and location relative to the jet stream, as reviewed by

Ruddiman and Wright (1987). Global similarities in environmental changes at that time included jumps or discontinuities, so that there were few or no clear uninterrupted trends in climate or biota, although stream discharges do seem to have been reduced over large areas. Vegetation became more zonal in distribution, rather than mosaic in form. However, in some regions, summer precipitation may have increased, leading to woodland replacement of grassland, whereas in other regions winter precipitation increased, at the expense of summer precipitation, leading to more erosion and more open habitats.

7.1.5.   The fossil evidence: does it indicate die-offs?

Large mammoth or mastodont accumulations appear to have been most abundant in unglaciated North America during the Late Glacial interval; prior to that interval, other taxa such as *Camelops* dominated the megafaunal record (G. Haynes and Stanford 1984). In the western and southern parts of the continent, the dominant megamammal taxa at the time of the Glacial Maximum or immediately after often were *Equus* and *Camelops*, genera that became much rarer compared with *Mammuthus* just at the time when one would expect the opposite trend – *Camelops* especially should have been capable of coping with seasonal drought better than did *Mammuthus* and should have survived longer, and yet the big accumulations of mammoth bones postdate the largest late Pleistocene *Camelops* bone accumulations, such as those at American Falls (Idaho), China Lake (California), Tule Springs (Nevada), and Rye Patch (Nevada) (G. Haynes unpublished data).

Camelops, like modern *Camelus*, undoubtedly was not as dependent on water sources as were proboscideans such as *Mammuthus* or *Loxodonta*. Modern camels roaming wild in the vast central Australian semideserts rely mainly on browse vegetation or succulent groundcover, not on water sources, for moisture intake during dry years or dry times of the year (K. Johnston 1987 personal communication; O. Williams 1987 personal communication; I. Cawood 1987 personal communication; R. Bryan 1987 personal communication; J. Heucke and B. Dörges 1988 personal communication). Camels do not continually need free water to drink. Giraffes in Africa survive drought the same way, by depending on browse, even while elephants are dying off locally (G. Haynes unpublished data).

Different big-game taxa react differently to a given set of climatic conditions. For example, free-roaming, unmanaged populations of camels (*Camelus*) in the central Australian desert (much of which is vegetationally steppe and savanna) are slow to react to severe drought conditions; their behavior is practically unaffected for many months, even while range cattle and wild horses are dying. Eventually they mob up and seek out pumped water sources, after browse has failed or can no longer be reached. Camels are competent to survive arid conditions (as, presumably, was *Camelops*) – they have evolved special features of their renal, behavioral, physiological and metabolic systems adapted to hot, dry conditions (Gauthier-Pilters and Dagg 1981; Maloiy 1972; Schmidt-Nielsen 1960). Wild horses facing the same conditions cope behaviorally – they travel long distances to find food, and return to drought refuges for water (G. Haynes unpublished data; Mohr 1971).

Like horses, elephants are "tethered" to water sources in hot, dry seasons (as, presumably, were *Mammuthus* and *Mammut*). However, camels are tethered to water sources only after drought conditions have become so severe that horses (and, of course, elephants, if they are present) are already dying at drought refuges. Therefore, it might be expected that during the late Pleistocene, the bones of *Mammuthus* and *Equus* at drought refuges should have underlain or been commingled with *Camelops* bones.

Camelops and *Equus* bones that were deposited before the Pleistocene interval, from which large mammoth bone accumulations are relatively abundant, show no direct evidence of drought-related die-offs. Many limb bones in these assemblages had been scattered, chewed, and broken by carnivores (G. Haynes and Stanford 1984) in a manner that does not suggest carcass abundance such as that seen around modern-day drought refuges; see G. Haynes (1980a, b, 1982, 1984b) for a general discussion of carcass use by carnivores, and Conybeare and Haynes (1984) and G. Haynes (1985a, 1988a) for more specific discussions of carcass use by scavengers at die-off locales. There is no unambiguous circumstantial, indirect, or direct evidence of human hunting of *Camelops* and *Equus* in the pre-Clovis-age deposits (G. Haynes 1984b; G. Haynes and Stanford 1984; G. Haynes unpublished data).

Under seasonally unstable climatic/environmental conditions, large mammals shift their feeding ranges and also fission social groups and local populations into separate, thinly distributed, and relatively small localized feeding bands that behave territorially around water sources. When water resources and forage vegetation are more readily available year-round, large mammals live in higher local densities of socially cohesive groups with smaller home ranges (McHugh 1972; McKnight 1969; Moehlman 1979). Short-term concentrations of animals into the last well-watered portions of their Late Glacial ranges may have accounted for the abrupt appearance of large numbers of mammoths in the Late Glacial fossil record, especially following upon an interval whose paleoenvironmental evidence reflects conditions that would have fostered proboscidean increase and range expansion, after the numbers of *Camelops* and *Equus* had declined, because of possible increases in snow depths and lengths of winter cold periods, and perhaps relatively low quality of the woody cover.

In times of drought (either seasonal or longer-term), elephants suffer mass deaths at water refuges. Animals cluster at remaining water sources and must walk for hours to find adequate forage. They lose condition and become weak and sometimes inattentive, such as when nursing cows ignore their begging calves. Fights over access to water are common and can result in injuries and breakdown of the social structure. The phenomenon of isolated cows associating habitually only with nursing calves, never with other adults, is not especially common under normal conditions, but that is often seen during droughts. Individual adolescent males are also commonly seen.

At some modern sites in southern Africa, serial die-offs have been recorded in which drought mortality *en masse* has been repeated several times, after the passage of years between episodes (see Chapter 4). The current cycle in southern Africa and central Australia seems to be about a decade (perhaps two) of poor precipitation, repeating every twenty years or so, with overall a cumulative drying trend (G. Haynes unpublished data), as described in unpublished records on file at Hwange National Park, Zimbabwe, and the Conservation Commission of the Northern Territory, in Alice Springs, Australia. In North America, any kind of cycle is unclear, but poor years seem to repeat about every ten years or so (G. Haynes unpublished data). These are very rough generalizations based on limited personal observations. The general characteristics of die-off sites in North America, Africa, and Australia include the following: (1) the presence of many bones of many individuals of different ages (and different taxa); some bones are articulated, and some are buried; (2) the presence of piled bones or clustered bones, either buried or on the surface; (3) the presence of broken limb bones lying on the surface (trampled), associated with buried unbroken ribs, or vice versa; (4) a representation mainly of subadult animals of either sex, and also of adult females; (5) the presence of scratched, fractured, and flaked bones in varying proportions, depending on the frequency of site visitation by big game. In elephant die-off sites, broken tips and medial fragments of tusks are abundant in the deposits.

Some of these characteristics are also seen in Late Glacial bone accumulations. For example, the age profiles for several Clovis mammoth bone assemblages are dominated by subadults (Saunders 1977a, 1980). At Blackwater Locality No. 1 (the original Clovis site) and at the Murray Springs Clovis site, it is possible that elephant-dug wells in a stream channel were foci of mammoth activity (C. Haynes 1987; Warnica 1966); another terminal Pleistocene site in Nevada's Black Rock Desert, the Wallman mammoth site (Dansie, Davis, and Stafford 1988; Stout 1986), contains the bones of an animal that may have died at a well dug about 11,000 years ago in search of water. Trample marks and bone fracturing are in evidence at several terminal Pleistocene assemblages, such as Lamb Spring, Colorado

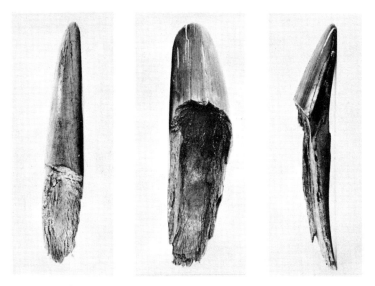

Figure 7.1. Three tusk tips from late Pleistocene *Mammuthus primigenius* at the Hofstade site, Belgium. Photographs courtesy of Mietje Germonpré; the specimens are from the collection of the Institut Royal des Sciences Naturelles de Belgique.

(the mammoth level is dated 13,000–11,000 years B.P.). Broken tusk tips have been found in several elephant bone assemblages around the world, such as Hofstade, Belgium (Figure 7.1), Old Crow, Yukon, and the Torralba and Ambrona sites in Spain (Figure 7.2), but have not been widely reported from the United States. In time-averaged deposits, tusk tips may have accumulated very slowly as a result of occasional fights between bulls; but in other water-hole deposits, tusk tips may have been broken in relatively higher numbers during dry periods, as a result of more aggressive interactions between animals seeking water.

The tip of one tusk of the Taimyr mammoth (collected in 1948, curated at the Zoological Institute in Leningrad) had been broken off in life, leaving flake scars that were wearing smooth at the time of death (Figure 7.3). The tip of the one surviving tusk of the Beryozovka mammoth had been broken in life, also leaving a flake scar that was in the process of being worn smooth at the time of the animal's death. Several other woolly mammoth tusks curated at the Zoological Institute of the USSR Academy of Sciences in Leningrad had been broken in life as well. Vereshchagin and Baryshnikov (1982) suggested that these tusks broke during attempts to lever or chip off pieces of ice, which the mammoths would have swallowed for water. Scratches and abrasions on tusk shafts may have resulted from use as ice-

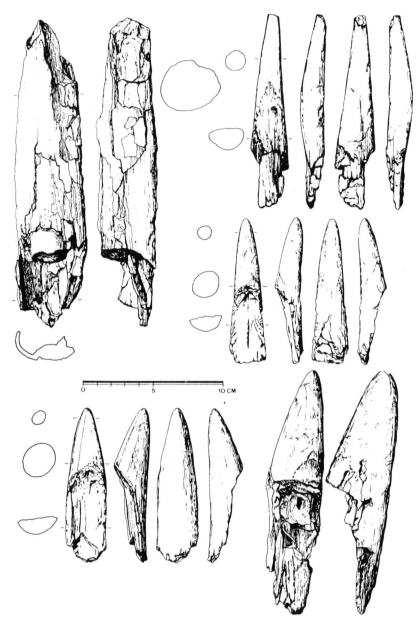

Figure 7.2. Tusk tips from the Ambrona middle Pleistocene site, Spain. (From Howell and Freeman 1983.)

Figure 7.3. The tip of the right tusk of the Taimyr mammoth (*Mammuthus primigenius*) (see Section 2.12). The tusk was broken in life, creating the flake scars that at the time of death were in the process of being rubbed smooth to reform a conical tip.

breakers, although Flerov (1931) and Garutt (1964) thought that the scratches resulted from the tusks being used to sweep snow off the ground to uncover vegetation. I propose the added possibility that the tusk tips broke off during aggressive interactions over access to open water during the Pleistocene; the effects of ice crusting and late-winter refreezing following a period of warmth and a thaw would have been similar to the effects of hot drought: Water would have been difficult to find, and mammoths would have fought over it.

7.1.6.   Conclusion of Part One

Even if the mammoth and mastodon bone accumulations resulted from mass mortality processes very similar to those documented in Africa, North America, and Australia, it is important to remember that these fossil deposits need not necessarily record the final *events* of extinction. They do provide a complex but readable record of a process whose outcome is clearly known. These deposits may not record the death of the very last animal, but they are powerful documenting evidence of major mortality processes that affected all big-game animals in the local populations, even those that died elsewhere.

Paul Martin (1967, 1973, 1984) has ably marshaled the facts and arguments that weaken the case that climatic changes alone could have accounted for the extinction of megafauna. In the first place, not all extinctions of the same or different genera were synchronous throughout the world, although the major climatic episodes of glaciation and deglaciation seem to have been closely synchronous in northern latitudes (Flint 1971; Shackleton and Opdyke 1973). Second, mammoths and other large mammals had survived earlier repetitive glacial–deglacial cycles affecting climates and had not become extinct. Temperature changes during at least one of those earlier cycles appear to have been more extreme than during the Pleistocene/Holocene transition (Owen-Smith 1987; Shackleton et al. 1983), although the rate of change may have been less rapid, and yet many fewer mammalian genera became extinct during that earlier cycle than at the beginning of the Holocene. A more rapid rate of change could have contributed to the many extinctions during the last cycle, although we lack the fine detail from the earlier deglaciations that would allow close comparisons. To these and other points raised by Paul Martin, I would add several others based on observations of the fossil record.

Some of the terminal Pleistocene mammoth bone assemblages appear quite similar to modern, nonculturally created elephant bone assemblages in Africa, and yet I have found no records of similar assemblages dating to the earlier glacial cycles in North America or Eurasia, especially to the periods of climatic stress that would have paralleled the stress at the end of the Pleistocene. Such assemblages may have existed, but been subsequently destroyed by cycles of erosion and basin flushing. However, it can be reasoned that in fact some earlier bone accumulations have survived to be discovered in modern times, and so the existence of die-off accumulations seems unlikely if none have been found after decades of searching. Judging from the lack of die-off evidence from earlier times, it might appear that the late Pleistocene period was by far a time of more severe or more abrupt stress for mammoths than were earlier climatic reversals, such as that at the end of the Illinoian glacial interval (140,000 years B.P.)

Paleoenvironmental evidence (including fossil bones, of course) from the earlier climatic oscillations is not as well preserved or as easy to find as is evidence from the end of the last Glacial period. But such evidence as exists suggests climatic changes no less severe than those that marked the end of the Pleistocene. Thus, if the late Pleistocene mammoth bone accumulations indeed resulted from climatic stresses on mammoth populations, there should have been earlier mammoth die-offs resulting from the earlier climatic reversals. Apparently there were none. Clearly, something exceptional happened at the end of the Pleistocene. The possibility remains that far harsher climatic shifts left no isotopic or foraminiferal evidence (Shackleton et al. 1983). The North American evidence has been interpreted as reflecting abrupt changes in climates and biota at the Pleistocene/Holocene boundary (COHMAP members 1988), but many researchers in Europe think that the climatic shifts and megamammal extinctions there were staggered over a lengthy span of several millennia. Could the added presence of humans have made it impossible for the mammoths to survive the already difficult conditions of changed climates, whereas previously the mammoths had been able to surive such climatic conditions?

7.2. Part Two: Overkill?

> *The big animal was soon made to look like a porcupine as the*
> *barbed spear thrown with unerring aim found an easy billet in*
> *his huge body.... These men seldom lose a wounded elephant.*

—E. Lewis, *Trader Horn* (1927)

> *They hunt, and eat and drink until they are ready to burst. Then*
> *they sleep.... They have dances, and make ready their poisoned*
> *arrows and knives of stone or bone. Then they hunt again.*

—H. Rugg and L. Krueger, *Nature Peoples* (1936)

At some time around 11,000 radiocarbon years ago, according to the archeological evidence, a distinctive kind of human culture became established throughout the Americas, from the northern Brooks Range in Alaska to the southern tip of South America. That time period was coincident with the extinction of 70–80 percent of the New World's Ice Age mammalian taxa (P. Martin 1984). During that time interval, and for a few thousand years before it, huge accumulations of mammoth bones had been deposited in the Old World. The origins of those deposits may have been cultural (kills made by human hunters) or, alternatively, natural (die-offs due to environmental stress). If cultural, the assemblages may have resulted either from episodic, successional kill events or from single mass slaughters. Mammoth bones may have come to rest where the animals were slain, or they may have been discarded in places other than where the animals had been killed. Many sites have been subject to conflicting interpretations, because the evidence is ambiguous.

The archeological record has provided an abundance of finely made stone implements in direct association with mammoth bones in large sites in Asia and Europe. Many Old World sites contain bones from dozens of proboscideans, as well as evidence for a developed human artistic tradition, the use of semisedentary or seasonal home bases, and an appearance of a relatively comfortable existence late in the Pleistocene, in spite of colder temperatures and the assumed harsher climate of the Late Glacial interval.

It seems clear from the fossil record that Paleolithic human groups in the Old World sometimes utilized proboscideans for food and other resources, and that source of supply apparently provided adequate raw materials for shelter, clothing, art items, fuel, weaponry, and tools. It has been suggested in the literature that the human subsistence orientation toward the mammoth may have contributed to its extinction in both the New World and Old World. I favor a slightly different interpretation of the human role in destroying mammoths. I propose that humans opportunistically spread their worldwide range in northern regions only after they had found a resource (mammoths) that was becoming more and more vulnerable because of environmental stress, such as harsher winters or seasonal drought. In other words, late Pleistocene hunting-gathering groups were not mammoth specialists who spread into the New World and caused the disappearance of mammoths because of their superiority as big-game hunters. Rather, the rapid spread of humans was in response to their awareness that mammoths were clustering in certain identifiable regions and habitats, where die-offs were resulting.

If people did indeed come to the New World because they discovered that game animals were easy prey for part of the year, those people may have reacted to the growing scarcity of their preferred subsistence base by increasing their mobility, to seek out the dwindling localized big-game populations. Greater mobility would have led to further explorations and rapid colonization of unoccupied ranges where mammoths and perhaps other large mammals still survived. To decide the plausibility of this and related hypotheses, we need to know whether the disappearance of the mammoth was abrupt or a lengthy, drawn-out process. If human mobility did increase so greatly at the end of the Pleistocene, then contact with other unfamiliar human groups may also have increased. Earlier people certainly were not vegetarians, but they may have been less oriented toward big game than were their descendants (such as Clovis people). The wide and rapid spread of North American projectile points at the Pleistocene/ Holocene transition could possibly indicate wide movements of people, or alternatively a peaceable and far-flung network of cooperation and reciprocal exchange.

7.2.1. Late Pleistocene mammoth bone accumulations

Clovis projectile points are found distributed virtually throughout the unglaciated (and some of the deglaciated) parts of North America. Possibly related forms occur as far south as the tip of South America. Of the thousands of reported Clovis finds and sites, most have been only surface occurrences. About two dozen buried sites have been chronometrically assigned ages that are acceptable to most prehistorians. Clovis points apparently are all of terminal Pleistocene age, roughly 11,000 years B.P. (C. Haynes 1987), evidently deposited within a narrow range of time. And wherever Clovis points are found directly associated with preserved animal bones, megafaunal taxa are present, especially *Mammuthus*. In fact, nearly all Clovis sites that have bones have mammoth bones although at least one site contains mastodon (*Mammut*) (Table 7.1). It is important to note that Clovis or "related" points appear abruptly in the archeological record; nearly as abruptly, and at apparently the same time, the megamammals *Mammuthus* and *Mammut* disappear from the fossil record. At first glance, it would therefore appear that the late Pleistocene people of North America were big-game specialists, oriented toward the biggest of game – mammoths, and perhaps mastodons as well. C. V. Haynes (1980, 1987), among others, has pointed out the similar material characteristics of many Clovis assemblages and Old World Paleolithic assemblages.

However, the Old World and New World cultures that utilized mammoths were quite different in important respects. As pointed out in Chapter 6, there were no late Paleolithic cultures anywhere in northern Eurasia whose subsistence can be characterized as mammoth specialization (G. Haynes 1990; Hoffecker 1987; Soffer 1985). A few individual sites may have resulted from mammoth-centered hunts and bone scavenging. Archeological evidence indicates that all known culture complexes appear to have specialized in reindeer (*Rangifer tarandus*), horse (*Equus* sp.), bison (*Bison priscus*), or, in many cases, smaller game. According to faunal counts and intersite comparisons, as summarized in Table 6.4, mammoth utilization by humans in far northern Eurasia appears to have been situationally opportunistic, not indicative of habitual preference for mammoth. The very few mammoth-dominated site assemblages may represent no more than human use of noncultural deaths.

According to my reading of the paleontological and archeological records

Table 7.1.   *Clovis sites containing proboscidean remains*

Site (& reference)	Geomorphic setting	[14]C date	MNI mammoths/ mastodonts
Blackwater Draw (New Mexico) (1)	Lacustrine & alluvial (spring-fed depression)	11,170 ± 360 (A − 481), 11,040 + 500 (A − 490), 11,630 ± 400 (A − 491)	6 (many locales),
Colby (Wyoming) (2)	Alluvial (but dry at time of bone-bed deposition)	11,200 ± 200 (RL − 392)	7[b]
Dent (Colorado) (3)	Colluvial sand	11,200 ± 500 (I − 622)	13
Domebo (Oklahoma) (4)	Alluvial/ lacustrine	Many dates, including 11,045 ± 647 (SMU − 695)[c]	1
Dutton (Colorado) (5)	Gleysol/ lacustrine	11,710 ± 150 (SI − 2897), underlying Clovis	?
Escapule (Arizona)	Alluvial/ colluvial	No date	1
Hiscock (New York) (7)	Spring-fed, lacustrine	11,250 ± 140(Beta − 16736) 10,930 ± 90(Beta − 16175)	6
Lange-Ferguson (North Dakota) (8)	Lacustrine spring pond	Not directly dated	1
Lehner (Arizona) (9)	Alluvial/ spring-fed stream	15 dates, average 10,900 ± 60	13
Leikem (Arizona) (10)	Alluvial/ streams	No date	2
Lubbock Lake (Texas) (11)	Lacustrine/ alluvial	11,100 ± 80(SMU − 263)	2(?)
Miami (Texas) (12)	Windblown/ lacustrine	No date	5
Murray Springs (Arizona) (13)	Alluvial/ lacustrine	11,300 + 500 (A − 8056), 11,150 ± 450 (A − 805a), 8 dates average 10,900 ± 50[e]	2
Naco (Arizona) (14)	Alluvial	No date	1
Rawlins (Union Pacific) (Wyoming) (15)	Lacustrine/ alluvial spring run	11,280 ± 350 (I − 449)	1
Kimmswick (Missouri) (16)	Alluvial/ colluvial	No date	2–5

[a]Age-profile types: A, normal herd, catastrophic shape; B, die-off type, selects mainly young (G. Haynes 1985); C, prime-age-selective (G. Haynes 1985); Ad., refers to single adult animal; Juv., refers to single juvenile animal.
[b]Graham (1986).
[c]Stafford et al. (1987).
[d]C. Haynes (1981).
[e]C. V. Haynes (1988 personal communication).

Table 7.1 (*cont.*)

No. Clovis points	Limb bones broken (+ = yes)	Anatomical order/articulation (+ = yes)	Cuts or chops (+ = yes)	Age profile[a]
11	+	+	−	C
4	−	+	−	B
3	−	?	−	B
3	−	+	−	Ad.
1	−	?	−	?
2	−	+	−	Ad.
3 (1 under bone)	+	?	+	?
3 (not in bone bed)	+	+	− / +	Ad. / Juv.
13	−	+	−	B
1 (not with bones)	?	?	?	Ad.
0	+	+partly	+	?
3	?	?	?	B
7[d]	−	+	−	C(?)
8	−	+	−	Ad.
0	+	+	+	Ad.
3(?)	+	−	?	?

Sources: 1, Agogino (1968), Hester (1972); 2, Frison (1976, 1978), Frison and Todd (1986); 3, Figgins (1933); 4, Leonhardy (1966); 5, Stanford (1979), Stanford (1982 personal communication); 6, Hemmings and Haynes (1969); 7, Jacobs (1989); 8, Hannus 1984); 9, Haury et al. (1959); 10, Saunders (1980); 11, Warnica (1966), E. Johnson and Holliday (1985); 12, Sellards (1952); 13, Hemmings (1970), C. Haynes (1981, 1982); 14, Haury (1953); 15, H. Irwin (1970); 16, Graham et al. (1981).

in northern latitudes, the prey species that figured in human diets reflected the local relative abundances of herbivores. In some subregions of northeastern Asia and North America, the proportions of species in both archeological and nonarcheological assemblages are broadly similar, indicating that humans were dominant predators, selecting game to hunt according to the local availability of different taxa [compare the results reached in other studies of big-game predation, such as that by Tilson, von Blottnitz, and Henschel (1980), a study of prey selection by spotted hyenas]. In subregions where humans were sympatric with other large predators, there probably were human preferences for specific prey taxa. However, studies such as that by Carbyn (1974) have shown that predators (*Canis lupus* in Carbyn's study) have clear preferences for certain taxa in geographic areas of high prey diversity, well out of proportion to the relative densities of the available species, even under conditions of little or no competition from other predators. I suggest that some predators, such as humans, may or may not show strong preferences for killing certain species, irrespective of competition from the other predatory taxa.

Humans may have been the dominant predators in many late Pleistocene ranges of the northern latitudes, and perhaps the only predators on mammoths and mastodonts. No carnivorous taxa of the late Pleistocene appear to have been predators on adult mammoths, although the Friesenhahn Cave (Texas) assemblage may have resulted from scimitar-toothed cats preying on young proboscideans. In comparable modern faunal communities, no carnivores have evolved that habitually elect to kill elephants, although lions and spotted hyenas in Africa occasionally have been recorded preying on young male elephants that have left the natal group.

No modern human hunters specialize in elephants. In fact, no human hunters except some arctic peoples specialize in only a small number of prey species or in one species. Prey selection by recent human hunter-gatherer groups outside the arctic in areas of prey diversity appears to reflect the abundance of prey taxa in the local environment – in other words, human hunters take medium-sized prey as they find it, in proportion to the relative densities of different species (Lee 1979; Marks 1976; Silberbauer 1981; Tanaka 1980). They do not deliberately specialize in particular taxa, although in general most human groups do prefer medium-sized animals over very small or very large ones, because of considerations of cost (time and energy required to find and procure game), the expected return, and the predictability of procuring animals.

Of course, this is not an argument (by ethnographic analogy) to refute the possibility that Pleistocene human hunters preferred to kill mammoths and mastodonts. The theoretical principles of predation are difficult to sort out when applied to social predators, especially humans, who have the capacity to invent new (cultural) adaptations to circumvent biological limits. Humans may optimize their hunting decisions in ways that are not clearly understandable to modern-day archeologists; see the papers in the volume by Winterhalder and Smith (1981). If the benefits or returns from the hunting of proboscideans were rich and abundant enough, the potentially high costs of hunting them may have been found acceptable.

The costs involved in mammoth hunting by people armed with spears would have been quite different than for people equipped with high-powered rifles or bows and arrows.

7.2.2. Potential ways of hunting mammoths and mastodons

Scholars are not agreed on Paleolithic hunting techniques. According to some writers, at the sites where bones of several mammoths are present, the individual animals most likely had been killed on separate occasions (e.g., Frison 1978, 1986; Haury et al. 1959). On the other hand, Saunders (1977a), 1980) suggested that Clovis hunters sometimes confronted and killed mammoth family groups as units, a suggestion that has met some resistance (Agenbroad 1980; G. Haynes 1985b), as well as some cautious acceptance as a useful hypothesis that may or may not prove correct (C. Haynes 1980, 1982). C. V. Haynes (1966) sought to explain one mass occurrence (the Dent site) by suggesting that Clovis hunters had stampeded a mammoth herd off a bluff, and Sellards (1952:23) thought that some animals at another mass occurrence (Miami) had died of disease, starvation, or drought and then had been scavenged by Clovis people. Some other mammoths may have been killed while they were in an enfeebled state.

On the basis of studies of drought-stressed African elephants (G. Haynes 1985b, 1987c), I have proposed that the evidence from some Clovis mammoth sites may indicate that noncultural mammoth die-offs accounted for the apparent family groupings of individuals found in certain sites. Presumably, drought or harsh winters would have affected mammoth behavior, and such stress would have been especially serious in the terminal Pleistocene, when extinction occurred. Mammoths clustering at water sources would have suffered and eventually died from malnutrition, dehydration (due to competition for dwindling water sources), and greatly increased vulnerability to external stresses such as low temperature.

In this hypothesis I imagine a different Clovis style of "specialization" in mammoth procurement: Instead of boldly confronting mammoth families, surrounding them with spearmen, and then killing one – or, more ambitiously, containing the group after slaying the herd leader (the matriarch), and eventually killing other family members (Saunders 1977a, 1980) – Clovis hunters were merely spectators at die-offs, perhaps occasionally dealing the killing blow to an already dying animal, or perhaps just waiting for animals to drop dead before cursorily butchering them. Elephant behavior during die-offs is described in Chapter 4 and in the earlier publications cited there. That behavior is predictable, and it would be appealing for meat hunters to exploit that predictability. Furthermore, the age profiles resulting from die-offs are similar to the age profiles from mass bone sites, such as Lehner (Arizona), Dent (Colorado), Miami (Texas), and Colby (Wyoming), all of which are Clovis sites (Graham 1986a; Saunders 1977a, 1980). I suggest not only that Sellards (1952) was correct in his explanation for the Miami site but also that his explanation will work just as well for many other Clovis occurrences.

It would seem worthwhile to look at the Clovis mammoth sites more closely and compare those sites' characteristics with the few available ethnographic descriptions of recent elephant hunters' sites and noncultural sites. The Clovis sites and the recent sites are not identical, and the differences may be significant in our attempt to find meaning in the record.

7.2.3. Characteristics of Clovis and modern sites

Few Clovis sites have been fully described in the published literature. Some buried Clovis sites had already been partly destroyed by erosion when discovered. Detailed information is at hand for less than two dozen buried sites, most of them located in the western United States. Table 7.1 presents summarized data from fifteen sites, but at least three of those sites may not be primary Clovis–mammoth/mastodont associations: The Lange-Ferguson, Dutton, and Rawlins (Union Pacific) sites. The Clovis points at Lange-Ferguson (South Dakota) were not found directly in the bone bed, but 10–15 m away in the same stratigraphic position as the bones; no Clovis points were found at the Dutton (Colorado) and Rawlins (Wyoming) sites, although the radiocarbon dates are near to or within the Clovis interval defined by C. V. Haynes (1987).

The notations in Table 7.1 for geomorphic setting are gross generalizations and probably obscure the fine details regarding sediments and the nature of the deposits containing the bones and Clovis points. For example, the fact that bones were enclosed by alluvium does not mean that they were originally deposited in moving water. There is little doubt that each of these sites at the time of bone deposition had been low ground at or within water sources, such as springs, ponds, or streams (C. Haynes 1980, 1984). Some site sediments may have been redeposited. At Dent, for example, the Clovis points and mammoth bones may have been redeposited after the Clovis occupation (Cassells 1983); the bones lay in three distinct alluvial-colluvial units (C. Haynes 1988 personal communication). Todd and Frison (1986) acknowledged that fluvial transport moved some of the Colby mammoth bones after their original deposition, which may have been as meat caches.

Some sites listed in Table 7.1 contained the bones of a single mammoth, whereas others contained the bones of numerous individuals. In four cases, some limb bones had been fragmented prior to final burial. However, in none of these examples were the limb-bone fragments directly associated with Clovis points or other stone tools, although they may have been situated in identical stratigraphic contexts (see Table 7.1 for references). Broken and possibly cut mammoth bones at Blackwater Draw (New Mexico) came from several different loci, but their association with Clovis implements is not always clear (Hester 1972; Saunders et al. 1990; Warnica 1966). The Rawlins site contained broken bones and several stone tools, but no Clovis points; unfortunately, the site has not been fully described in the literature. At Dutton, one or more limb bones may have been broken, but clear Clovis association is lacking. The Kimmswick site contained Clovis tools and broken mastodont bones, but the bone preservation was extremely poor, and much of the breakage probably can be attributed to postdepositional processes (R. Graham 1987 personal communication). At Lange-Ferguson, the Clovis points were not found within the bone bed.

7.2.4. Broken limb bones: why or why not?

If we could determine that limb bones were rarely or never broken in the other (less ambiguous) cases of Clovis–mammoth association, such as Lehner, Naco, Dent, Murray Springs, and Colby, we might be able to explain why there were differences among the sites. In fact, a few bones had indeed been broken at some sites (although most fragments have never been excavated or reported), judging from the occurrence of a bone "shaft wrench" (C. Haynes and Hemmings 1968) and a tibia fragments (Hemmings 1970) at Murray Springs, bone "foreshafts" (Hester 1972) and sectioned ivory (Saunders et al. 1990) at Blackwater Draw, and bone foreshafts at Sheaman (Frison and Stanford 1982) and at Anzick (Lahren and Bonnichsen 1974). These kinds of specimens are exceptionally rare; the broken bones in most assemblages had generally been fully shaped and finished into implements. Unmodified fragments of broken-up limb bones are very rare (or rarely reported). The scarcity of unshaped bone fragments in Clovis sites seems unusual when we compare their abundance in ethnographically documented big-game processing sites.

Unfortunately, searching the ethnographic literature to find examples of the human use of megamammal carcasses is a disappointing undertaking. Ethnographic analogy as a tool to explain prehistoric behavior has been a subject of lengthy debate in archeological theory; reliance on analogy to explain the archeological record may or may not be wise or appropriate (Wylie 1985). References to ethnographies here are not intended to be definitive "proofs" of what happened in the prehistoric past. Yet descriptions of how humans behave in the present may add a valuable perspective to our search for meaning in the archeological record.

Although most of the ethnographic literature ignores the details and particulars of what happens to bones when humans butcher large game such as elephant, a few documentary sources make it clear that the fresh bones of large carcasses often suffer heavy damage. According to the stories and tales of many travelers in Africa, utilization of elephant carcasses by native people typically was thorough; for example, see the reports by J. Wood (1868) and F. C. Selous, cited in Lydekker (1908 : 16).

When the Ba-Mbuti (a Pygmy people of Zaire) slew an elephant, they thoroughly sectioned the carcass, often chopping with axes or slicing with heavy iron spear blades, then stripped off meat, chopped open leg bones to eat the oil-rich cancellous interiors, and finally cooked, carried off, and scattered much of the skeleton; see Duffy (1984), Turnbull (1961), and K. Duffy's motion picture "Pygmies of the Rain Forest." Recent observations of other African people (Crader 1983, 1984; J. Fisher 1986 personal communication) have verified that thorough use is made of elephant carcasses. Chopped, cut, or broken bones may be abandoned at the kill/butchery sites, and much of the skeleton is disarticulated.

What people get from the fractured limb bones is a thick, fatty substance, called marrow when found in the "hollow" limb bones of ungulates. Most elephant limb bones have no extensive medullary cavities full of marrow; instead, the interiors are packed with trabecular bone in three-dimensional latticework, the hollows of which contain the marrow material. This substance tastes good, is usefully greasy, and is in fact healthful, because fat usually is in short supply in the muscle meat of wild animals (Allen et al. 1976; P. Brooks 1978; P. Brooks, Hanks, and Ludbrook 1977; FAO 1977; Speth 1983, 1987; Talbot et al. 1965). Jochim (1976, 1981) and Speth (1983, 1987) surveyed the ethnographic literature and found that many hunter-gatherers throughout the world make special efforts to seek out sources of fat, a biologically sound strategy, because lean-meat diets offer relatively poor sustenance to humans with limited access to carbohydrates. Elephants, like other wild animals, have relatively little body fat compared with domesticated species, so their limb bones are potentially rich sources of necessary nutrients; see Speth (1983) and Speth and Spielmann (1983) for more details on the conditions under which fats are beneficial in human diets. Aside from enjoying the flavor and texture, Pygmies and other elephant hunters simply were eating right whenever they broke open elephant limb bones for the marrow. However, not all elephant hunters behaved the same way; Crader (1983 : 127) noted that although the Zambian Bisa, sedentary horticulturalists, highly prized animal fat, they did not intentionally break open elephant limb bones seeking it, although they did eat fat from the feet.

Stanford et al. (1981a) argued that in sites containing broken proboscidean limb bones (virtually all of which are non-Clovis associations), humans were not seeking marrow but were manufacturing cutting, scraping, or chopping tools made of thick cortical bone. Hannus (1983, 1984, 1986) classified some of the broken bones from the Lange-Ferguson site as retouched and utilized butchering tools, a diagnosis of fractured mammoth bones that has also been applied to a number of other site assemblages in North America, as reviewed elsewhere (G. Haynes 1981). It is impossible to find recent ethnographic records that hunter-gatherers broke proboscidean bones for the manufacture of butchering tools; if megamammal limb bones were broken by primitive people, they would seem to have been further modified, shaped, and decorated before use. Occasional reports of minimally modified bone "tools" in archeological assemblages are not considered here to be proven examples (Frison 1970, 1974; E. Johnson 1978, 1985; E. Johnson and Holliday 1985). North American plains Indians may have used unmodified bones from fresh carcasses as aids in butchering, but there are no clear records of them breaking bones to use without further shaping. Appeals to ethnographic analogy cannot provide guidelines for explaining broken limb bones as tool sources in elephant sites.

At any rate, the Clovis sites do not contain nonnegligible quantities of broken bones. Because modern elephant hunters usually break bones for the marrow fat, and because hunting people put a premium on animal fats, it would seem that Clovis hunters often behaved in an unexpected manner.

A couple of possible explanations come to mind: If the diet of Clovis people was rich in carbohydrates, they would have had little or no need to break open bones and eat the oil-rich cancellous interiors (Speth 1983, 1987). Or if the mammoths utilized by Clovis hunters were carrying much greater quantities of body fat than modern elephants and other megamammals, once again there would have been no need to break limb bones. Neither of these possible explanations for the lack of broken bones seems satisfactory.

The amount of carbohydrates in the Clovis diet undoubtedly was limited in the American west, just as it was for the bison-hunting Plains Indians or any other human group dependent mainly on game animals and living in open habitats where sources of fruits, nuts, and wild seeds are scattered, rare, and seasonally unavailable. People living in open habitats routinely broke limb bones, as do nearly all recorded hunter-gatherers of arctic, temperate, and tropical latitudes. The fossil record has not preserved evidence that Late Glacial plant resources provided appreciable carbohydrates to Clovis diets. For example, seed-grinding tools and parched seeds range from extremely rare to nonexistent in Clovis sites. Sellards (1952) reported a grindstone from the Clovis level at Blackwater Draw, but the scarcity of such implements probably indicates that there was little or no seed-grinding activity during Clovis times. Seed-grinding implements tend to be heavy and repetitively used, often showing signs of reuse or rejuvenation; amorphous grindstones may be expedient tools made to utilize local resources, part of a toolmaking repertory adjusted mainly to exploit other types of resources (M. Smith 1986).

As for the idea of mammoths having abundant body fat, it may find some degree of support from published descriptions of frozen Siberian woolly mammoth carcasses, but there have been few such reports, and they are open to some question. Northern mammoths (*M. primigenius*) were well insulated by a dense coat of hair, but the thick layer of subcutaneous fat mentioned in the literature (Augusta and Burian 1964; Garutt 1964; Pfizenmayer 1939) may in fact partly have reflected the great thickness of the skin itself. Thick subcutaneous fat, though an effective insulator, is a heavy burden for animals to carry, and in fact such an insulating system relying on fat is not seen in modern arctic land mammals (L. Irving 1966; Scholander et al. 1950). Of the animals indigenous to northern regions, only polar bears and marine mammals possess thick subcutaneous fat over the entire body as an adaptation to life in cold waters. Migrating or cyclically starving ("hibernating") mammals such as bats and bears do put on body fat, both subcutaneously and intraperitoneally. Other northern mammals put on fat mainly along the dorsal part of the body, but this functions as an energy reserve for the hard times of winter, not so much as an insulating layer, because it is depleted over the course of fall and winter.

All frozen partial carcasses of mammoths recovered in northern Asia were decayed to different degrees (Garutt 1964; Herz 1902; Howorth 1887; Pfizenmayer 1939; Shilo et al. 1983; Stewart 1977; Tolmachoff 1929; Vereshchagin 1981b, 1982), and even the best-preserved specimens, such as the Beryozovka mammoth and the Magadan calf (Dima), were desiccated, distorted, partly eaten or partly rotted carcasses that had lost mass and volume to scavenging carnivores, anaerobic decomposition, or aerobic decay, or some combination of those factors. Pfizenmayer (1939: 103, 163) stated that both the Beryozovka and Sanga-Yuryakh mammoths had 3–3.5-inch-thick layers of well-preserved fat below the hide, as well as plenty of fat in the body. How that fat survived autolysis and decay is not easily explained in view of the decomposition of skin and viscera. In my own examinations of preserved mammoth tissue and carcasses in the Zoological Institute and Paleontological Institute of the USSR Academy of Sciences (in Leningrad and Moscow, respectively), I did not find any subcutaneous fat, but I did see several masses of tendons, ligaments, and other soft tissue enclosed in adipocere, a firm kind of fat that is produced by postmortem hydrogenation and hydrolysis of soft tissues in the body. However, I acknowledge that the measurements of fat thicknesses reported by Herz and Pfizenmayer suggest that the fat was actually seen, although it was not preserved in the museum collections.

Columbian mammoths (*M. columbi*) were generally larger than the northern or woolly type; a thick layering of subcutaneous fat would have been relatively inefficient insulation for them, as it is for modern arctic mammals, aside from being perhaps less necessary in temperate latitudes. No doubt fat was put on seasonally around the viscera and laid down along the back, where large mammals commonly store energy reserves for the winter. Columbian mammoths probably had a greater potential than did woolly mammoths for putting on body fat, because more woody plants were present in columbian habitats (Guthrie 1982; Shilo et al. 1983). Studies of modern African elephants have shown that during seasons of nutritional stress (such as winter or very dry periods), elephants that have access to browse are in much better nutritional states than those that are mainly grazers (B. Williamson 1975a, b, 1976). Elephants opt to eat the bark of particular trees because those taxa (such as *Acacia*) contain relatively higher concentrations of fatty acids (Guy 1976).

The elephants studied in southern Africa by B. Williamson (1975a, b, 1976) had much greater proportions of browse in their diets than did other samples of elephants from east Africa (Laws et al. 1975) and were in better condition during dry seasons. Although subcutaneous fat layers were present in healthy animals, the fat was never very thick. Modern African elephants in some ranges are faced with the need to survive cold winter weather, when nighttime temperatures drop below freezing for one to three months each year. Elephants living in the high interior plateaus of south central and east Africa, where the weather is seasonally cold, do not insulate themselves with thick layers of subcutaneous fat (G. Haynes unpublished data). They do put on body fat, but most of it is around the kidneys, the stomach, and other internal organs. The elastic mass of fatty tissue and collagen in each foot that acts to cushion the step does not change volume seasonally. During winter, fat reserves are rapidly depleted because plant foods are at the lowest level of nutritiousness.

A conflicting argument can be made concerning thicker subcutaneous fat during the Pleistocene, based on studies of the bioenergetics of migrating Recent vertebrates (Orr 1970: 75–85). Heavy body fat (both subcutaneous and intraperitoneal) is put on by birds, fish, seals, sea lions, and bats that migrate seasonally or that enter winter torpidity and do not migrate. Body-fat deposition is a predictable characteristic of adult organisms that engage in seasonal movements; besides providing effective insulation because of reduced vascularity, it provides a sizable quantity of water as a by-product of metabolism. However, this water may not represent a net gain to the body, because fat oxidation requires increased oxygen, which usually is provided by greater ventilation in the lungs, resulting in water loss in respirated air (Schmidt-Nielsen 1960).

At this time it is impossible to prove that mammoths had enough subcutaneous body fat to satisfy Clovis hunters, such that there would have been no need to break open bones for marrow. Perhaps the research of D. Fisher and his associates (D. Fisher 1987, D. Fisher and Koch 1983; Koch, Fisher, and Dettman 1989), attempting to determine the seasons of death for proboscideans, will contribute to a resolution of the issue by revealing whether Clovis mammoths died in seasons when body fat would have been at its maximum or minimum level. D. Fisher, in studying some mastodonts in the Great Lakes region of the United States, has found that surficial marks ("cut marks") and bone breakage often accompanied a late autumn death (as deduced from thin-section analysis of teeth and tusk dentine). However, the association of possible butchering traits with autumn death is not without exceptions (Kapp et al. 1990). Although I doubt that all of the autumn deaths in fact reflected butchering, it is worth noting that in the view of Fisher and associates the mastodonts that lacked possible traces of butchering died in late winter or early spring, when their body fat would have been at its minimal level.

7.2.5. *Mass kills: meat and fat procurement in large quantities*

Saunders (1977a, 1980) suggested that Clovis hunters occasionally were successful in killing whole groups of mammoths, which, if true, would have resulted in a great quantity of body fat salvageable from the entire assemblage, enough fat to satisfy small human groups, except when such animals were in poor condition (as in late winter or early spring). If five to ten animals were killed together, relatively small amounts of remaining fat from viscera, muscle masses, and subcutaneous layers could have been gathered together into a sizable sum total, making it unnecessary for limb bones to be split open for marrow.

Speaking on the basis of experience with elephant herds encountered on foot, Frison (1986) has doubted that spear-wielding Clovis hunters could ever have contained and killed mammoth groups. My own personal experiences, like those of Frison, were with modern African elephants, and I have another point of view. I acknowledge that judging mammoth behavior strictly on the basis of African elephant behavior may be a little like judging black rhinoceros behavior on the basis of observations of only white rhinoceroses. But lacking indications to the contrary, my reasoning here is as supportable as any other. I do not think that Clovis people would habitually have slain mammoth groups, but I do think that *they would have tried, and succeeded (at some cost), when it was necessary.* D. Johnson,

Figure 7.4. Nineteenth-century artist's conception of an elephant hunt in Africa. From *Histoire Naturelle*, extracts from Buffon and Lacépède (1868), published by Alfred Mame et Fils (Tours).

Kawano, and Ekker (1980) and C. Haynes (1980, 1982), among others, have provided ethnographic examples of elephant hunts, some of which, if true, indicate that primitive people using only spears or swords often confronted and killed elephants under conditions that would be unthinkable to twentieth-century writers (Figure 7.4).

For example, J. Wood (1868 : 492) described a situation in east equatorial Africa in which 500 African people held in containment a herd of eighteen elephants by building a large encircling fire in the vegetation, then showered spears into the stupefied group after the fires had burned down. Apparently all were slain where they stood.

If that description is true, it lends plausibility to the proposition that Clovis hunters could have contained family groups of mammoths. But to have brought a group to bay and then to have killed some or all of them, a large number of spearmen would have been needed to surround the mammoths, relying on an encircling fire or other kind of impenetrable enclosure to prevent the animals from bolting free. An enormous quantity of spears would have had to have been thrown at the mammoths. Note how imaginary these kinds of reconstructions are.

If mammoth groups were killed by Clovis hunters in this way (and any other way is difficult to envision, although the strong tendency of elephants to run from humans today may in fact be a learned response to recent hunting pressures), the large numbers of carcasses would have provided fat in quantities sufficient to make bone breaking unnecessary. Most of the many spears that would have been required could have been salvaged by the hunters, and any spears tipped with bone and wood that were never salvaged probably would not have been preserved in the fossil record. Surprising numbers of stone points were left behind at some Clovis sites.

J. Wood (1868 : 134–8) described another kind of elephant hunt somewhere in the southern part of Africa. In that example the strategy was quite different from that in the example of the mass kill. Several men first carefully observed elephants and picked out one that had the most desirable tusks; the men noted the peculiarities of that animal's footprints and then tracked it, starting at a water hole, where the soft ground preserved the different spoor of individual elephants. On encountering their animal, the men hurled spears into it. The elephant ran off into the bush, pursued by the men and their dogs. The dogs distracted the wounded elephant and harried it while the men threw more spears at every opportunity.

Wood most likely recounted that type of hunt from a reliable source, because he supplied details about the dogs, certain ceremonies, and other aspects of the hunt that give his description the appearance of trustworthiness. Wood pointed out that iron-tipped spears thrown at an elephant rarely, if ever, penetrated the thick skin to strike vital organs. As a result, the chase was always lengthy and arduous, because wounded elephants can run for miles (G. Haynes unpublished data). The elephant would be encouraged to run, in order to tire it and hasten its end.

Wood's two examples of elephant hunts show us two different styles of procurement. In the first, a family group was contained by fire, and spears were thrown by hundreds of people. In the second, the single animal selected was not held at bay but was pressed to flee by a small number of spearmen. Either strategy could have been employed by Clovis people.

It is fruitless to express doubts about the wisdom of confronting mammoth herds when one is armed only with spears. Not all elephant herds will stand at bay, and not all will break in different directions. The argument about group kills cannot be solved by invoking ethnographic or egographic analogy. Perhaps a competing hypothesis will give us a new perspective. I suggest here another possible reason that we find assemblages of animals whose bones were rarely or never broken for fat. The animals may have been starving to death, and their limb-bone marrow may simply have been too poor for eating.

7.2.6.   Bone marrow: fat content and appeal

One factor affecting the fat content of marrow is the age of the animal. Bone marrow functions in young animals as an active factory and storehouse for red blood cells; in adults, the marrow has similar functions, but is less active in manufacturing red blood cells, and thus is more fatty. The marrow of rapidly growing animals is more pinkish and less solid than the thick, yellowish, fatty marrow of full-grown healthy adults (de Calesta, Nagy, and Bailey 1977; Hanks et al. 1976). These generalizations about bone marrow and ages of individuals have been verified in elephants (Albl 1971). In fact, many Clovis mammoths were subadults (Saunders 1977a, 1980), and their limb-bone marrow would have had relatively little appeal, even if they were healthy.

Another important factor potentially affecting the appeal of marrow has to do with the overall condition of the animal. Ungulates rely on body fat to get through periods of nutritional stress, and they mobilize fat in a patterned sequence, beginning with fat around the kidneys before progressing to bone-marrow fat (Albl 1971; P. Brooks et al. 1977; Cheatum 1949). Elephants or any other animals that have lost condition and whose fat reserves are depleted have poor-quality bone marrow, consisting of pink and runny matter, which has no appeal to anyone seeking fat.

If Clovis mammoths had been starving to death, the marrow in their limb bones would have been useless as a fat source for humans eating the animals. And because the marrow in other bone elements, such as the skull, scapulae, ribs, innominates, and vertebrae, normally contains less fat than does limb-bone marrow, starving animals' bones other than the long bones would not have been worth breaking. Because subcutaneous and perinephric (kidney) fat in herbivores is depleted before the fat in bone marrow, a human hunter who wishes to make a quick check of the condition of his prey need only make an incision into the carcass's skin or visceral cavity, either through the abdominal wall or between the distal ribs, to observe the kidney fat. Low stores of kidney fat will mean that limb-bone marrow fat probably also will be low.

The possibility of mammoths starving to death in Clovis sites should be kept in mind; the following sections present additional arguments and evidence in support of this proposition and also offer possible explanations for grouped mammoth remains.

7.2.7. Articulated skeletons

Note in Table 7.1 that all but a few of the fifteen sites contained articulated or nearly articulated bones, including large portions of mammoth skeletons (Figure 7.5–7.7). For the sites that have not been documented as containing articulated skeletons, hasty salvage excavation, erosion, or postdepositional disturbances could account for the jumbled disposition of bones.

Actualistic studies show that skeletons deposited on the surface of the

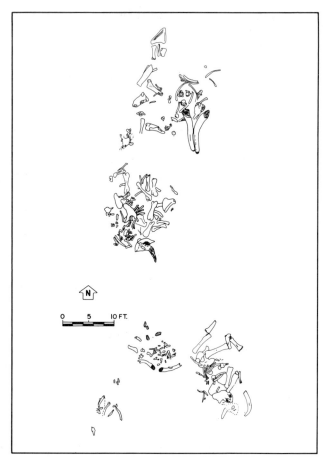

Figure 7.5. Bones of mammoths (*Mammuthus columbi*) (at least four) at the Blackwater Locality No. 1 Clovis site, New Mexico. Redrawn from Warnica (1966). Note the close clustering of each individual's bones. Numerous artifacts were recovered within the bone bed, and a "well" was found in the deposits about 2 m from the mammoth skeleton at the botton right.

earth, even in arctic environments, normally are disarticulated and scattered in a matter of days or weeks, because of a variety of different agencies (Haglund, Relay, and Swindler 1988; G. Haynes 1980a, 1982, 1988a; A. Hill 1979; Toots 1965; Voorhies 1969). Articulated skeletons are preserved during burial only when ligaments, tendons, muscle masses, or skin are still in place, holding the bones and body parts in anatomical order. Partial or whole carcasses that are buried relatively quickly by windblown, alluvial, or lacustrine sediments will be preserved associated or in

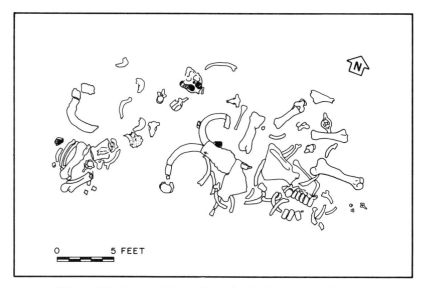

Figure 7.6. Mammoth bones from the 1936 Mammoth Pit excavations at Blackwater Locality No. 1. Redrawn from Hester (1972).

articulated order. The closely associated Clovis mammoth skeletons must have been abandoned unsectioned for the bones to have been preserved in such tight clusters. Note that the bones of some Clovis mammoth carcasses apparently had been dispersed to different extents.

Unique explanations have been proposed to accommodate the aggregation or dispersal of bones at different Clovis sites. For example, Frison (1976, 1978; Todd and Frison 1986) proposed that some parts of the mammoths at the Colby site had been stacked to protect meat caches, although he recognized that moving water had also influenced the distribution of elements. Irwin (1970:125) suggested that the Rawlins mammoth had been killed (or died when mired) in a bog, and the bones of only its left side had been preserved after sinking into the mud. However, he inferred that some bones had been "hacked off and dragged to [a] gravel bar" by the butchers, presumably to escape the sticky mud. Warnica (1966) attributed bone breakage and scattering at Blackwater Draw to either moving water or human action, processes that affected the various mammoth carcasses to differing degrees. He noted that the bones of the mammoths were not in identical states of preservation, an observation also made by Frison (1986) regarding the Colby mammoth bones, perhaps indicating different times of deposition for the various carcasses. Hemmings and Haynes (1969) thought that the Escapule mammoth had not been butchered, because the site yielded no butchering tools or artifacts of any

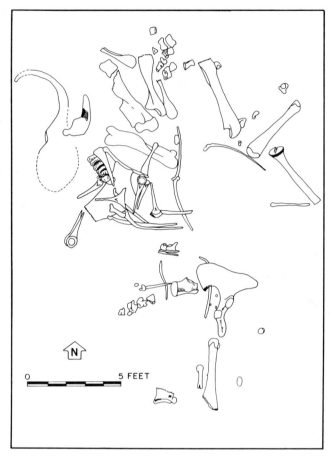

Figure 7.7. Mammoth bones from the Domebo Clovis site in Oklahoma. Redrawn from Leonhardy and Anderson (1966). Three Clovis points were excavated within the bone bed.

kind aside from two Clovis points, no cultural features such as hearths, and no characteristics of the bone bed suggestive of human utilization. Haury et al. (1959) believed that the wide bone dispersal for some Lehner-site mammoths and less scattering for others having articulated elements resulted from spaced kills that had been subjected to varying degrees of scavenging by carnivores and trampling by mammoths and other animals.

These and other potential explanations for bone dispersal or aggregation are plausible, but sometimes not satisfying, at least to me, primarily because the agencies that are invoked to explain bone dispersal at one site may be called on to explain bone aggregation at another. For example, the butchers at Colby are alleged to have stacked bones, some in articulated units; the butchers at Domebo are believed to have flipped individual bones short distances from each other; the butchers at Blackwater Draw (various loci), Lehner, and Naco are said to have moved bones only minimally (Naco may have been an escapee from Clovis hunters); the lack of patterned bone placements at Escapule is interpreted as indicating an absence of human actions.

Most Clovis-site bones have been too poorly preserved to show clear cutting marks or scars from butchering tools. Assemblages that show such damage usually are ambiguous cases of Clovis–mammoth associations or are ambiguous cases of actual cut-marking. Saunders et al. (1990) found fabrication marks on a mammoth tusk from Blackwater Locality No. 1 and possible cut marks on several bone fragments. In some cases, marks that appear to be cuts may have resulted from trampling, gnawing, or even the excavation process. However, the absence or scarcity of tool marks is not a reliable guide for understanding the degree of butchering of any mammoth carcass (Crader 1983). Because conarticulated elephant elements can be separated without the need for wedging them apart or cutting into periosteal bone on limb shafts, epiphyses, or articular surfaces (G. Haynes 1987b), tool marks would hardly be expected anyway. The cartilage that covers the articular ends of limb bones is quite thick, and once it has decayed, even seemingly deep cut marks will have disappeared, because they rarely penetrate to cortical bone surfaces. The chopping damage caused by heavy knives and axes might have been expected routinely to have left permanent scars on bone surfaces, but that type of damage has almost never been observed in Clovis mammoth bone assemblages, possibly because most Clovis tool kits seem not to have included any prepared heavy chopping implements. The large bifaces in the Simon and Anzick Clovis assemblages would have served well as choppers (C. Haynes 1988 personal communication), although in fact they may have been large cores or blanks. The whittled and pecked tusk segment described by Saunders (1985; Saunders et al. 1990) may indicate that the sectioning of heavy bones and ivory was possible, but not commonly done by chopping, and was rarely carried out in Clovis mammoth sites. In place of chopping tools, flakes or bifaces could have been used to section ivory or bones.

Crader's descriptions (1983, 1984) of the degree to which the Bisa people fully sectioned elephants (in order to remove meat) are similar in many ways to Duffy's observations (1984) among forest Pygmies. A copyrighted photograph by a person identified only as Bailey in an article by Redford and Johnson (1985:43) shows an elephant being butchered by Efe Pygmies, suggesting full utilization of the carcass. Likewise, Binford's descriptions (1978, 1981) of Nunamiut Eskimo utilization of caribou, Marks's description (1976:121–2) of buffalo butchering by the Bisa, three accounts (Lee

1979:222–3; Silberbauer 1981:213–14; Tanaka 1980:32–3, 35) of how San groups utilize big-game carcasses, and O'Connell's observations (1987) of Hadza butchery and processing, though different in itemized details and fine points, all show a similar pattern, or a comparable utilization "paradigm" – when individual large animals are to be eaten or processed for later consumption, the body is sectioned into units that often include a bilaterally split thoracic cage, limbs or parts of limbs, and head/neck units, with the sectioning done at joints, for the most part, followed by further subdivision, meat stripping, and usually bone breaking. Much of this cross-cultural paradigm derives from the need to divide up large-game carcasses into parcels that can be carried to a home camp some distance away. However, in many instances when large game is killed by hunters such as the Dobe San or Mbuti, the home camp itself is moved to the carcass site (Duffy 1984; Lee 1979; Turnbull 1961). It is clear that extensive carcass sectioning has as much to do with reducing the food into parcels conveniently sized for preparation, consumption, or redistribution as with reducing the carcass to portable fractions.

Clovis people seem to have left behind whole or nearly whole mammoth carcasses, some of which were unsectioned, an odd pattern of carcass utilization when seen in the light of recent ethnographic descriptions documenting the habitually heavy use of elephant carcasses. I point out that I do not claim that heavy use of proboscidean carcasses is unimodally "programmed" into all human hunting cultures; obviously, human uses would have varied, for a number of different reasons. For example, the Bisa of Zambia did not habitually break elephant bones for marrow. The point being examined here is that modern hunting people are not wasteful of big-game carcasses, and their nutritional return from elephant carcasses is regularly maximized. It is difficult to visualize a hunting people who would *not* maximize their returns, although it could perhaps be proposed that hunters entering a new continent and encountering naive, never-hunted megamammals would have found mammoth killing unstoppably easy.

At this point it would be useful to examine the whole general concept of carcass utilization by carnivores, including humans, and see what the meaningful variables are, because a structured appreciation of these attributes might reveal more meaning in bone assemblages. In this discussion, the butchery actions of elephant-hunting humans are considered to be reflections of economic motivation, not mere "ethnic" preferences. Thus, human uses of elephant carcasses are, in my discussion, comparable to models of carcass use by other mammalian carnivores.

7.2.8. Carcass utilization by carnivores

The degree of carcass utilization by carnivores varies according to the conditions under which food is obtained. Hungry carnivores eat more than well-fed ones, of course. The cause of the hunger may be difficulty in finding prey, difficulty in killing it, or a general increase in metabolic levels resulting from greater activity, such as extended traveling (G. Haynes 1980a, 1982). A very low prey density in a carnivore's catchment area would lead to greater mobility of the carnivore, plus greater utilization of prey carcasses when obtained (Blumenschine 1986, 1987). A very low prey vulnerability to predation or other causes of death would also result in much greater carcass utilization when a carnivore finally did gain access to a dead prey animal.

On the other hand, when prey animals are highly vulnerable to predation or other causes of death, such as drought-related stress, carcasses are lightly utilized by carnivores (Mech and Frenzel 1971; Pimlott, Shannon, and Kolenosky 1969). For example, during unusually severe winters in North America, wolf packs find it relatively easy to kill typical prey animals such as moose (*Alces alces*), bison (*Bison bison*), and deer (*Odocoileus* spp.). In these instances, when hunting is easier, prey carcasses often are abandoned in one to six units of articulated body parts (G. Haynes 1982:271–2): Lower legs tend to survive uneaten, articulated, and sometimes still attached to the thoracic/pelvic units. The skin usually remains on lower limbs, head and neck, and most of one side of the body (sometimes both). Some bones will be exposed, including ribs, whose proximal articular ends will still be attached to vertebrae, but whose shafts may be broken. Teeth marks may be visible on innominates, proximal femora, proximal humeri, and ribs and vertebrae, but the marks generally are separate, shallow furrows, scratches, pits, and grooves, rather than deep gouges or heavy networks of scars in cortical bone. Light marks result from the stripping of soft tissue, and deeper marks result from actual bone chewing. Heavily chewed bones result from competition for access to soft tissue and may provide one direct indicator of carnivore hunger at the time of feeding. Heavily chewed bones are also found at carnivore homesites, where animals gnaw bones both out of hunger and as an activity separate from feeding.

When heavily damaged individual bones are found at kill sites or sites of scavenged carcasses, the sequencing of damage to different big-game bones is rather patterned (G. Haynes 1982). Another patterned aspect of carcass utilization is carnivore separation of body parts and individual bones. When carcasses are more fully utilized, legs are detached from the body, the hide is eaten or pulled off the bones, and long bones within the limb units are themselves separated. Sometimes these limb bones are then fractured, after epiphyseal ends are eaten off. Skulls show tooth marking and breaks resulting from feeding. Elsewhere (G. Haynes 1980a, 1982) I have provided more details on the differences among light, full, and heavy utilization of North American big-game carcasses.

Blumenschine (1986), Brain (1981), and Richardson (1980), among a few others, have observed carcass utilization by African carnivores such as lion (*Panthera leo*), leopard (*P. pardus*), spotted hyena (*Crocuta crocuta*), and other taxa, and it is clear from their studies that the degrees of carcass use vary for reasons similar to those discussed for North America.

When times are tough for carnivores, virtually all the meat on a carcass is consumed, most skin is eaten, parts of bones are chewed and broken, and skeletons are dispersed and scattered. But when prey is easy to kill or easy to find dead or dying, such as during harsh climatic conditions, carcass use by carnivores is light: Part of the muscle meat is eaten, few bones are broken or gnawed, and skin is left covering the limbs and part or all of the axial skeleton. Carnivores, in effect, only nibble at carcasses, because fresher remains can be found nearby.

Wherever North American wolves find hunting easy, they often wander away from prey herds and carcasses to explore other parts of the landscape. They may choose to risk losing contact with prey and even voluntarily abandon uneaten meat because they have the security of knowing that if more meat cannot be procured, then at least they can return to the old carcass sites (G. Haynes 1982:278; Oosenbrug, Carbyn, and West 1980a, b). Spotted hyenas in southern Africa have the same kind of security during very dry seasons, when ungulates are weakened by drought conditions and starvation, hyenas are content to nibble at carcasses of elephants (Conybeare and Haynes 1984; G. Haynes 1987a, c, 1988a), as a result of which many carcasses remain nearly intact.

The available ethnographic studies of people who kill single elephants reveal that utilization usually fits into the category of full or heavy (Crader 1983, 1984; Duffy 1984; Turnbull 1961): Skin is removed, although it may not be eaten or otherwise used; limb bones are separated and often broken open; and body parts are thoroughly sectioned. Chopped and stripped bones are abandoned, except for tusks. Full utilization of elephant carcasses is, of course, not a genetically unchangeable mode of human behavior. J. Fisher (1988 personal communication) saw considerable variation in the thoroughness of elephant carcass use by Efe Pygmies. As is the case with other carnivores, human use of elephants may vary with situational contexts. However, I expect that full use of carcasses occurs more often than not, because humans, as a rule, find elephants very difficult to hunt and rarely attempt to procure them for food. When an attempt is successful, most of the returns from the hunt are fully utilized.

In Clovis sites, the degrees of utilization are difficult to decipher clearly, because bone assemblages in some sites will have undergone postdepositional disturbances, in the form of erosion, redeposition, or weathering destruction. In instances where postdepositional disturbances have been minor, and where Clovis mammoths have been examined conscientiously, Clovis mammoths have been found strikingly intact, with few or no clear instances of artifactually damaged bones or widely dispersed body parts. These carcasses must have been lightly utilized, or perhaps they were unutilized, as suggested by Hemmings and Haynes (1969) in the case of the Escapule mammoth.

To be fair, I acknowledge that some, or a few, or many skeletal joints may have been disconnected in the skeletons at the various Clovis sites and that all sites have contained bones that have not been in anatomical order. But a look at the maps of the excavated levels should convince anyone that large parts of individual mammoths have remained together at many sites (Figures 7.5–7.7).

It is impossible to factor out every noncultural and cultural agency that could have affected bones at every site. Todd and Frison (1986) evaluated a range of possible factors that could have accounted for disturbance of the Colby mammoth bones, and they decided that deliberate stacking by people and secondary stream displacement or transport had affected the mammoth skeletons. No in-depth studies of other sites are available.

I suggest that in the absence of cut marks, chop marks, and clear scars left by deliberate bone fracturing intended to section mammoth carcasses (features that have been ethnographically recorded in modern assemblages of butchered elephants), it is possible that in the different Clovis sites most of the skeletons represented by varying numbers of bones in varying states of aggregation and dispersal may originally have been deposited as articulated and unsectioned carcasses that were subsequently jumbled around, scattered, and otherwise modified by moving water, trampling animals, and scavengers to different degrees in the various sites. Each site is recognized as possessing its own unique geomorphic and biotic history, and so the end effects of noncultural processes should also be unique. The underlying assumption for Clovis sites is that the mammoth carcasses originally were abandoned by humans in a minimally modified state – each mammoth underwent light butchering and subsequently was disturbed by nonuniform events that differed from site to site. This hypothesis pictures every mammoth carcass starting its entry into the fossil record in a high state of integrity. Some kinds of sediments preserved the carcasses nearly unmodified following Clovis utilization, whereas others subjected the carcasses and skeletons to diverse and multiple processes of disturbance.

If the Clovis mammoths were killed (or found dead) in winter, the carcasses could have frozen before people or other carnivores had time to consume them completely. North American wolves often abandon partially eaten bison carcasses in winter because their inclination to seek out live prey is greater than their satisfaction from eating solidly frozen meat (G. Haynes 1982), although if they need to, wolves will return to frozen carcasses and continue eating them. Early elephant-hunting humans may have behaved similarly.

Early humans, however, would have had an advantage over wolves – their mastery of fire could have extended the period of utilization of a solidly frozen mammoth. Yet if mammoths were extremely vulnerable in winter, and much easier to hunt then, as are big-game mammals in North America (G. Haynes 1982; Oosenbrug et al. 1980a, b) and elephants in Africa (A. Conybeare 1983 personal communication; G. Haynes unpublished data; Till 1936–7), then perhaps elephant-hunting humans, behaving much like wolves or hyenas, only "nibbled" at the carcasses until the meat froze hard, and then went away to find fresher, steaming victims.

Winter freezing may have encouraged light utilization; but I propose another set of circumstances to explain the light utilization of Clovis mammoths: The circumstances are the same as those experienced by spotted hyenas, jackals, lions, and other African carnivores present in elephant ranges during drought-caused die-offs. I refer to the scenario mentioned earlier in which I envisioned Clovis hunters witnessing elephant die-offs.

Elephants congregate at water sources in dry seasons, and a drought year effectively limits their movements to remaining water refuges. An extended hard freeze or lack of snow cover during late Pleistocene winters would have had the same effect on mammoths that hot droughts have on modern elephants. The affected mammoths would have sought open water in warm springs or would have tried to break stream or pond ice to reach the unfrozen water below it. And because all elephants need considerable amounts of moisture daily, taken in the form of water or derived from their solid food, in times of water shortages elephants can quickly weaken and become highly vulnerable to predation. The Late Wisconsin winters of North America may have been times of severe water shortage, especially in grassland habitats where there were no large masses of live roots, dormant buds, or bark layers that could provide moisture as in forests and woodlands.

Relatively light utilization of elephant carcasses by humans is seen today when mass kills are carried out by teams of elephant killers using high-powered rifles, as part of officially authorized wildlife management operations (see Chapter 6). During recent reductions of elephant herds in Zimbabwe, groups of five to seventy-five elephants were shot dead as units in very brief time spans, usually seconds, and then butchered in the field by experienced laborers. Each carcass in the shot herds was abandoned in one to three articulated units, with one or both forelimbs detached but lying next to the rest of the carcass. The only elements regularly removed from the sites were lower jaws and tusks still rooted in socket bone that had been chopped off the skulls.

At these cull sites, utilization was much less heavy than it is at sites where single elephants are killed by Pygmies, Bisa, or other human groups hoping to subsist on the carcasses. It is important to note that the relatively light utilization at these cull sites resulted not because each day's abundance of meat, skins, and ivory discouraged further butchering. Carcass utilization was quite patterned and efficient and closely directed so as to allow optimal use of body by-products that had commercial or economic value – all of the skin, about 80 percent of the meat, all the ivory, and sometimes the limb bones. The butchering of herd groups was quickly performed and completed, before the midday heat could affect the quality of skins and meat. The meat that was not removed would have taken many more hours to strip, because it was left in small quantities around joints, clinging to bones, or in other situations requiring much more time and care in cutting.

The culling is mentioned again here as an example of "assembly-line" utilization, when carcasses are abundant and the time available for butchering tasks is tightly structured. If Clovis hunters killed family groups as units and were highly efficient, rapid butchers of the massed carcasses, then each dead mammoth could have been abandoned as an articulated skeleton, lacking most of the meat, and perhaps skin too. After rapid burial, the articulated bone units would have retained the same tight organization that results when whole bodies are buried.

If mammoths were seriously fat-depleted as a result of environmental stress, then other megafaunal taxa probably were under some stress as well. Thus, their body fat and bone marrow would have been inferior. If mammoths were unable to find adequate moisture in their forage, then other grazing animals also would have suffered from lack of moisture and would have centered their movements closely around dwindling water sources (as evidenced with free-roaming wildlife populations anywhere in the world). If mammoths and other megafauna were under *no* environmental stress, the time and effort required for humans to contain and slay several proboscideans together at water holes would have been relatively excessive; groups of healthy proboscideans could not have been contained without the efforts of large numbers of spearmen. If it is ever proved that Clovis hunters killed mammoth herds, then we might be justified in speculating that their abundance of food encouraged a much lower degree of carcass utilization than one sees among recent elephant hunters in Africa.

7.2.9. Age profiles in Clovis sites, modern die-offs, and hunted samples

Several other patterned features characterize verified modern archeological sites where single elephants were killed, butchered, and utilized, as distinguished from prehistoric elephant bone assemblages whose origins may or may not have been cultural. In some modern cases, the age profiles for elephant kills by hunting people over the course of several years are biased toward adult males, simply because ivory tusks are valued commodities (D. Wilson and Ayerst 1976), and male tusks are much larger than female tusks (Elder 1970). In Chapter 6 I discussed some recent descriptions of elephant hunting in Africa by the Bisa in Zambia and the Efe Pygmies in Zaire. The age profiles for their kills most likely are dominated by adult males, because ivory is the main target of procurement.

The Clovis mammoth locales in western America often are dominated by subadult animals and small adults that may be females. In the two largest Clovis mammoth assemblages, Lehner (Arizona) and Dent (Colorado), 61 and 83 percent of the mammoths were subadults (Saunders 1977a, 1980). However, other multi-animal archeological assemblages, such as Murray Springs (Arizona) and Blackwater Draw (New Mexico) – the latter including several different loci – have provided age profiles dominated by adults (Saunders 1980). The fact that the age profiles are so variable in Clovis sites requires some thought. Could the assemblages dominated by subadults have been produced by meat hunters, with the other assemblages having been produced by ivory hunters?

That is probably a rejectable explanation, because most adult mammoths in Clovis assemblages still had their tusks. The differences conceivably could have resulted from Clovis hunters attacking mixed herds (mostly adult females and their immature offspring) in some cases, and all-bull bands in other cases. If Clovis hunters were attacking mammoths whenever they were encountered, as opposed to preselecting their victims, then sometimes one type of group or the other would have been confronted at water holes.

Mixed herds of elephants are made up of sexually mature females, their offspring, and the offspring of their offspring. This type of herd is much more vulnerable to predation than is an all-bull herd, because it contains younger animals; all-bull herds contain sexually mature individuals. Also, besides being less able to defend themselves against large carnivores, young elephants (or mammoths and mastodonts) are much more vulnerable to environmental stress, such as drought. Older adult females are also relatively vulnerable to drought, unlike adult males. Hence, during hard times, mixed herds of mammoths would have been much easier hunting for predators, such as prehistoric human groups, than would all-bull herds, although it would have been easier to sneak up on a solitary bull. Archeological assemblages with high proportions of young animals and adult females may reflect herds that were attacked by humans because they were already suffering from drought conditions. In that case, they had been set up for the kill by factors outside the control of the culture utilizing them. Thus, we have another potential "explanation" for the different representations in bone assemblages, one with major ramifications for understanding both human behavior and animal ecology.

7.3. Can the present explain the past?

The evidence recovered from Clovis mammoth sites can be used to support contradictory conclusions. For example, some say that the evidence indicates that Clovis hunters slew whole herds; on the other hand, it has been argued that Clovis people did not kill, but merely scavenged die-off locales. It is possible that Clovis people found abundant fat on the mammoth carcasses and did not bother to break long bones, unlike modern elephant killers; but I envision Clovis people as eating starved mammoths with no fat on their bodies or in their long-bone marrow.

Although it has not been fully axiomatic in the archeological literature that Clovis people were exclusively big-game hunters, specializing in mammoth and bison, the conventional wisdom seems to be that they focused their hunting attention on the biggest of big game. Jennings (1974:81, 91), while referring to Clovis people as big-game hunters, also speculated that mammoth kills were the exception, and that much smaller game was more often sought. Willey (1966:37, 38) stated his belief that smaller game undoubtedly was taken by Clovis people as opportunities arose. Yet neither of those authorities nor the students who learned from their writings seem to entertain major doubts that the famous Clovis mammoth sites, such as Blackwater Draw, Lehner, and Domebo, were anything other than true kill sites.

The nearly universal acceptance that all Clovis-point associations with mammoth bones represent kill sites apparently is a metaphysical decision based on the simplicity and plausibility of an *a priori* theory: That humans who kill elephants leave a few tools behind, tools that could hardly be expected to be found with the bones of animals if those animals had not been slain or utilized in some way. All the facts concerning Clovis sites and the fossil record have been expected to fit into this theory, and if they have not fit by a natural process they sometimes have been forced into place by paradoxes and subtleties – to paraphrase Howorth (1887:xi), in discussing the scientific refusal to acknowledge that the biblical flood was the most likely cause for the extinction of mammoths. The paradoxes in the literature concerning Clovis sites arise because the mammoth carcasses do not look utilized when judged against ethnographic examples; the subtleties involve numerous post hoc accommodative arguments about the peculiarities of individual sites.

In some instances, when analysts are dealing with fossil proboscidean assemblages that are difficult to interpret, the emphasis is placed on similarities between certain attributes of modern elephant bone specimens and fossil samples; in other cases, analysts speak of the uniqueness of their fossil assemblages, or the deviations from what is expected after studying modern elephants and recent primitive people. The degree of nonconformity becomes a springboard from which increasingly imaginative versions of the past are propelled.

The Clovis samples clearly are different in many ways from modern elephant bone assemblages, but they are also similar in other ways. Their greatest similarities are to modern noncultural bone assemblages. They simply do not look like ethnographic examples of butchered elephants.

Besides the differences already described, the Clovis samples are distinct in that they contain well-made stone projectile points bedded with mammoth bones. No modern elephant bone sites that were culturally produced contain discarded or lost butchering tools, no doubt because of the high value that modern people place on iron and steel tools. We must not presume that Clovis points were less valuable simply because they were left behind with the underutilized mammoths. Perhaps the Clovis points were deliberately left as "offerings," in thanks for the abundance of mammoth meat obtained with such little effort. That kind of idea about the past has as much verifiability, using the available evidence, as does the idea that Clovis points were mislaid or accidentally dropped into pooled blood, piles of stomach contents, or muddy water, to be lost from sight forever.

Modern ethnographic studies have never documented elephants being slain at water holes. Modern sites are located in the bush, too far from water for burial by alluvial or lacustrine sediments. Every Clovis mammoth seems to have died at a water source (unfortunately, the actual proportion of upland Clovis mammoths is unknowable, because bones in such sites weathered away long ago). Probably none of the modern sites of elephant butchering will be preserved as a fossil deposit, because the bones in such sites are subject to a full range of purely subtractive processes, such as subaerial weathering, scavenger feeding, and trampling destruction. But the Clovis mammoths had been relatively well protected from subtractive processes, at least until first burial. Modern elephants dying in drought are also found at water sources, where the postmortem disturbances to carcasses vary from minimal to heavy, the burial rate is rapid even in the absence of flowing streams (Conybeare and Haynes 1984; G. Haynes 1985a), the carcass additions are virtually continuous over long periods of time (at least fifty-five years, documented in Hwange National Park), and the modifications to bones and skeletons vary widely within extremely restricted spatial areas (G. Haynes 1988a).

Of course, some features are shared by Clovis sites and sites where modern elephants have been slain and utilized. Both fossil and recent archeological sites contain hearths or fire features (i.e., "hearths" that lack rock linings). The fire features at recent sites may be shallow, scooped-out basins in ground surfaces, from which ashes and much of the burnt wood have been blown or kicked away. Discarded meat, gristle, and bone may have been removed by scavengers after the human abandonment of the site. Several Clovis sites had fire features that contained bones.

Bones at modern butchery sites (and Clovis sites) may be aggregated into apparent activity areas or loci of discard behavior (Crader 1983, 1984; J. Fisher 1986 personal communication). Such bones may be fully fragmented or relatively unmodified.

Overall, I find great differences between Clovis sites and the ethno-archeologically documented sites of elephant kills. The differences are sufficient to require that we accentuate the contrasts rather than explain them away when considering analogies to aid in reconstructing the past. Clovis sites contain items of material culture, features such as burnt areas or hearths near bone beds, articulated skeletons together with disarticulated bones and body parts, few or no cut-marked and chopped bones, few or no fractured limb bones directly associated with stone tools, and good evidence of water-lain sediments. Modern single-elephant butchery sites contain no artifacts, occasional fire features, articulated body parts together with disarticulated elements, relatively few cut-marked bones but frequent chopped, fractured, and burned bones, and evidence of windblown but not water-lain sedimentation around or over the bones prior to or following subaerial weathering. Modern multi-elephant butchery sites (cull sites) are similar to the single-animal sites, except that the skeletons are much less sectioned; there are no artifacts discarded in the carcasses, but there are several fire features around the periphery where meat snacks were cooked or tea was prepared by the butchers, and there are no fractured limb bones or cut-marked bones. Such sites are never located at water holes, although mass poaching sites in other parts of Africa may very well be near water sources. Modern sites of noncultural mass deaths of elephants (the drought die-off locales) contain both separated and "stacked" carcasses, skeletons, or articulated body parts, mixed weathered and unweathered bones, no cut-marked or chopped bones, but many elements that trampling elephants have scratched and broken in spiral fractures; see P. Andrews and Cook (1986), Behrensmeyer, Gordon, and Yanagi (1986), Conybeare and Haynes (1984), Fiorillo (1984), G. Haynes (1988a), and G. Haynes and Stanford (1984) for descriptions of cut-mark mimics. Noncultural sites may contain large accumulations of bones, although some may contain the bones of only one or a few elephants. Fire features usually are present at the peripheries of such sites, places where campfires were made by travelers such as Bushmen, scientists, and patrolling park rangers. These camps often have been found to contain scattered artifacts spanning the time from the Middle Stone Age through the late twentieth century, but having little to do with the elephant die-offs.

The age profile for a cull site is reflective of that for a typical herd – 30-50 percent subadults, very few or no adult males, and the rest of the group made up of females over twelve years old. The age profile for a noncultural mass death site (a die-off) typically contains 60–80 percent subadults, plus adult females, but no adult males until the last stages of very severe die-offs

(G. Haynes 1987c). Age profiles for modern single-elephant assemblages have never been cumulatively recorded, and the sample is quite small; I would guess that such assemblages contain mainly adult male elephants.

The large quantities of meat recovered from proboscidean carcasses may or may not have been worth the effort needed to kill such huge beasts. The meat could have been preserved and stored, or eaten quickly. Perhaps the recovered ivory was so valuable as a trade item or scarce commodity that it made proboscidean hunting a worthwhile speciality. Perhaps the oils and fat recovered from the viscera and bones of carcasses were especially prized.

Questions about the assemblages are without self-evident answers from recent ethnographic observations. I suggest that the costs involved in hunting proboscideans normally were far too great for such an economic activity to be routine in the Pleistocene and Holocene epochs. But I also suggest that during the last millennia of the Pleistocene, proboscideans would have been under environmental stress due to changing climates, and their appeal as prey would have gone up significantly at that time. Their predictability as a resource, their aggregated and immobile ("tethered") locations in local ranges, and their increased vulnerability to predation would have made them an optimal resource.

I believe that the large Clovis mammoth assemblages are the remains of die-offs exploited by the people who made fluted points. The mammoths were dead or dying when encountered. This is not to say that Clovis hunters were incapable of killing healthy mammoths. But healthy mammoths killed by Clovis people probably would have been much more fully utilized than die-off carcasses; they might have been killed in the uplands some distance from water holes, as are modern healthy elephants killed by African spearmen (who are poachers), because elephants run away when attacked. Modern elephant kills that do not involve high-powered rifles invariably result from pursuit and harassment, not from containment. The sites where healthy mammoths would have been slain and butchered most likely would have been situated in upland microenvironments where preservation and fossilization were improbable, and as a result the remains never entered the fossil record.

The sites that did enter the fossil record must be regarded as biased samples of the types of decisions and the range of behaviors of Clovis people. These sites provide evidence concerning the life and times of terminal Pleistocene big-game specialists, but it is not evidence of extraordinary human perserverance or boldness. That evidence has not been preserved. Clovis sites give us a view of opportunistic people who learned to exploit a resource that was predictable and abundant. The end effect of their opportunism was the rapid exploration of the New World by people who existed always at low densities over huge ranges, but who managed to maintain a cultural identity by sharing similar emblems, such as (perhaps) tool types and projectile points (Weissner 1983; Wobst 1977). The technological and stylistic characteristics of the widespread Clovis assemblages varied considerably, as did economies and settlement patterns, similar to what we see with modern-day Pygmies, Australian Aborigines, and the San in Africa, and yet overall those Clovis peoples could have recognized themselves as being one by means of identifiable artifact styles. The new, rapidly changing environments of North America were unstable and unpredictable, and early hunters must have allowed each other access to resources throughout wide areas, resulting in a certain homogeneity of material culture (Gamble 1985; Yellen and Harpending 1972). Clovis projectile points may have functioned on one level as emblems of affiliation (Weissner 1983), allowing communication with socially and spatially distant people (Wobst 1977).

7.4. Conclusions

Pleistocene people increased their utilization of mammoths in both the Old World and the New World during the same Late Glacial time interval in which mammoths were dying off because of noncultural causes (G. Haynes 1987c; Vereshchagin 1977; Vereshchagin and Baryshnikov 1984). The die-offs were outcomes of environmental stress, resulting from unstable and harsh climatic conditions in the Late Glacial interval (16,000–10,000 years B.P.). Severe warming and drying trends, in combination with heightened seasonal weather differences (L. Martin and Neuner 1978), led to (1) periodic or sustained crowding of mammoths around water holes, resulting in loss of condition (as nearby food was exhausted), dehydration as competition for decreasing water supplies became more and more exhausting, and increased injuries due to fighting (Conybeare and Haynes 1984; G. Haynes 1985a, 1987a), (2) increased subadult mortality rates, and (3) increased susceptibility to predation, disease, and accidental death.

Late Glacial human groups opportunistically took advantage of the mammoths' vulnerability and availability: In the Russian plain, they gathered the bones of dozens of mammoths that had died natural deaths *en masse*, and they made dwellings from the bones, possibly because wood was in short supply (Hoffecker 1987). In northern Asia, people also used the abundant bones and ivory of dead mammoths, and perhaps the meat as well. In North America, Late Glacial people followed mammoths to their dry-season or winter water refuges, by traveling heavily used game trails, but they made only meager use of the abundant carcasses they found there. The Clovis culture may have spread so rapidly because it was adapted to finding and exploiting die-offs. A cause, rather than an effect, of the rapid Clovis spread was the mortality process that eventually led to mammoth extinction. During the time of that process, specialists in big-game hunting would have found the pickings easy, because much of elephant behavior is predictable in droughts. Clovis people would have had a great deal of security, a safety net of tremendous meat availability, which could have encouraged wide exploration into new territory where more mammoths could be found dying off periodically. Yet if vulnerable mammoths turned out to be scarce in the new territory, then wandering Clovis groups could have backtracked to the places where prey had last been found dying.

The traits of utilizing mammoths (or mammoth carcasses) and manufacturing Clovis points, and a few other associated characteristics, could have spread by actual migration or, alternatively, by diffusion among resident pre-Clovis populations. I personally doubt the reality of pre-Clovis occupation, but this is a disputable point, in spite of vigorously argued opinions to the contrary. It matters little to the supportability of the scheme proposed here whether or not the Clovis culture was the first occupation of the Americas.

A sweeping conclusion that would indisputably pin the blame for proboscidean extinction on Clovis hunters would be a literary triumph but a scientific impossibility. The ultimate impact of the Clovis culture is difficult to gauge. Even before the Clovis point and associated traits appeared, Late Glacial mammoth and mastodont populations probably were in overall decline throughout the last few millennia of the Pleistocene, although some intervals would have been more favorable than others. The decline left interesting bone assemblages behind, many quite similar to those created recently by starving *Loxodonta* populations.

Yet in spite of recurring heavy mortality during drought, *Loxodonta* populations have been able to bounce back and maintain healthy growth rates. This ability to recover following serious die-offs in Africa is striking and has made me think again about the effects of late Pleistocene climatic stresses on mammoths and mastodonts. Recall that in each of the worst drought years in the 1980s, at least 20 percent of the elephants present at the Hwange seeps died, including all calves and a high proportion of individuals under twelve years old. If mammoth and mastodont reproductive rates were lower than those of modern *Loxodonta*, would they likewise have been able to recover from the serious environmental changes indicated by paleoenvironmental evidence from the end of the Pleistocene? They *had* recovered from similar changes during earlier deglacial intervals.

The question, reworded, is this: Were they killed off, or did they die off? It is tempting to explain mass extinctions by finding unique and irresistible agents that clearly are not part of the ordinary ebb and flow of biotic change on earth. For example, in the case of the Cretaceous-boundary extinctions, the possibility of extraterrestrial bodies catastrophically impacting the earth has been eagerly accepted by some scholars as a robust explanation for the sudden disappearance of so many dinosaur species. In the case of the

Pleistocene/Holocene extinctions, every empirical comparison I can make between my actualistic data and the fossil data seem to provide support for the die-off scenario rather than for the kill-off scenario, therefore making Clovis hunting unnecessary to explain extinction.

But Clovis hunters perhaps made an already bad situation worse for megamammals: While climatic changes were driving proboscideans to die off, the added stress of Clovis hunting drove them to *die out*. In the absence of Clovis hunters, mammoths and mastodonts (and other megafauna) should have been able to survive the changing conditions of terminal Pleistocene environments. What I am leading up to is my inability to reject the prospect that Pleistocene proboscideans were in fact *killed off*. I cannot discard this possibility in spite of the fact that the best evidence I can find does not contradict the thesis that drought or some other serious environmental stress accounted for the terminal Pleistocene accumulations of mammoth and mastodont bones.

The most important lessons to be learned from studies of modern elephant die-offs have to do with proboscidean responses to climatic change and with proboscidean resilience, allowing recovery from nearly any environmental stress *except human overhunting*. Illegal elephant hunting with high-powered rifles has halved Africa's elephant numbers in only a decade; yet the past twenty years of range loss, terrible droughts, and population compression into overused habitats did not lower the numbers anywhere near as much or as rapidly.

Clovis hunters did not have rifles, and their ability to kill as quickly as modern poachers cannot be demonstrated. But the Clovis hunter is still a leading candidate to fill the role of the unique and irresistible agency behind extinction, at least until evidence for more abrupt and much more disastrous climatic change everywhere at the end of the Pleistocene is found.

8

Final words

*The Mammoth and Mastodon will ever be the most interesting
of the Pleistocene animals, partly because they are so different
from any now found in North America, partly because they
have become extinct so recently that it is entirely possible
that they were contemporary with early man.*

—F. A. Lucas, *Animals before Man in North America* (1902)

Like most archeologists, I began learning about mammoths and mastodonts long before gaining any personal experience with their bones. The
earliest sources of my knowledge were archeological interpretations, all of
which were published as dependable facts. Those "facts" had been created
using classical formulas for concocting past events from an examination of
scattered objects such as artifacts and bones.

But during the course of this study I had to seek help from other kinds of
sources, and I came to realize that most of the alleged facts about extinct
proboscideans are fabrications whose foundations are exceptionally shaky.
I have had to slog – increasingly irritated – through hundreds of published
descriptions, interpretations, and explanations, where the conclusions were
based on nothing more substantial than mere plausibility or hearsay. The
end point of my wandering education has been to find that the partial
information available from past events can be readily manipulated to
develop fictions of almost any sort, often quite elaborate and thus giving the
appearance of a conclusiveness founded in reality. Such "realities" are
constructed and framed by mentally transforming ambiguous information
into unequivocal statements of meaning. People who do this share certain
conventional beliefs about the meaning of the fossil record, and because of
their similar points of view they usually reach widespread agreement about
the stories told to accommodate fossil data. But the conventional beliefs
may be faulty.

I did not write this book to threaten anyone else's mental well-being, but I
have not hesitated to doubt many of the stories that archeologists tell about
the past. Prehistorians generally either admire or despise disbelieving and
disputatious colleagues. Whatever the personal judgement about this book,

318

I hope that a critical reevaluation of proboscidean sites will lead us to seek more reliable methods for finding meaning in fossil assemblages. My motives are to stimulate rethinking about mammoths, mastodonts, and elephants in Quaternary prehistory.

To do so, to conceive better study techniques, we must first recognize that there is more ambiguity in the fossil record than is generally acknowledged, ambiguity that modern analytical and intuitive techniques cannot yet eliminate or nullify. The next step will be to separate the facts from the fabrications, the certainties from the inventions. The business of interpreting the past has been far too much mythologized, in the name of science, to such an extent that popular novelists can do it with the same believability and authority as Ph.D. archeologists.

Of course, all studies of the past rely on analogical thinking, proxy data, or theoretical probabilities, because these are the only available bases for reasoning when dealing with events for which there are no eyewitnesses. In spite of caution about our conclusions, we are always going to be at a disadvantage in that we have to think long and hard about the reliability of our statements. We are unable to run laboratory experiments that could prove them or could unmistakably separate the meaningful signals from the noise preserved in the fossil record.

Perhaps the closest we can come to experimental tests are actualistic studies. Such studies will educate us about the ways in which fossil bone assemblages either preserve their original meaning (the signal) or alter it through many processes of ambiguation (adding the noise, which typically overwhelms the original signal).

My observations of living elephants were designed to be actualistic problem-solvers, and their future usefulness as sources of deductions about the behavior of mammoths and mastodonts must be decided by paleoecologists. Their true test may come from new actualistic studies of elephants and other megamammals, part of a growing scientific awareness concerned with middle-range theory, in which the goal is to demonstrate decisively the behavioral meanings of preserved fossils, artifacts, and sediments.

I have spent some time following wild animals and am grateful for the opportunity I have had to watch them, especially the elephants that found strength by drinking at the Hwange seeps. Sad to say, the chances for carrying on more such studies are shrinking rapidly, because human beings may soon kill off all the wild elephant populations left on earth. In spite of the fact that most of the human race no longer thinks of wild animals as prey to be hunted for meat, elephants are surely going to be hunted down and destroyed during the lifetimes of many people reading this book. Elephants in Africa and Asia will vanish by attrition, as more and more die at the hands of poachers or farmers needing more land, or crowd into game reserves surrounded by rural villages and farmland. Even in protected lands they are now under great pressure from humans.

These dwindling elephant populations may be our best bet for understanding long-gone proboscideans of the past, such as mammoths and mastodonts. Very soon, perhaps within twenty years, no more studies of free-roaming animals will be possible as elephant populations disappear, shot for ivory or meat, and become more and more closely managed to guarantee that the species will live on, severely limited in their movement and behavior. Soon, in a last attempt to save some of them, no more elephants will be allowed to die natural deaths in regions uninhabited and unvisited by human beings. Before long, the last remaining wilderness areas of Africa and Asia will be developed, manicured, and managed, and wild elephants, like cattle on ranches, will be carefully herded and supervised to ensure their survival.

Hundreds of wild elephants died during my African fieldwork, but I could see that they fought strongly to stay alive. It was – and is – a losing battle, but their struggles have the power to disclose to us how mammoths and mastodonts also faced death thousands of years ago. If we thoughtfully record these last remaining decades in the existence of free-roaming modern elephants, perhaps we shall be enabled to see how the end came also to mammoths and mastodonts.

Appendix: Methods for determining age in proboscideans

A1. Elephant dentition

The dental formula is $\dfrac{1.0.3.3}{0.0.3.3}$ in living elephants (*Loxodonta* and *Elephas*), as well as in *Mammuthus*. In *Mammut*, the formula may vary by the presence of a lower incisor, the mandibular tusk.

The cheek teeth of living elephants are molarlike, but deciduous, and throughout life they are continually being worn and progressively replaced by new teeth erupting from behind in the jaws. All cheek teeth are similar in overall morphology, from the earliest appearing in young animals to the last in place in the jaws of old animals, but they differ greatly in size and several key attributes, as described later.

Emerging from the mouth and rooted in the upper jaw are two modified incisors called tusks. They are round or oval in cross section and usually have conical tips shaped by wear from digging, bark stripping, and fighting with other elephants. Very young elephants have a deciduous pair of tusks, which drop out at about one year of age and are followed by the permanent pair developing behind in the alveoli. The tips of permanent tusks are covered by cementum over an enamel layer, which wears off within a few years (Deraniyagala 1955; Miller 1890a,b). In *Loxodonta*, both males and females have tusks, although some elephants will lack one or both tusks because of accidental breakage or because they were born without them. Usually, modern *Elephas* females do not have tusks, and even when they do the tusks are rarely visible outside the lip.

The permanent tusks have persistent pulps and open rooted ends, so they grow continuously throughout life (Bland-Sutton 1910; Brown and Moule 1977a, b). The pulp cavity is relatively longer and deeper in males than in females (Elder 1970), especially after sexual maturity is reached. The pulp cavity in adult females (and possibly also in sterile males) becomes smaller with increasing age. R. Martin (1978) measured tusk growth rates as high as 15 cm/year in Zimbabwe elephants. Colyer and Miles (1957) found 17 cm of growth on the tusk of an African elephant one year after an injury to a tusk; Colyer and Miles also referred to observations by Humphreys (1926)

321

regarding an Asian elephant's broken tusk that probably grew 16 cm in a year.

In *Mammut,* two additional (and smaller) incisors may be found emerging from the front of the lower jaw; these are the mandibular tusks. I have found these in the jaws of either sex, and they appear to have been shed or worn completely down by middle age.   '

The molariform teeth in the living taxa and in *Mammuthus* are hypsodont, but the roots are not continuously open throughout life. As a result, the maximum height of each tooth is limited. Teeth have several parallel-arranged cores made up of enamel enclosing dentine, attached to each other by cementum. At the beginning of the use-life of teeth, the roots are open at the bottom and are filled with pulp. There are nearly as many roots on buccal and lingual sides as there are enamel cones, although the most anterior cones may share one large root. In mandibular teeth the roots are oriented nearly perpendicular to the occlusal surface, but over time they move forward in the bone at slower rates than the occlusal surface, causing a posterior-directed bend at the lower ends. The number of enamel cones (which appear as loops–called laminae or lamellae when viewed on the occlusal surface) varies from tooth to tooth during the life of an elephant, but generally increases with each replacement tooth. The loops, when worn, are diamond-shaped in the African elephant, but appear as parallel ridges in the Asian elephant and mammoth. These loops are formed by the wearing down of enamel to form rims separated by depressions made up of softer dentine (see Figure 1.3). The repeating rims and depressions act to clip and shred vegetation during occlusion.

Members of the living taxa, as well as *Mammuthus* and *Mammut,* normally grow a total of twenty-four cheek teeth in their lives (de Blainville 1845; Sikes 1971). Six successive molariform teeth develop in each quadrant of the mouth, although occasionally a seventh (supernumerary) tooth may be found in one or more quadrants. A seventh tooth is always smaller than the preceding sixth tooth, reversing the trend of progressive increase in size (Eltringham 1982:9; Laws 1966; Warren 1855:83–90).

As each tooth wears, it moves forward in the jawbone, so not only is its size reduced but also its position changes in the bone relative to other anatomical features. As the worn teeth move forward, replacement teeth erupt from behind and begin to wear, also moving forward. For *Loxodonta, Elephas,* and *Mammuthus* there is only one brief period in life (around two years of age) when more than eight teeth at a time are in wear in the entire mouth. Thus, usually no more than two teeth are in wear at the same time in each side of each jaw.

However, Osborn (1936) and Saunders (1977b) noted with *Mammut* dentition that several times during life three teeth functioned together in each quadrant of the mouth. These times were at about two years of age, at about twelve to fourteen years, and between twenty and twenty-five years of age, according to my comparisons of descriptions of the dentition-to-age-determination criteria worked out for modern elephants.

The phylogenetically earlier taxa of proboscideans also had more than two teeth in function in each quadrant of the mouth (Osborn 1936, 1942). For example, Tassy and Pickford (1983) found that the three molars of an early Miocene mammutid, *Eozygodon morotoensis*, were in wear simultaneously. *Mammut* was an earlier phyletic branching than *Mammuthus*, *Elephas*, and *Loxodonta* (Coppens et al. 1978; Maglio 1973; Tassy 1982), and hence would be expected to have retained primitive characteristics such as more cofunctioning teeth.

Eventually, worn teeth are reduced to relatively small, forward-positioned vestiges with shallow roots. These small remainders either are swallowed or drop out of the mouth. Molariform tooth progression may be an effect of continual remodeling of the jawbone around tooth roots, rather than being the cause of it, in reaction to the forces applied during chewing (Sikes 1971). The evidence for bone remodeling propelling teeth forward in response to chewing pressure is admittedly circumstantial. For example, one lower jaw that I found in Zimbabwe had a transversely split final tooth on one side and an unbroken tooth on the other side. The rear part of the broken tooth was not as worn as the forward half and had not moved ahead in the bone to keep pace with the forward half.

The evolutionary development of this unusual system of dentition may have been caused by the lengthiness of an elephant's growth period and the great increase in size each animal undergoes during growth (Eltringham 1982:7). Young animals are so much smaller than adults that their jaws cannot accommodate teeth of the size needed by adults. Yet, small teeth would be quickly outgrown in a few years as the jaws enlarged. Elephants have therefore developed an "assembly line" of larger and larger teeth that move through the growing jaws.

The dentition of African elephants has been studied and described by numerous researchers, such as Bourlière and Verschuren (1960), G. Craig (1983 personal communication), Fowler and Smith (1973), Frade (1955), Hanks (1979), O. Johnson and Buss (1965), Laws (1966), Morrison-Scott (1947), Perry (1954), and Sikes (1971). The terms used to refer to the successive teeth vary somewhat among different writers. The first four molariform teeth in some earlier fossil proboscideans were in fact replaced by teeth growing from below rather than from behind in the bone, and therefore are deciduous premolars. Modern paleontologists therefore designate the first teeth in *Loxodonta*, *Elephas*, *Mammuthus*, and *Mammut* as Dp2 (or dM2), Dp3 (or dM3), and Dp4 (or dM4); Dp1 is present in some fossils, but not in the taxa of interest here. Coming into function after these teeth, and posterior to them, are the molars M1, M2, and M3.

Table A1.   *Tooth terminology*

Paleontological nomenclature (Saunders 1970)	Common field designation (Laws 1966)	(Sikes 1971)
Dp2	M1	I
Dp3	M2	II
Dp4	M3	III
M1	M4	IV
M2	M5	V
M3	M6	VI

However, field biologists rarely, if ever, refer to modern elephant teeth in these terms. As Laws (1966:4) pointed out, for sake of simplicity the teeth of living elephants usually are designated M1 to M6, because *all* teeth are deciduous (in the sense of being shed), not just the first three. Thus, in Laws's scheme [and the one adopted by most modern field workers, such as Roth and Shoshani (1988) and Sikes (1971)], "M1" is equivalent to Dp2, "M2" is equivalent to Dp3, "M3" is equivalent to Dp4, "M4" is equivalent to M1, "M5" is equivalent to M2, and "M6" is equivalent to M3. I favor the simplicity of this scheme, which numbers teeth according to their appearance, not according to their ancestry (Table A1). Sikes (1971) designated teeth with Roman numerals (I–VI); in her system, Arabic notation was used only to distinguish the number of plates in each tooth.

A2.   Tooth wear and age determination

Elephants grind their food between upper and lower teeth that move in a backward–forward motion (Maglio 1973). This motion shears the food into smaller, more easily digestible bits. However, the grinding rarely reduces the food to an undifferentiated mass, and elephant dung usually contains plainly recognizable mashed wood, compressed grass leaves, and other plant parts sheared within the mouth.

Because teeth continually wear down, there is an adaptive advantage to having very high tooth crowns. Upper-cheek teeth are higher-crowned than lowers, because the skull has more room to accommodate a larger tooth mass than does the lower jaw. A result of the grinding of larger upper teeth on smaller lower teeth is that lower teeth wear down faster than uppers. Some of that faster wear has been avoided by the evolutionary development of an angle formed by the enamel loops and the occlusal surface (Eltringham 1982; Maglio 1973). Still, in the living elephants and *Mam-*

muthus, the lower teeth often show more wear than uppers at any one time. This fact must be kept in mind when one attempts to determine the life-ages of elephants by reference to teeth (the commonest means of age determination). The upper teeth often will appear younger than the lower teeth.

Saunders (1977b:52) reported similarities in tooth eruption and epiphyseal fusion in an Arizona mammoth and Asian elephants described by Deraniyagala (1955). Because eruption, wear, and progression of cheek teeth are similar in modern elephants and extinct taxa, researchers have applied the detailed criteria for determining ages of African and Asian elephants to the tooth assemblages from mammoths and mastodonts (e.g., C. Haynes 1967; C. Haynes unpublished data cited by Saunders 1977a; G. Haynes 1985b, 1990; Roth and Shoshani 1988; Saunders 1977a, b, 1980). My experiences with the teeth of fossil and living proboscideans support the validity of determining ages for fossil animals based on criteria worked out with modern taxa. I have examined teeth in the upper and lower jaws of several thousand African elephants, most of which had been killed during controlled population-management operations. I have also examined jaws and teeth of fossil proboscideans in large collections (the U.S. National Museum of Natural History, the American Museum of Natural History, the University of Alaska Museum, the University of Nebraska, and the USSR Academy of Sciences Zoological and Paleontological Institutes, among others).

The anatomical process of epiphyseal fusion is quite scheduled during the life spans of mammals: Epiphyses fuse to certain elements in a specific sequence and at specific ages. Therefore, the age-specific correspondence of both tooth eruption and epiphyseal fusion in *Elephas* and *Mammuthus* (Saunders 1977b:52) is a strong indicator that the extinct mammoths developed through life in stages closely similar to those of Asian elephants. Section A4.5 describes observations on the scheduling of epiphyseal fusion in the postcranial skeletons of modern African elephants. I have calibrated the fusion scheduling to stages of tooth eruption and wear, and I suggest that a similar calibration would be valid for extinct *Mammuthus* and *Mammut*.

The method for determining the ages of modern elephants requires that a complete quadrant of the mouth be examined. The preferred quadrant is one side of the lower jaw (which in elephants is a fully fused, single bone throughout life), because teeth in upper jaws wear differently and have not been so thoroughly studied. The wear on the two sides of the lower jaw may be asymmetrical (G. Haynes unpublished data; Roth 1989; Roth and Shoshani 1988; Sikes 1971), but it is impractical to determine if one side or the other is consistently more worn within any population of animals. Therefore, it may not matter whether the left or right side is examined, as long as the dentition is complete in the examined side.

The most difficult aspect of jaw evaluation is distinguishing the identity of each individual tooth within the progressing series of M1 through M6 (Roth and Shoshani 1988; Saunders 1970). M1 teeth are quite small and distinctive; M2 teeth are also relatively easy to identify correctly. Unfortunately, M3, M4, M5, and M6 often can be confused with each other. The M6 tooth is always the largest to erupt in any individual's mouth, but some individuals have much larger teeth than others, and so size alone is not an adequate standard by which to distinguish it. Roth and Shoshani (1988) found among Asian elephants that teeth M3 through M5 were also difficult to identify correctly. Morrison-Scott (1947) reviewed earlier formulas for identifying teeth and introduced the "laminary" index and "enamel-loop" index. Even using these indices and measurements of length, width, and number of loops, M4 and M5 are quite difficult to distinguish from each other. Besides measurements of length, width, and number of enamel loops, qualitative observations of morphology and eruption are necessary for accurate identification.

Saunders (1970) measured teeth of *Mammuthus columbi* and concluded that individual teeth were, on the whole, distinguishable. Graphed data (Saunders and Graham n.d., cited in Graham 1986b: Figures A2.2 and A2.3) showed good separation of small and large teeth (M1 and M6), but relatively poorer isolation of presumed M4 and M5 teeth. There was a great deal of overlap in both length–width plotting and length–plate-number plotting. Judging from my experiences with modern elephants, some variations in tooth measurements result from gender differences (the teeth of males usually are much larger than the teeth of same-age females). Teeth also change lengths, widths, and plate numbers through normal wear (Roth and Shoshani 1988). The maximum lengths of entire teeth (not grinding surfaces) are reached before much wear occurs. Roth (1989) realized the potential for great morphological variation in elephant teeth as a result of the many forces (mastication, eruption, progression) working on them throughout the lives of individuals.

Each tooth gains its maximum number of loops before wear begins on all of them and before complete eruption, but after all enamel plates ("lamellae") have fused. Often the first and final enamel cones, which are relatively smaller than the others, are worn away by contact with adjoining teeth.

After a tooth is in function, its maximum length decreases as the anterior enamel structures are worn down. Also, the forwardmost enamel loop may be worn away by contact against the tooth in front, and the last loop may be lost (before coming into occlusal wear) by contact against a tooth or bone behind. it. Major decreases in length occur as entire enamel loops are worn completely away. Roth and Shoshani (1988) and Deraniyagala (1955:52) noted variations in width and length resulting from the degree of wear, a potential source of error in measuring the maximum dimensions of any tooth.

The occlusal width actually increases for a time after wear begins as wider and wider parts of the loops become the occlusal surface; eventually, width decreases over time.

Individual variations also account for some differences in tooth sizes and measurements, a result of sex distinctions as well as phenotypic idiosyncracies (Roth 1989). I have recorded great differences in plate numbers and maximum sizes of African elephant teeth, in some cases as great as the differences that are thought to reflect the taxonomic separation of different species such as *Mammuthus imperator* and *M. columbi* (Graham 1986a; Kurtén and Anderson 1980; Maglio 1973).

A3. Life span: maximum expectable age

There is still disagreement in the literature over the maximum life spans for members of proboscidean taxa, including the living genera *Loxodonta* and *Elephas*. This disagreement, of course, has had a profound bearing on the issue of age determination based on tooth progression and wear. The teeth of some allegedly "known-age" animals in museum collections are anomalous when compared with the teeth of other "known-age" animals. But because elephants are so long-lived (sixty years or more), the state of the teeth in "known-age" animals in zoos, circuses, or museum collections has figured prominently in all attempts to develop criteria for age determination. Fieldworkers simply cannot mark wild calves of known age and check their teeth over long life spans. Laws (1966:17) discussed this problem and noted that a few published sources attribute ages of up to sixty-seven years to Asian elephants; see Flower (1943, 1947) and citation of a study by Macnaughten. Laws (1966), in his study of the African elephant, assumed that its longevity was similar to that of the Asian species, and therefore he set sixty-five years as an average upper limit. Although that figure seems to be generally agreed on, occasionally the potential upper limit may be estimated higher in the literature. For example, Kurtén and Anderson (1980:349) placed the upper age limit for modern elephants at seventy years. Sikes (1971:170, 175) was certain that the life span in hunted populations of elephants living in large forested regions was much greater than the life span in "semiprotected" populations living in grasslands. She believed that some elephants lived eighty to one hundred years. Many other writers have also expressed their belief that elephants can live past the century mark (Deraniyagala 1955:74, 132; Kunz 1916:191, 221).

Patterns of dental eruption, wear, and progression are similar for *Loxodonta*, *Elephas*, *Mammuthus*, and *Mammut*. And because the various taxa do not differ substantially in terms of body size, ontogenetic processes, and dental development, their life spans probably are also similar. Of course, that remains to be proved conclusively. Wide individual variations in longevity are possible within any elephant deme, and environmental differences will affect life spans.

In a study of *Mammuthus* tusks recovered in northern Asia, Vereshchagin and Tikhonov (1986) observed what appeared to be annual growth lines in the dentine. Osborn (1936) also noticed alternating dark and light rings on the tusks of an American mastodont; similar rings were known from living elephants, both *Elephas* (Osborn 1936:183) and *Loxodonta* (Kunz 1916:232). The dark and light rings appear to reflect the rhythmic increments known to be laid down in mammalian teeth and hard tissue (Neville 1967). The highest number of these tusk lines found in the sample analyzed by Vereshchagin and Tikhonov was seventy-three, leading the authors to conclude that the maximum life span of the arctic Siberian woolly mammoth (*Mammuthus primigenius*) was seventy-five to eighty years, over 20 percent longer than the life span generally accepted for African elephants. Furthermore, based on an increase in spacing of the rings, those authors tentatively suggested that sexual maturity for mammoths was reached at about age eighteen to twenty, meaning that the mammoths lived more than one-quarter of their lives in a sexually immature state. As discussed in Chapter 3, modern African elephants may reach sexual maturity as young as age eight, although twelve to fifteen years is more common in healthy populations that are not subject to unusual conditions such as crowding.

It does indeed seem possible that proboscideans could live as long as eighty years if their teeth held out. It is not merely a question of the teeth surviving the wear that results from occlusion, because the scheduling of eruption and progression also has a lot to do with the age at which the individual teeth are worn down and therefore become ineffective. As mentioned earlier, studies of a few "known-age" animals (though not providing solid evidence) indicate that sixty-five years is old for African elephants, and records for captive Asian elephants also do not indicate a possible longevity much past sixty-five years, if that.

A possible explanation for the seemingly excessive number of growth rings on the arctic mammoth tusks might be that the separate rings are not annual increments. Growth layers in teeth and hard tissue may reflect rhythmic metabolism synchronized with daily, monthly, or annual events in the environment (such as day–night cycles, lunar cycles, or season-to-season cycles) (Klevezal and Kleinenberg 1967). Thus, several rings on the tusks could have been laid down over the course of a single year. Numerous studies of wild cervids have found extra annular layers not laid down in a regular yearly way, but apparently deposited during intra-annual phases, such as the rutting season. Therefore, multiple layers may represent a single year of life. Armstrong (1965), who noted "false" annuli in the root cementum of molar teeth from bison and bighorn sheep, suggested that annuli in tooth cementum were not necessarily laid down in response to purely seasonal changes in an animal's metabolism or general physiological state, but could also be laid down during abnormal times, such as periods of physical stress. Later studies of periodic growth increments have verified that the rhythm of incremenal additions may not always be synchronized with environmental changes or astronomical events; incremental additions produced by individual physiological states may be superimposed over increments affected by astronomical/biological events (Dean 1987).

The tusks of mammoths and elephants are indeed teeth, although they are exceedingly unusual and specialized for functions other than the mastication of food. Tusks are made up mainly of dentine covered with a very thin cementum layer attaching them to their bony sockets in the skull. Their dense mass is constructed in the manner of cones within cones, which may or may not develop in regular, annual rhythms. Cementum is a slowly forming hard tissue, and its widely spaced growth increments may be seasonal, monthly, or annual. Dentine forms more quickly and may record rhythms having shorter periodicity; see Dean (1987) for a literature review. No doubt some growth lines record individual states that do not parallel environmental events.

Until the particular problems connected with age determination from tusks can be clarified, I continue to accept sixty to sixty-five years as the upper age limit for modern elephants, mammoths, and mastodonts. The overwhelming bulk of research on age determination has been based on acceptance of this limit. It is important to note that body-size scaling indicates a theoretical life span of about sixty years for animals the size of proboscideans. In scaling from body size, such life-history variables as age at sexual maturity, gestation time, interbirth interval, and life span can be shown to increase in regular ways as body size increases (Western 1979).

A4. Age determination

Most of the data discussed here were taken from studies conducted by G. Colin Craig (1983 personal communication, 1987 personal communication, 1989 personal communication) in Zimbabwe. Providing the basis for the age-determination criteria developed by Craig were related studies by Jachmann (1985, 1988), E. Lang (1980), Lark (1984), Laws (1966), and Sikes (1971), all of whom studied African elephant samples, the latter two from east Africa. Fatti et al. (1980) pointed out that when Laws's criteria (hereafter referred to as "the Laws criteria") were applied to other samples of elephants, they often resulted in anomalous age distributions. Hanks (1979) also observed this with a sample of teeth from Zambia. For example,

Table A2.   *Assignment of ages using different scales (Laws, Craig)*

Age class	Median age assigned by Laws	Approximate equivalent age assigned in this study
I	0.25	0.3
II	0.5	0.5–0.7
III	1.0	~ 1.7
IV	2.0	~ 2.5
V	3.0	~ 3.3
VI	4.0	~ 4.3
VII	6.0	~ 5.0
VIII	8.0	~ 6.7
IX	10.0	~ 8.0
X	13.0	~ 9–10
XI	15.0	~ 10–14
XII	18.0	~ 14–15
XIII	20.0	~ 15.5
XIV	22.0	~ 16
XV	24.0	~ 17.7
XVI	26.0	~ 19–21
XVII	28.0	~ 23
XVIII	30.0	~ 24–27
XIX	32.0	~ 29
XX	34.0	~ 31–33
XXI	36.0	~ 33–35
XXII	39.0	~ 36–38
XXIII	43.0	~ 39–42
XXIV	45.0	~ 42–45
XXV	47.0	~ 47–48
XXVI	49.0	~ 50
XXVII	53.0	~ 52–53
XXVIII	55.0	~ 56
XXIX	57.0	~ 57–58
XXX	60.0	~ 60

certain age categories may be entirely missing, or disproportionate numbers may be assigned to certain other age classes.The sample of elephants that Craig studied was far larger than any previously studied, and Craig applied more information available from studies of known-age animals, as well.

I present here some of the major points gleaned from Craig's study, but I am not intending to declare them once and for all the ultimate truths about elephant age determination. All age-determination criteria have been developed in a context of considerable inference, assumption, and inadequate hard data. I prefer the Craig system because of its greater detail and apparent theoretical and empirical strengths, such as the less anomalous age distributions. Table A2 shows age equivalences for two systems of age determination. Craig's system differs in important ways from those of Laws and Sikes; for example, it has more age categories. Also, the Craig system assigns ages that may be up to six to seven years different from ages that the Laws criteria would assign for the same state of tooth wear. As Table A2 shows, the two systems often assign identical ages, but also often diverge.

A4.1. Individual teeth: identification of uppers and lowers, lefts and rights

As stated earlier, maxillary (upper) teeth are wider, higher, and larger overall than corresponding mandibular (lower) teeth. Upper teeth also tend to have more enamel loops than corresponding lower teeth. When upper teeth erupt from the jawbone, their roots are oriented approximately 0–30° to the occlusal surface, whereas with lower teeth the roots form a less acute angle with the wear surface, from about 45° to 90°.

When an upper tooth that is free from bone is oriented in proper anatomical position, a single anterior root will be aligned in the same plane as the bulk of the tooth; another root behind it, as large as the front root, diverges off to the lingual side.

When an upper tooth is viewed on the anterior (front) surface, with the occlusal surface downward, the second (or internal lingual) root diverges in the same direction (to the viewer's right) that the tooth occupies in the mouth (the right side of the jaw). As the tooth moves forward in the jaw during normal progression, the distal end of the anterior root seems to stay nearly in place, causing the root to curve back from its connection with the occluding surface. Eventually the anterior roots are resorbed, because the tooth continues moving forward to positions in the bone that do not leave enough room for the roots.

A lower tooth generally has a single unforked anterior root; however, the second root and succeeding roots may be forked, with a branch descending from each side (buccal and lingual). Hence, after the first (anterior) root is resorbed during the normal tooth movement over time, the forked roots may then be in anteriormost positions. Although this happens late in the life of the tooth, it may cause some confusion when loose teeth from bone assemblages are being identified, because the forking may be mistaken for the root division of an upper tooth. Note the similarities in upper and lower teeth: The first and second roots are separate, they diverge from each other, they are of nearly equal sizes (viewed from the front), and they are attached to separate enamel loops.

The forward (anterior) part of the occlusal surface in either an upper or lower tooth is always more worn than the rearward parts.

When viewed on its occlusal surface, a mandibular tooth is concave on its buccal aspect. To identify whether a lower tooth comes from the right or left side of the mandible, first hold the tooth with its wear surface up. Next, orient the tooth so that worn loops are forward and unworn loops are toward the back when you look down on the wear surface. If all loops happen to be worn, then position the largest roots rearward, which is the same as positioning the most curved roots to the front. When you look at the tooth, the concave side of the wear surface will correspond to the side of the jaw from which the tooth came. If you reposition the tooth in your hand so that you are looking at the rear end, you will see that the distal ends of the roots curve off to one side or the other; if they curve toward the left side of your body, the tooth is from the left side of the jaw. Maxillary teeth do not curve along their lengths as noticeably as do mandibular teeth.

A4.2. Distinguishing M1, M2, M3, M4, M5, and M6

Modern African elephants show great variation in the number of loops on each tooth. For example, most of the lower M6 teeth examined in the Zimbabwe sample had nine or ten loops, but some had as many as fourteen. Similarly, the average number of loops on M5 was seven or eight, but variations up to twelve loops were also seen. Table A3 shows the observed ranges of loops counted in the Zimbabwean sample, the sample studied by Sikes (1971), and the sample studied by Laws (1966) from western Uganda.

The numbers of loops in the teeth of Asian elephants also are subject to variation. Far fewer studies have been carried out on age determination for Asian elephants. Table A4 shows the counts of numbers of loops per tooth made by Roth and Shoshani (1988).

The teeth of *Mammuthus* and those of *Mammut* are quite different. The teeth of *Loxodonta* and *Elephas* differ less. The teeth of *Elephas* are somewhat similar to those of *Mammuthus columbi*, but *Mammuthus primigenius* teeth usually are distinctive. Hence, it is necessary to detail the characteristics of each taxon's teeth. Table A5 shows the ranges of loop counts for the teeth of two species of *Mammuthus*, as derived from studies by Osborn, Madden, Saunders, Maglio, Garutt, and Vereshchagin.

Morrison-Scott (1947) described parameters that help to identify individual teeth of *Loxodonta*, such as length, width, and total length of all complete loops divided by the number of measured loops (the laminary index). Osborn (1942) and Whitmore et al. (1967) applied similar measures to fossil taxa. Saunders (1970, 1977a, 1980) concluded that in addition to the number of loops, the length, width, and height are sufficient measurements to identify teeth. Data from Osborn (1936, 1942), Laws (1966), and Saunders and Graham (Graham 1986b) showed that "separation" was possible when tooth widths and lengths were plotted against each other, using samples from modern African elephants and *Mammuthus*. No comparable measure-

Table A3. *Ranges in numbers of lamellae counted on mandibular teeth in three samples of L. africana*

Tooth	Zimbabwe sample	Sikes (1971)	Laws (1966)
M1	2–4	5	3–4
M2	5–7	7	6–7
M3	7–10	10	8–10
M4	6–10	10	7–10
M5	8–12	12	9–12
M6	9–14	13	10–14

Table A4. *Ranges in numbers of lamellae counted in a sample of E. maximus*

Tooth	Mandible	Maxilla
M1	5	4–5
M2	6–9	7–8
M3	11–14	11–15
M4	14–17	14–17
M5	16–21	17–21
M6	21–29	20–26

Source: Roth and Shoshani (1988).

ments were made for teeth from the large Zimbabwe sample, except in cases in which there were questions of identity; those measurements were not systematically collected and therefore are not presented here. A more convenient check on field identification of teeth from the Zimbabwe sample involved comparisons of shoulder heights, a method that proved quicker and much more reliable than measuring width or lengths of teeth. As described later, the degree of epiphyseal fusion of long bones is also a reliable check on the correctness of tooth identification. The final molar, M6, is perhaps most easily identifiable by its posterior taper and usual lack of posterior wear.

Each tooth in the series present within the mouth of an elephant begins and ends its use-life at specific ages. For example, the M6 begins its wear at age twenty-seven, although it is present inside the jaw for several years before that time. The M5 tooth begins to wear at about age fourteen and a half years and is gone from the mouth at about age forty. Proboscideans from earlier geologic periods did not have precisely the same methods and age-specific sequencing of tooth replacement, but the system described here is appropriate for *Loxodonta, Elephas, Mammuthus,* and *Mammut.*

Table A5A.   *Ranges in numbers of lamellae counted in different samples of M. columbi*

Tooth	Madden (1981) Max.[a]	Mand.[b]	Saunders (1990) Max.	Mand.	Osborn (1942) Max.	Mand.
M1	4	4	3+	—[c]	—	—
M2	6–9	6–8	6–10	14	—	—
M3	9–11	9–11	13	9–12	—	11.5
M4	11–13	11–15	11–13	12–15	12	12.5
M5	13–16	12–16	13–17	11–15	13–16	12.5+
M6	18–24	18–23	18–20	15–18	18–19	15–16+

[a] Maxillary.
[b] Mandibular.
[c] No data available.

Table A5B.   *Ranges in numbers of lamellae counted in different samples of M. primigenius*

Tooth	Maglio (1973) Max.[a]	Mand.[b]	Garutt (1984) Max.	Mand.	Osborn (1942) Max.	Mand.	Vereshchagin (1977) Mand.
M1	—[c]	—	5	4–6	4	4	5
M2	8	8	7–10	8–10	8	8	8–10
M3	9–13	10–11	9–15	9–15	12	12	8–15
M4	12–14	11–15	11–16	11–15	12	12	10–17
M5	15–17	15–16	14–21	15–21	16	16	12–20
M6	20–27	20–27	20–29	20–28	24–27	24–27	23–26

[a] Maxillary.
[b] Mandibular.
[c] No data available.

Table A6 shows data on the ages during which the different teeth are in wear, derived from a large sample of Zimbabwean elephants (G. Craig unpublished data), with comparative data from E. M. Lang (1980), Laws (1966), and Sikes (1971), as well as the Roth and Shoshani (1988) study of Asian elephants. My data are not entirely comparable to other data reported in the literature because many other reports stated the ages at which teeth "appeared" and were lost, rather than when they began wear and later were lost. Loose teeth in bone assemblages usually are fragmentary, and so it is nearly impossible to determine an animal's age from the size or appearance of the tooth. Too much is going to be missing – for

Table A6. *Age estimation for loose teeth*

	Tooth					
Age span when in wear	M1	M2	M3	M4	M5	M6
This study	0.1–2.0	0.3–4.1	2.3–13	5.5–21	14.5–40	27–61
E. Lang (1980)	0–2	Lost at 4 yr	Lost at 9–10 yr	Lost at 9–25 yr	?	?
Laws (1966)	0–1.5	0.5–4.5	2.3–14.5	7.5–25	19–43	31–60
Sikes (1971)	?–<2	~2–~4	4–5<10	~10–~24	~25–~41	>30–?
Elephas (Roth and Shoshani 1988)	0–1.5	1.5–5+	2–10	6–25	13–38	~30–67+

Note: At the lower age limits, teeth are beginning to wear; only one cusp may show wear, while others may be unerupted, or their lamellae unfused. At the upper age limits, occlusal surfaces usually are very smooth, chipped in front, or vestigial.

Table A7.   *Percentage of tooth life for each tooth in wear*

Tooth	Number of years in wear	Percentage of total time all teeth in wear
M1	1.9	2
M2	3.8	4
M3	10.7	12
M4	14.5	16
M5	25.5	28
M6	34.0	38
Total	90.4 use-wear years	

example, with broken teeth it is impossible to determine how many lamellae were developed, or how many were fused together, which is an important criterion in determining an age category. However, even with fragmentary teeth, often it is possible to determine if wear had begun and progressed beyond the first stage. Thus, simply knowing the age at which wear begins and having a vague idea of the tooth's identity based on overall size will allow an analyst to estimate absolute minimum age for the animal from which the tooth came, which can then be further checked against stages of epiphyseal fusion.

The total span of years that the teeth are in wear will be more than the maximum life span of the animal, because there is overlap in the wear-lives of the individual teeth. I have added the wear-lives of the six teeth in *Loxodonta* and come up with 90.4 years, which I call "tooth life." Table A7 shows the proportion of that total span during which each individual tooth is in wear.

A4.3.   Age determination by reference to teeth: problems summarized

In elephants, there is no permanent set of teeth on which to measure yearly wear, as can be done with ungulates such as bison. The molariform teeth progress and replace each other in sequence. The chronology of the sequence has never been well studied with a large sample of known-age animals, because elephants live so long. The size and shape of each tooth change through its use-life, and individual teeth also differ in size and shape.

Individual variations in tooth wear and progression are not major problems, according to the studies that have been completed thus far, although there are important differences in tooth sizes reflecting the possible variations in body sizes of individual animals.

After about five years of age males and females of the same age show considerable size differences, and the teeth of some males are far larger than those of same-aged females. Because it is generally assumed that the replacement teeth in any elephant's mouth become progressively larger and larger, and that small teeth are comparatively young, there is a risk that older female teeth may be mistaken for younger male teeth.

A tooth begins its use-life with only a small portion of the occlusal surface in function. The wear increases over time, and as it does so the measurable wear surface becomes much wider and longer, and the number of wearing enamel loops increases. Generally, when the number of loops in wear reaches the maximum possible, the tooth's occlusal surface is at its widest and longest. Over its life of wear, there is a continuing loss of loops (enamel cones) at the front, and the gradual loss of one enamel cone at the back of the tooth. If one counts the loop numbers and measures tooth widths and lengths for one tooth over its use-life, one will end up with a range of quite different measurements.

A4.4. Age determination: The stages of tooth progression and wear

Table A8 shows an approximation of ages correlated with tooth wear and progression, based on data from G. Craig (unpublished data) and G. Haynes (unpublished data). This table has been checked against known-age animals described in the published literature and has also been compared with Table 7 of Roth and Shoshani (1988), which provides similar information about Asian elephant teeth. The ages that the Laws criteria (Laws 1966) would assign are found in the far left column; the wear stages are in the next column, and the Craig ages are in the next column to the right. The descriptions in the other columns are rough guides that have been "averaged" from thousands of typical and atypical cases. Examples of individual variations may not fit neatly into this scheme, which is currently being checked and amended whenever unconformable cases are seen in the field.

If an analyst examining mandibular teeth and jaws of *Mammuthus columbi* first identifies individual teeth correctly (see Table A5) and then applies the number from Table A5 to the correct columns of Table A8, a relatively precise age can be determined for the animal that lost the teeth.

The information in Table A8 is perhaps most applicable when entire lower jaws (or complete dentition in at least half the lower jaws) are available. However, it can be seen that even in cases in which part of the dentition is missing, the proportion of wear on the remaining teeth may in fact provide enough information to allow an age determination. This proportion is determined by comparing the number of loops remaining and

Table A8. *Age estimation of dentition in lower jaws*

Laws AEY[a]	Laws class	Craig AEY	Tooth in wear					
			M1	M2	M3	M4	M5	M6
0–0.25	I	0.1	No wear					
0–0.25	I	0.2						
0.75	II	0.5–0.7						
1.0	III	1	100% wear	30% in wear				
1.5	III	2	Smoothing	100% wear				
3.0	V	3	Gone	100% wear	15–20% wear			
6.0	VII	5		Smoothing & chipping off	75% wear			
8.0	VIII	6		Gone	100% wear, smoothing front			
11.5	IX	8			75% left	20% wear		
13.0	IX–X	10			30% left	> 50 % wear		
15.0	XI	12			Smooth	75% wear		
18.0	XII	14			Gone	100% wear		
22.5	XIV	16				100% front smoothing	10% wear	
		18				75% left, smoothing.	50% wear	
						30% left or less	80–90% wear	

			Gone or very smooth	> 90% wear	
26.5	XVI	20			Lamellae unfused
		22	> 90% wear		Lamellae fused
30	XVIII	24		100% wear	20% wear
32.0	XIX	28		50% left, smoothing, chipping	
33.0	XIX	30		< 50% left	20–30% wear
34.0	XX	32–33		30–40% left	40% wear
37.0	XXI	35		30% left	50% wear
41.0	XXII	37–38		20–25% left	65% wear
		40–42		Very smooth or gone (hole in jaw)	80% wear
45.0	XXIV	45–46		No hole left	90% wear
47.0	XXV	46–48			100% wear, front smoothing
50.0	XXVI	50			65% left, jawbone behind calcifying
53	XXVII	52–54			> 50% left
56	XXVIII	56			50% or less left
57.0	XXIX	56–58			30% left
60.0	XXX	60			nearly smooth
Dead	Dead	Dead			

[a] African elephant years.

Table A9.   *Possible combinations of teeth in wear in lower jaw*

Stage	Description	Age span[a] (years)	Laws (1966) years (approx.)
Animals sexually immature, growing rapidly			
A	None in wear, or M1 only	0.0–0.5	0.0–0.5
B	M1 and M2 in wear	0.5–2.0	0.5–1.5
C	M1, M2, and M3 in wear	2.0–2.0 (+ 3–4 months)	1.5–2.5
D	M2 and M3 in wear	2.5–4.0	2.5–4.5
E	M3 in wear	4.0–5.5	4.5–8.0
F	M3 and M4 in wear	6.0–13.0	8.0–14.5
Animals sexually mature, some limb-bone epiphyses fused or fusing			
G	M4 in wear	13.0–14.5	14.5–18.5
H	M4 and M5 in wear	14.5–22.0	18.5–28.0
I	M5 in wear	22.0–27.0	28.0–30.0
Animals reaching end of growth, or fully grown and in prime			
J	M5 and M6 in wear	27.0–40.0	30.0–42.0
Animals passing prime, or in old age			
K	M6 in wear	40.0–60.0	42.0–60.0

[a]G. Craig (unpublished data).

the maximum number possible (Tables A3–A5). And even in cases in which it is simply impossible to define how much wear has occurred on isolated or loose tooth crowns that are not in jawbones, analysts still can place the age at which those teeth were in the mouth, by reference to Table A6. Thus, any correctly identified tooth in the series M1-M6 will provide a valid age or age range, and by factoring out uppers from lowers, and rights from lefts, a useful age profile based on minimum numbers of individuals can be determined with confidence. Such a profile will lack fine detail, but still will be valuable.

Table A9 provides information that can be used as another "check" on tooth identification: Data on age spans and the teeth that would be expected to be in wear during those ages.

Table A8 has been checked against many lower jaws of *Mammuthus columbi* and *M. primigenius;* there are some minor differences in wear rates. For example, the period of life when both M5 and M6 are in wear together is from about age twenty-four-plus to age forty-two in *Loxodonta;* M5 is lost when M6 is at about 90 percent of its wear. However, in *M. primigenius,* M5 is lost when M6 has reached only 50 percent of its wear. Before the loss of M5 in *Loxodonta,* about ten years of life will pass before the wear on M6 will increase from 50 percent to 90 percent. The same transition in *M.*

primigenius (from 50 percent wear to 90 percent wear on M6) may take place in less than ten years, because once the M5 is lost, wear increases greatly on the remaining tooth. Hence, the M6 may wear faster in the absence of the M5. If this accelerated wear does indeed occur, then the relatively greater amount of wear on the final tooth (M6) in *Loxodonta* when it alone remains in the jaw need not indicate that *Loxodonta* has a shorter life span than did *M. primigenius*, whose M6 was alone in the jaw with much more wear-life remaining.

It is also possible that the transition from 100 percent of the tooth in wear to about 65 percent remaining (from wear class XXV to class XXVI) may require more time than the four to six years cited in Table A8.

A4.5. Age determination: epiphyseal fusion

During my field studies in Africa, I collected postcranial bones from fresh carcasses of elephants that had been shot in the cull or had died in drought (see Chapter 4). The culled animals were objects of scientific study, and extensive data were collected on each individual, including information such as age (based on tooth eruption and wear), sex, shoulder height, tusk girth, quantity of kidney fat (a measure of condition), number of placental scars in females, and the presence or absence of lactation in females. Data from the studies are on file at Hwange National Park. I collected limb bones from a large sample of carcasses in order to measure growth rates and the scheduling of epiphyseal fusion. As for the drought-killed elephants, I determined age (based on dentition) and sex (when possible) and measured selected limb bones to monitor growth and stature.

Measurements of limb elements were compared to measures of the stages of dental wear and shoulder heights to gain an idea of how closely they might correlate with ages of animals. The sample was modest in size, and individual variations in limb lengths during growth were to be expected. Also, male and female differences were great, and the sample was weighted toward females. The data on shoulder heights are currently under analysis in Zimbabwe, but my impression is that there are two distinct female growth curves, which are themselves distinct from the male growth curve. The male curve is characterized by a longer growth period and, of course, much greater total growth; the two female curves show different rates of growth and different adult statures. One "type" of female grows slowly and attains a relatively small adult shoulder height; the other "type" grows more quickly to a taller stature. Both types end their growth at about the same age interval (M. Jones 1986 personal communication; A. Conybeare 1985 personal communication). I emphasize that these are impressions, and they may have resulted from sampling quirks, data glitches, or unjustified inferences. However, these observations may prove to be supportable;

Table A10.   *Femur measurements from sample of 41 elephants shot in cull and 26 that died in drought*

Age[a] (years)	Sex	Shoulder height[b] (cm)	Diaphyseal length[c] (cm)	Fusion stage[d] Proximal	Distal
57	F	—[e]	89.0	5	5
57	M	—	112.0	5	5
5?	F	—	90.0	5	5
50	F	—	86.0	5	5
47	F	262	92.1	5	5
47	F	279	93.0	5	5
47	F	—	90.5	5	5
47	F	—	85.0	5	5
47	F	279	93.0	5	5
43	F	—	86.5	5	5
43	F	282	91.5	5	5
40	F	267	92.5	4	4
32	M	—	110.0	1	4
31	F	259	87.0	3	4
31	F	250	88.0	5	5
28	M	—	102.5	(chewed off)	
28	F	—	90.0	5	5
27	M	—	98.0	0	4
25(?)	F	245	80.7	0	1
25	F	261	94.0	4	5
25	F	—	85.0	3	5
22	F	245	84.5	5	5
22	F	265	93.0	2	5
22	M	315	102.0	0	1
20	F	247	87.6	0	4
20.	F	260	87.0	0	3
20	F	250	79.0	0	0
20(?)	F	261	93.5	5	5
20	F	245	87.5	5	5
19	F	254	85.7	1	3
19	F	253	87.5	2	4
18	M	—	102.5	0	0
18	F	—	85.0	0	1
18	F	245	84.5	2	4
18	F	255	85.0	0	1
18	F	266	91.0	2	4
17	F	242	78.7	0	3
17	F	248	84.5	0	4
16	F	251	85.0	0	0
16	M	—	95.0	0	0
15	M	272	91.0	0	0
15	F	241	85.0	0	3
15	M	255	86.5	0	0
15	F	248	85.0	0	0

Table A10. (*cont.*)

Age[a] (years)	Sex	Shoulder height[b] (cm)	Diaphyseal length[c] (cm)	Fusion stage[d] Proximal	Distal
14	F	—	78.0	0	0
13	F	258	85.1	0	0
13	F	249	83.2	1	2
13	M	270	95.3	1	1
12	M	254	85.0	0	0
11 +	F	—	76.0	0	0
10	F	251	88.0	0	1
9	F	230	77.5	0	0
8	F	207	71.0	0	0
8	M	234	79.5	0	0
8	M	223	75.0	0	0
8	M	243	80.7	0	0
8	F	202	69.0	0	0
7	M	—	75.5	0	0
7	M	—	72.0	0	0
7	M	222	74.5	0	0
6	M	190	67.5	0	0
6	F	—	65.5	0	0
6	M	—	70.5	0	0
5	F	—	61.5	0	0
4	M	—	58.0	0	0
4	M	—	57.5	0	0
4	M	—	55.0	0	0
2	?	—	42.5	0	0

[a] Using Craig criteria, determined from mandible.
[b] Measured along front limb.
[c] Measured from epiphyseal fusion line, distally and proximally.
[d] Unfused = stages 0, 1, and 2; fusing = stages 3 and 4; fused = stage 5.
[e] Data unavailable.

perhaps the two types of female growth result from different ages at first ovulation and first pregnancy, or perhaps from maturation during drought years *versus* years of good rains.

Table A10 shows measurements of femora from a sample of forty-one culled elephants and twenty-six elephants that died during drought (1982–7). This table shows each animal's sex, age (as determined from dental stages), shoulder height (for culled animals only), and "diaphyseal length" of the femur. The diaphyseal length was measured with a tailor's cloth tape laid out along the anterior side of the bone: The distance from the fusion line of the distal epiphysis to the fusion line of the proximal trochanteric epiphysis at approximately the midpoint between femoral neck and greater trochanter. If the proximal trochanteric epiphysis was fused to the shaft, its thickness was considered part of the diaphyseal length, because usually it was impossible to find a fusion line or suture. This part of the epiphysis may be 10 mm thick or thicker, which may be up to 1 percent of the total diaphyseal length. If the distal epiphysis had fused, the location of the suture generally was found to remain visible for several years following synostosis and thus could be used as a measurement point. If the suture was not visible, its location was estimated, as based on past experience with its placement relative to the trochlear rims and other anatomical features. However, the estimation procedure probably had the potential to add an error of ± 5–10 mm to the diaphyseal length measurement on any specimen with a fused distal epiphysis, or 0.5–1 percent of the measurement.

Table A10 also shows the degree of epiphyseal fusion for each measured femur. The grading system, from 0 to 5, follows the example of earlier publications, such as Roth (1984) and Thorington and Vorek (1976), but with an additional increment added in order to gain slightly more detail. A 0 indicates no attachment of epiphysis and diaphysis, and relatively uncomplex fitting surfaces on diaphysis and epiphysis; the numbers 1 and 2 indicate increased complexity of the conjoining surfaces of diaphysis and epiphysis, that is, more roughening and detailed sculpting of the surfaces, such that they fit together more precisely. The epiphysis is still unfused. The number 3 indicates the early stages of fusion, although sutures are still open, and the epiphysis sometimes can be pulled off the diaphysis, although with some difficulty. The attachment is beginning to ossify. The number 4 indicates that fusion has progressed to the point that the sutural opening is ossifying, although still visible as a "seam." The number 5 indicates full sutural closure and obliteration of the fusion line.

I should emphasize that the ages assigned to the elephants whose femora were measured are not necessarily the true ages of the animals, but are instead the apparent ages determined by tooth wear and progression. The animals were analytically placed in certain age classes, which were assumed to correspond with ages in years, but these ages may be subject to revision if future studies indicate errors in the system.

Table A11 shows the breakdown of age and sex correlations with epiphyseal fusion of the proximal femur; note that the degree of fusion is not so complexly defined as in Table A10. Table A12 shows correlations of age and sex with distal epiphyseal fusion of the femur. Epiphyses on females fuse at younger ages than do those on males.

Table A13 shows data on femor lengths and circumferences from the same sample; note that there are no significant differences in midshaft circumferences for male and female femora having similar diaphyseal lengths until the lengths exceed 85–90 cm. Femora with diaphyseal lengths

Table A11. *Fusion of proximal femoral epiphysis*

Age (years)	Male				Female			
	N	Unfused[a]	Fusing[a]	Fused[a]	*N*	Unfused	Fusing	Fused
0–2	0	—[b]	—	—	0	—	—	—
2–4	1	1	—	—	0	—	—	—
4–6	3	3	—	—	1	1	—	—
6–8	3	3	—	—	2	2	—	—
8–10	3	3	—	—	1	1	—	—
10–12	1	1	—	—	0	—	—	—
12–14	0	—	—	—	3	3	—	—
14–16	3	3	—	—	4	2	2	—
16–18	1	1	—	—	1	—	—	1
18–20	1	1	—	—	7	3	4	—
20–22	0	—	—	—	4	—	—	4
22–24	1	1	—	—	2	—	1	1
24–26	0	—	—	—	3	1	—	2
26–28	1	—	—	1	1	—	—	1
28–30	1	—	—	1	1	—	—	1
30–32	0	—	—	—	0	—	—	—
32–34	0	—	—	—	2	—	—	2
34–40	0	—	—	—	1	—	—	1
40–50	0	—	—	—	6	—	—	6
50–60	1	—	—	1	3	—	—	3
Total	20				42			

[a]Unfused = stages 0, 1, and 2; fusing = stages 3–4; fused = stages 4–5 and 5.
[b]No recorded specimens.

of 85 to 95 cm (see Table A10 and Figure A1) and with fused epiphyses therefore belong to females: femora with shafts longer than 85 cm and having unfused epiphyses belong to males.

The sample of measured bones other than femora was small. Table A14 organizes the observations on different elements. I present the data here as a preliminary guide to understanding epiphyseal fusion scheduling. The work of Roth (1984) should be consulted for more discussion of recent elephant growth patterns and how they correlate with developmental stages in other proboscidean taxa. Roth demonstrated that the state of skeletal fusion is positively correlated with dental stage in *Loxodonta* and *Elephas*, and the apparent sequence of epiphyseal fusion is also similar for a sample of *Mammuthus columbi* and the dwarfed mammoth *M. exilis*. Deraniyagala (1955:54, 57) attempted to calibrate tooth wear and ossification for Ceylonese elephants, but his measurements and observations lacked detail and were limited in number, although the data are

Table A12. *Fusion of distal femoral epiphysis*

Age (years)	Male				Female			
	N	Unfused[a]	Fusing[a]	Fused[a]	N	Unfused	Fusing	Fused
0–2	0	—[b]	—	—	0	—	—	—
2–4	2	2	—	—	0	—	—	—
4–6	2	2	—	—	1	1	—	—
6–8	3	3	—	—	2	2	—	—
8–10	3	3	—	—	1	1	—	—
10–12	1	1	—	—	0	—	—	—
12–14	0	—	—	—	5	4	1	—
14–16	3	3	—	—	3	2	1	—
16–18	1	1	—	—	1	—	1	—
18–20	0	—	—	—	7	3	3	1
20–22	1	1	—	—	3	—	2	1
22–24	1	1	—	—	1	—	—	1
24–26	1	1	—	—	5	1	—	4
26–28	0	—	—	—	0	—	—	—
28–30	2	1	—	1	0	—	—	—
30–32	0	—	—	—	1	—	—	1
32–34	0	—	—	—	2	—	1	1
34–40	0	—	—	—	1	—	—	1
40–50	0	—	—	—	8	—	—	8
50–60	1	—	—	1	1	—	—	1
Total	21				42			

[a] Unfused = stages 0, 1, and 2; fusing = stages 3–4; fused = stages 4–5 and 5.
[b] No recorded specimens.

consistent with data from Roth's study and my own. There *are* some differences in fusion scheduling for the two surviving elephant genera, according to Roth (1984: Table 2), but they are relatively minor, and the overall sequences are quite similar (Roth 1984: "note added in proof"). It should be noted that data on the sexes of the animals that provided specimens for her study were not always available (Roth 1984: 127) and that some specimens may have been derived from captive animals, which often develop differently than do free-roaming individuals.

These possible causes of variability, plus individual variations and the limited sizes of elephant samples available anywhere for examination, make the close correspondence between my observations and those of Roth all the more gratifying, especially in light of the wee size of the Zimbabwe sample. Unless more studies can be carried out, it will be difficult to prove that generic differences, gender differences, individual differences, or strictly local differences in life conditions can account for distinctions in skeletal changes with age. See İşcan, Loth, and Wright (1987) for an example of a

Table A13A. *Femur measurements calibrated to age determined by reference to teeth*

Female		Male	
Diaphysis length (cm)	Age (years)	Age (years)	Diaphysis length (cm)
(42.5)[a]	(2.3)[a]	(2.3)[a]	(42.5)[a]
		3.8	55
		4.1	57.5
		4.5	58
61.5	5	6.3	70.5
65.5	6.3	6.3	67.5
		7.5	72
		7.6	75.5
		7.6	74.5
71	~8	~8	76
69	8.4	8.4	75
		8.4	79.5
		8.4	80.7
		11.3–12.8	85
88[b]	13.8		
77.5	13.8		
78	14.4		
83.2	14.9		
85.1	14.4		
80.7	15.5	15.5	86.5
85	15.5	15.5	95.3
		15.6	91
		~16	95
87	12.8–18.9		
84.5	17.7		
85	17.7		
85.6	17.7		
85	18.5	~18.5	102.5
85	18.9		
87.5	18.9		
87.6	18.9		
84.5	18.9		
90.5	~19		
78.8	19.5		
91	21.4		
85	23.3	23.3	102
87.5	23.3		
93	23.5		
84.5	25.2		
79	26.6		
93.5	~27		

Table A13A.   (*cont.*)

Female		Male	
Diaphysis length (cm)	Age (years)	Age (years)	Diaphysis length (cm)
90	27.2	27.2	98
94	27.2		
88	31		
87	32.1		
		32.8	110
86.5	38		
85	∼47		
90.5	∼47		
91.5	∼47		
92.5	47.5		
93	47.5		
86	50		
92.1	51		
90	∼56.5	56–57	112
89	57(+)		

Minimum differences: Male and female	Correlation of lengths and ages	
	Female lengths	Male lengths
0–6 yr = 2–5 cm	∼60 cm = 5 yr	60 cm = 5 yr
6–15 yr = 6–10 cm	60–70 cm = 5–8 yr	60–70 cm = 6–7 yr
15–25 yr = 8–15 cm	70–80 cm = 8–15 yr	70–80 cm = ∼7–9 yr
> 25 yr = 10–25 cm	80–90 cm = 15–60 yr	80–90 cm = 9–15 yr
	> 90 cm = > 21 yr	90–100 cm = 15–30 yr (some shorter animals possible). > 100 cm = > 18 yr

[a] Sex unknown.
[b] Pregnant.

study of racial differences in human skeletal changes; in that study, in spite of the full availability of data on the age and sex of each human individual examined, it still was impossible to distinguish between racial and social factors to find a main "cause" for differences in skeletal changes.

Note that earlier (G. Haynes 1985b) I mistakenly stated that all long-bone epiphyses in *Loxodonta* males are fused by age thirty-six years, and in females by thirty-two years. There is now unequivocal evidence that some long-bone epiphyses in healthy males remain unfused well into the fifth decade of life.

Table A13B. *Mid-diaphysis circumference, femur length, and sex*

Male		Female	
Length (cm)	Mean circum-ference (cm)	Length (cm)	Mean circum-ference (cm)
50–60	20	50–60	—[a]
60–70	22	60–70	22.2
70–80	25.3	70–80	25.8
80–90	27.7	80–90	28.6
90–100	33.5	>90	30.9
100–120	38.5	(None over 100)	

Note: Largest female circumference = 32.4; largest male circumference = 42.5 cm.
[a]No data available.

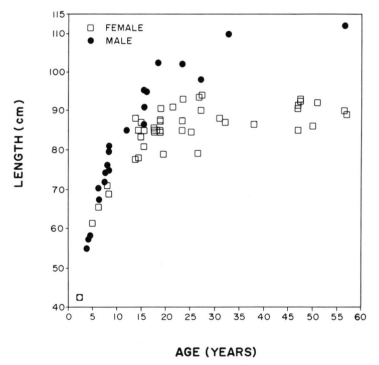

AGE (YEARS)

Figure A1. Graph of measured femoral diaphyseal lengths versus age in years for Hwange *Loxodonta*.

Table A14.   *Epiphyseal fusion scheduling in Hwange elephants (calibrated to tooth ages)*

	Age at which epiphysis has fully fused (years)	
Element	Female	Male
Femur, distal	17–23	26–29
Femur, proximal	25–32 (probably 19–25)	> 29 (often unfused into late 30s)
Tibia, distal	~ 19 (18–20)	~ 32?
Tibia, proximal	17–24 (probably > 20)	28–32 (possibly earlier)
Humerus, distal	< 19 (~ 13–14)	< 18
Humerus, proximal	Fusing 19–26	~ 32 (often unfused into late 30s)
Radius/ulna, distal	> 24	> 32 (probably late 40s)
Radius/ulna, proximal	Fusing at 19	< 32 (probably by 20s)
Innominate/ sacrum	Sacral vertebrae fused (fused < 19)	Sacral vertebrae fused (> 32)
Vertebral centra, plates	Some centra fused to arches; plates attached middle only (< 19)	Lengthy fusion
Innominate ischium ilium, pubis	Fusing ~ 8.0	Fused at 8.0

Note: Knee epiphyses fuse before hip or ankle. Elbow epiphyses fuse before shoulder or wrist.

Table A15 organizes the data on epiphyseal fusion in chronological order through the life of an elephant. The ages in Tables A14 and A15 should not be considered averages; they are either minima or maxima. Variations of a few years either way will be encountered in any comparative sample. Also, the caution stated earlier about future revision of ages assigned to elephants will be repeated here: The ages in Tables A11 and A12 were determined by using a system that is subject to change. If specific epiphyses were fused in any individual elephant's skeleton, the animal's teeth were examined to see in which wear stage the teeth belonged. These wear stages are thought to correspond to actual age.

Moss (1988) noted that some adult female elephants in her study area, Amboseli National Park in Kenya, seemed to stop growing in terms of shoulder height but continued growing in body length (several of the photographs in Moss's book are clear demonstrations of this kind of growth). The plates of some vertebrae in the *Loxodonta* skeleton do not fully fuse to the centra for several decades, perhaps allowing the centra to continue growing anteriorly–posteriorly (lengthwise) after statural growth ceases, in both males and females. In fact, fusion of centra and plates is not

Table A15. *Ages at epiphyseal fusion*

Laws class and age (years)	Craig age (AEY)[a]	Females: Element and epiphyseal state[b]	Males: Element and epiphyseal state[c]
IXa (9–12)	8	Three bones of innominate fusing	Three bones of innominate fused
XVIa (26)	18–19	Sacrum fused to innominate (19)[d]	
		Humerus distal fused (18–19)	Humerus distal fused (18–19)
		Humerus proximal fusing (19)	
		Radius/ulna proximal fusing (19)	
		Tibia distal fused (19)	
XVIa–XVII (26–29)	18–24	Tibia proximal fuses	
		Femur distal fuses (17–23)	
XVIIIa–XX (30–35)	25–32	Femur proximal fused or fusing	Femur distal fuses (~ 30)
		Radius/ulna fuses	Tibia proximal fuses (28–32)
			Femur proximal fuses (> 29)
			Radius/ulna proximal fuses (< 32)
			Tibia distal fuses (< 32)
			Humerus proximal fused (32)
XXII–XXX (39–60)	36–60		Radius/ulna distal fuses (> 40)
			Innominate fused to sacrum (> 50)

[a] AEY, African elephant years.
[b] Sample: 44 females (6–55 years old)
[c] Sample: 22 males (3–55 years old).
[d] 19 years.

complete until the sixth decade of life in both males and females, that is, into the fifties.

The sacrum remains unfused to the innominates in males into the fifties, and perhaps fuses a bit earlier in females.

The vertebral borders of the scapulae seen to be actively growing bone tissue into the fifties in males, as do the highest (dorsal) borders of each ilium, allowing continued natural growth after long-bone epiphyses have fused to diaphyses.

The rib–sternum articulating ends remain unfused into the late fifties in both males and females.

In males and females, the suture joining nasal and maxillary bones of the cranium does not fully fuse during life. The suture's unfused proportion probably is relatively greater in males, allowing for greater growth of tusk circumference.

These unfused or growing bones are the residual growth centers in the *Loxodonta* skeleton. Body growth, both lengthwise and staturewise, remains possible even in relatively old adults.

A comparative study of associated *Mammuthus* skeletons in North America and the Soviet Union has indicated similar sequences of skeletal maturation, epiphyseal fusion, and sutural ankylosis. Adrian Lister (in press, 1988 personal communication) has reported similar findings for European mammoth skeletons. Also, it seems that these changes occur at the same or very similar stages of tooth wear in *Loxodonta* and *Mammuthus*, reflecting similar life-ages in the different taxa.

A5.   A word about anomalies

At least two associated skeletons of *Mammuthus primigenius* do not fit neatly into the age categories defined earlier, based on criteria of tooth eruption/wear and skeletal maturation. One skeleton is the Beryozovka mammoth (see Chapter 2) found in Siberia as a frozen carcass in 1900, and currently on display at the Zoological Museum in Leningrad. The other skeleton is the Lyakhov mammoth found in Siberia during the nineteenth century and sold to the Paris Natural History Museum, where it is now curated (Vaufrey 1955). Both skeletons are nearly complete, both probably are males, and both are anomalies for the same reasons (Lister 1989).

In both specimens, erupting from the jaws are large, curved, molariform teeth that appear to be M6s. The mandibular examples in both specimens each have twenty-two lamellae (a minimum number), at least one more lamella than M5s are thought to have (see Table A5B). Judging from tooth eruption and wear, both mammoths could have been in their midthirties at the time of death, which is the opinion of Dr. V. Garutt (1987 personal communication) of the Zoological Institute in Leningrad and Dr. A. Lister (1988 personal communication) of Cambridge University. Yet the skeletons have many unfused long-bone epiphyses and rather small tusk circumferences (about 44 cm on the Beryozovka's left tusk). Their "skeletal" ages appear to be sixteen to twenty years, based on the *Loxodonta* scale. Both animals are rather small in stature, with the Beryozovka animal standing about 265 cm (G. Haynes unpublished data) and the Lyakhov mammoth standing about 250 cm at the shoulder (Vaufrey 1955 : 846). Could these animals be examples of pathological growth in which skeletal maturation had been arrested, perhaps because of nutritional or other stress? There are no clear pathological findings on any bones or the teeth, and there may or may not be microscopic traces of stress in the skeleton.

It has been shown in human populations (Vyhnanek and Stloukal 1988) that Harris's lines in bones may result from nutritional stress of disease during growth years, but that not all people who were hungry or ill during childhood growth have Harris's lines. However, when Harris's lines are present in bones, there is a high correlation with open sutures and unfused epiphyses. It seems quite possible that the two anomalous mammoths had been starved or sick during their earlier growth years and thus retained open sutures far beyond the age when those features disappear in healthier proboscideans; in view of that possibility, the bones of these mammoths ought to be examined for the presence of other stress indicators, such as Harris's lines, although there is no guarantee that elephants and humans will be comparable in the correlation of such attributes.

A related anomaly in the Zoological Museum in Leningrad is the mounted skeleton of an *Elephas maximus*, perhaps one given by the shah of Persia to the Russian tsar in the nineteenth century. This captive elephant may have been a carnival animal used in official parades, but little other information is available concerning its life (V. Garutt 1987 personal communication). Its tusk circumference is rather large (a 34-cm average for the two tusks measured at the alveolar insertion), and the teeth in wear are also large, appearing to be the M5 and M6, based on size measurements and numbers of lamellae (Table A4) (Roth and Shoshani 1988). Yet many long-bone epiphyses appear to be only partly fused, specifically the distal femora and tibiae, distal humeri, and distal ulnae and radii. Many skull sutures are also unfused. According to Tables A14 and A15, this Asian elephant should have been fourteen to twenty-one years old. According to Dr. Garutt, who carefully evaluated evidence from tooth eruption, the animal (a male) was in its middle to late thirties when it died.

There are several abnormalities on the bones, including very rough bone surfaces near or on joints, bony spicules and growths on vertebral spinous processes (perhaps a result of frequently carrying riders or heavy weights), and a strikingly shaped left femur, very constricted in its anteroposterior shaft thickness and unusually curved (perhaps a result of habitually being chained there in restraint). This animal may have been poorly fed or overworked, and perhaps kept in a very cold environment, and thus prone to joint inflammation. If the animal was in its midthirties at the time of death, then the unfused epiphyses and open sutures resulted from stresses during its earlier growth, and Harris's lines may be preserved in the bones.

Until in-depth studies can be carried out on these two mammoths and the Asian elephant (and specimens in other museums that seem anomalous), it is impossible to determine conclusively whether these animals were in their midthirties or were at least ten years younger than that.

References

Abramova, Z. A. 1979a. *Palyeolit Yeniseya: Afontovskaya Kultura.* Novosibirsk: Nauka.

1979b. *Palyeolit Yeniseya: Kokorevskaya Kultura.* Novosibirsk: Nauka.

1979c. K voprosu o vozraste aldanskovo palyeolita. *Sovyetskaya Arkheologiya* (1979) 4: 5–14.

Absolon, K., and B. Klíma. 1977. Předmostí, ein mammutjägerplatz in Mähren. *Fontes Archaeologiae Moraviae VIII* (Prague).

Adams, M. 1807. Relation d'un voyage à la mer glaciale et découverte des restes d'un mammouth. *Journal du Nord* 32 (Supplément): 633–40, 621–8 (pages misnumbered).

Adams, R. M. 1953. The Kimmswick bone bed. *The Missouri Archaeologist* 15: 40–56.

Agenbroad, L.D. 1980. Quaternary mastodon, mammoth and men in the New World. *Canadian Journal of Anthropology* 1(1): 99–101.

1984. New World mammoth distribution. In P. S. Martin and R. D. Klein, eds., *Quaternary Extinctions, a Prehistoric Revolution,* pp. 90–108. Tucson: University of Arizona Press.

1990. The mammoth population of the Hot Springs site and associated megafauna. In L. D. Agenbroad, J. I. Mead, and L. Nelson, eds., *Megafauna and Man: The Discovery of America's Heartland,* pp. 32–5. Hot Springs, South Dakota: Mammoth Site of Hot Springs.

Agenbroad, L. D., and R. L. Laury. 1984. Geology, paleontology, paleohydrology, and sedimentology of a Quaternary mammoth site, Hot Springs, South Dakota: 1974–1979 excavations. *National Geographic Society Research Reports (1975 Projects)* 16: 1–32.

Agenbroad, L. D., and J. I. Mead. 1987. Late Pleistocene alluvium and megafauna dung deposits of the central Colorado Plateau. In G. Davis and E. VandenDolder, eds., *Geologic Diversity of Arizona and Its Margins: Excursions to Choice Areas.* Arizona Bureau of Geology and Mineral Technology Special Paper No. 5, pp. 68–85.

1989. Quaternary paleoecology and distribution of *Mammuthus* on the Colorado Plateau. *Geology* 17: 861–4.

Agogino, G. A. 1968. Archeological excavations at Blackwater Draw Locality No. 1, New Mexico, 1963–1964. *National Geographic Society Research Reports (1963 Projects),* pp. 1–7.

354

Akersten, W. A., T. M. Foppe, and G. T. Jefferson. 1988. New source of dietary data for extinct herbivores. *Quaternary Research* 30: 92–7.

Albl, P. 1971. Studies on assessment of physical condition in African elephants. *Biological Conservation* 3(2): 134–40.

Alexander, R. M., A. S. Jayes, G. M. O. Maloiy, and E. M. Wathuta. 1979a. Allometry of the limb bones of mammals from shrews (*Sorex*) to elephant (*Loxodonta*). *Journal of Zoology (London)* 189: 305–14.

Alexander, R. M., G. M. O. Maloiy, B. Hunter, A. S. Jayes, and J. Nturibi. 1979b. Mechanical stresses in fast locomotion of buffalo (*Syncerus caffer*) and elephant (*Loxodonta africana*). *Journal of Zoology (London)* 189: 135–44.

Allen, C. E., D. C. Beitz, D. A. Cramer, and R. G. Kauffman. 1976. *Biology of Fat in Meat Animals*. North Central Regional Research Publication No. 234, University of Wisconsin College of Agricultural and Life Sciences.

Anderson, A. D. 1962. The Cooperton mammoth: A preliminary report. *Plains Anthropologist* 7(16): 110.

ed. 1975. The Cooperton mammoth: An Early Man bone quarry. *Great Plains Journal* 14(2): 130–73.

Anderson, J. F., A. Hall-Martin, and D. A. Russell. 1985. Long-bone circumference and weight in mammals, birds and dinosaurs. *Journal of Zoology (London) A* 207: 53–61.

Andrews, C. W. 1906. *A Descriptive Catalogue of the Tertiary Vertebrata of the Fayûm, Egypt*. London: British Museum of Natural History.

Andrews, P., and J. Cook. 1986. Natural modifications to bone in a temperate setting. *Man* 20: 675–91.

Anthony, H. E. 1949. Nature's deep freeze. *Natural History* 58(7): 296–301.

Anzidei, A. P., and M. Ruffo. 1985. The Pleistocene deposit of Rebibia [*sic*] Casal de'Pazzi (Rome–Italy). In C. Malone and S. Stoddart, eds., *Papers in Italian Archaeology IV. Part i: The Human Landscape*, pp. 69–85. Oxford: BAR International Series 243.

Armstrong, G. G. 1965. An examination of the cementum of the teeth of Bovidae with special reference to its use in age determination. Unpublished M.Sc. thesis, University of Alberta Zoology Department.

Augusta, J., and Z. Burian. 1964. *A Book of Mammoths*, 2nd ed. (translated by M. Schierl). London: Paul Hamlyn.

Austen, B. 1953. Report of assistant game warden for May and June, 1953. Unpublished manuscript on file, Hwange National Park, Zimbabwe.

1954a. Report of assistant game warden for June, 1954. Unpublished manuscript on file, Hwange National Park, Zimbabwe.

1954b. Report of assistant game warden for October, 1954. Unpublished manuscript on file, Hwange National Park, Zimbabwe.

Aveleyra Arroyo de Anda, L. 1955. *El Segundo Mamut Fósil de Santa Isabel Iztapan, México, y Artefactos Asociados*. Instituto Nacional de Antropologia e Historia, Direccion de Prehistoria Publicacion Nombre 1.

Aveleyra Arroyo de Anda, L., and M. Maldonado-Koerdell. 1953. Association of artifacts with mammoths in the Valley of Mexico. *American Antiquity* 18(4): 332–40.

Badgeley, C., and A. K. Behrensmeyer. 1980. Paleoecology of Middle Siwalik sediments and faunas, Northern Pakistan, *Palaeogeography, Palaeoclimatology, Palaeoecology* 30: 133–55.

Bahuchet, S., and H. Guillaume. 1982. Aka–farmer relations in the northwest Congo Basin. In E. Leacock and R. Lee, eds., *Politics and History in Band Societies*, pp. 189–211. Cambridge University Press.

Baker, R. R. 1978. *The Evolutionary Ecology of Animal Migration*. New York: Holmes & Meier.

Baker, S. W. 1866. *The Albert N'Yanza, Great Basin of the Nile, and Explorations of the Nile Sources*. London: Macmillan.

Bakker, R. T. 1980. Dinosaur heresy – dinosaur renaissance: Why we need endothermic archosaurs for a comprehensive theory of bioenergetic evolution. In R. D. K. Thomas and E. C. Olson, eds., *A Cold Look at the Warm-Blooded Dinosaur*, pp. 351–462. AAAS Selected Symposium 28. Boulder; Colo.: Westview Press.

1986. *The Dinosaur Heresies: New Theories Unlocking the Mystery of the Dinosaurs and Their Extinction*. New York: Morrow.

Barash, D. P. 1977. *Sociobiology and Behavior*. New York: Elsevier.

Barbour, E. H. 1925. Skeletal parts of the Columbian mammoth *Elephas maibeni*, sp. nov. *Nebraska State Museum Bulletin* 10: 95–118.

Barker, W. 1953. The elephant in the Sudan. In W. C. Osman Hill et al., eds., *The Elephant in East Central Africa: A Monograph*, pp. 68–79. London: Rowland Ward.

Barnes, R. F. W. 1982. Mate searching behaviour of elephant bulls in a semiarid environment. *Animal Behaviour* 30: 1217–23.

Barnett, R. D. 1982. *Ancient ivories in the Middle East*. Qedem 14 Jerusalem.

Barnosky, C. W. 1989. Postglacial vegetation and climate in the northwestern Great Plains of Montana. *Quaternary Research* 31(1): 57–73.

Barry, R. G. 1983. Late-Pleistocene climatology. In S. C. Porter, ed., *Late-Quaternary Environments of the United States. Vol. 1: The Late Pleistocene*, pp. 390–407. Minneapolis: University of Minnesota Press.

Baryshnikov, G. F., I. Y. Kuz'mina, and V. M. Khrabrii. 1977. Rezul'tati uzmerenii trubchatykh kostyey mamontov Berelyokhskovo "kladbishcha." In Ya. I. Starobogatov, ed., *Mamontovaya fauna russkoy ravniny i vostochnoy sibiri*, pp. 58–67. Trudi Zoologicheskovo Instituta 72.

Bate, D. M A. 1903. Preliminary note on the discovery of a pygmy elephant in the Pleistocene of Cyprus. *Proceedings of the Royal Society of London* 73: 498–500.

Bateman, A. J. 1948. Intra-sexual selection in *Drosophila*. *Heredity* 2: 349–68.

Bax, P. N., and D. L. W. Sheldrick. 1963. Some preliminary observations on the food of elephants in the Tsavo Royal National Park (East) of Kenya. *East African Wildlife Journal* 1: 40–53.

Behrensmeyer, A. K. 1975. The taphonomy and paleoecology of Plio-Pleistocene vertebrate assemblages east of Lake Rudolf, Kenya. *Harvard University Museum of Comparative Zoology Bulletin* 146(10): 473–578.

1978. Taphonomic and ecologic information from bone weathering. *Paleobiology* 4: 150–62.

1982a. Time sampling intervals in the vertebrate fossil record. In *Proceedings of the Third North American Paleontological Convention*, Vol. 1, pp. 1–5. Montreal.

1982b. Time resolution in fluvial vertebrate assemblages. *Paleobiology* 8(3): 211–27.

1988. Vertebrate preservation in fluvial channels. *Palaeogeography, Palaeoclimatology, Palaeoecology* 63: 183–99.

Behrensmeyer, A. K., K. Gordon, and G. Yanagi. 1986. Trampling as a cause of bone surface damage and pseudocutmarks. *Nature* 319: 768–71.

Bell, R. H. V. 1985. Elephants and woodland – a reply (letter to the editor). *Pachyderm* 5: 17–18.

Benedict, F. G. 1936. *The Physiology of the Elephant*. Carnegie Institution of Washington publication No. 474. Washington, D.C.: Carnegie Institution.

Beregovaya, N. A. 1960. *Palyeoliticheskiye mestonakhozhdeniya SSSR*. Materiali i Issledovaniya po Arkheologii SSSR No. 81. Moscow: Nauka.

1984. *Palyeoliticheskiye mestonakhozhdeniya SSSR (1958–1970 gg.)*. Leningrad: Nauka.

Berg, J. K. 1983. Vocalizations and associated behaviors of the African elephant (*Loxodonta africana*) in captivity. *Zeitschrift für Tierpsychologie* 63: 63–79.

Berger, J. 1983. Ecology and catastrophic mortality in wild horses: Implications for interpreting fossil assemblages. *Science* 220: 1403–4.

Berglund, B. E., S. Hakansson, and E. Lagerlund. 1976. Radiocarbon-dated mammoth (*Mammuthus primigenius* Blumenbach) finds in south Sweden. *Boreas* 5(3): 177–91.

Best, A. A., and W. G. Raw. 1975. *Rowland Ward's Records of Big Game (Africa)*, 16th ed. London: Rowland Ward.

Biberson, P., and E. Aguirre. 1965. Expériences de taille d'outils préhistoriques dans des os d'éléphant. *Quaternaria* 7: 165–83.

Biddittu, I., P. F. Cassoli, F. Radicati di Brozolo, A. G. Segre, E. Segre Naldini, and I. Villa. 1979. Anagni, a K-Ar dated lower and middle Pleistocene site, central Italy: Preliminary report. *Quaternaria* 21: 53–71.

Binford, L. R. 1978. *Nunamiut Ethnoarchaeology*. New York: Academic Press.

1981. *Bones: Ancient Men and Modern Myths*. New York: Academic Press.

1987. Were there elephant hunters at Torralba? In M. N. Nitecki and D. V. Nitecki, eds., *The Evolution of Human Hunting*, pp. 47–105. New York: Plenum Press.

Blair, P. 1710a. Osteographia elephantina (Part I). *Philosophical Transactions* 27(326): 51–116.

1710b. Osteographia elephantina (Part II). *Philosophical Transactions* 27(327): 117–168.

Bland-Sutton, J. 1910. The diseases of elephants' tusks in relation to billiard balls. *The Lancet* (November 26, 1910): 1534–7.

Blumenschine, R. J. 1986. Carcass consumption sequences and the archeological distinction of scavenging and hunting. *Journal of Human Evolution* 15: 639–59.

1987. Characteristics of an early hominid scavenging niche. *Current Anthropology* 28(4): 383–407.

Bodenheimer, F. S. 1957. The ecology of mammals in arid zones. In *Arid Zone Research VIII: Human and Animal Ecology*, pp. 100–32. Paris: UNESCO.

Bökönyi, S. 1985. Subfossil elephant remains from southwestern Asia. *Paléorient* 11(2): 161–3.

1986. Subfossile Elefantenknochen aus Vorderasien. In R. Hachmann, ed., *Bericht über die Ergebnisse der Ausgrabungen in Kānuid el-Lōz in den Jahren 1977 bis 1981*, pp. 187–9. Bonn: Dr. Rudolf Habelt.

Bond, G. 1948. The direction of origin of the Kalahari sand of southern Rhodesia. *Geological Magazine* 85: 305–13

Bonfield, W. 1975. Deformation and fracture characteristics of the Cooperton mammoth bones. In A. D. Anderson, ed., *The Cooperton Mammoth: An Early Man Bone Quarry. Great Plains Journal* 14(2): 130–73.

Bonnichsen, R. 1973. Some operational aspects of human and animal bone alteration. In B. Gilbert, ed., *Mammalian Osteo-Archaeology: North America*, pp 9–24. Columbia: Missouri Archaeological Society.

1979. *Pleistocene Bone Technology in the Beringian Refugium*. Archaeological Survey of Canada paper No. 89.

1981. Inference and illusion: Comments on R. D. Guthrie's "The First Americans? The Elusive Arctic Bone Culture." *The Quarterly Review of Archaeology* (June): 17–18.

Bourlière, F., and J. Verschuren. 1960. *Introduction à l'écologie des ongulés du Parc National Albert*. Brussells: Institut des Parks Nationaux aux Congo Belge.

Bower, B. 1987. Uncovering life by an ancient lake. *Science News* 131: 264.

Boyle, D. 1929. Height in elephants. *Journal of the Bombay Natural History Society* 33: 437.

Brain, C. K. 1981. *The Hunters or the Hunted?* University of Chicago Press.

Briggs, C., Jr. 1838. Report of C. Briggs, Jr., fourth assistant geologist, no. 4. In *First Annual Report of the Geological Survey of the State of Ohio*, pp. 71–98.

Brooks, A. S., and J. E. Yellen. 1977. Archeological excavations at ≠Gi: Preliminary report on the first two field seasons. *Botswana Notes and Records* 9: 21–30.

Brooks, P. M. 1978. Relationship between body condition and age, growth, reproduction, and social status in impala, and its application to management. *South African Journal of Wildlife Research* 8(4): 151–7.

Brooks, P. M., J. Hanks, and J. V. Ludbrook 1977. Bone marrow as an index of condition in African ungulates. *South African Journal of Wildlife Research* 7(2): 61–6.

Brown, G., and A. J. Moule. 1977a. The structural characteristics of elephant ivory. *The Australian Gemmologist* 13(1): 13–17.

1977b. The structural characteristics of various ivories. *The Australian Gemmologist* 13(2): 47–60.

Bryan, A. L. 1973. Paleoenvironments and cultural diversity in Late Pleistocene South America. *Quaternary Research* 3(2): 237–56.

ed. 1978. *Early Man in America from a Circum-Pacific Perspective*. Department of Anthropology, University of Alberta, Occasional Papers, No. 1.

Bryan, A. L., R. Casamiquela, J. M. Cruxent, R. Gruhn, and C. Ochsenius. 1978. An

El Jobo mastodon kill at Taima-Taima, northern Venezuela. *Science* 200: 1275–7.

Buchner, A. P., and L. J. Roberts. 1990. Reevaluation of an implement of "elephant bone" from Manitoba. *American Antiquity* 55(3): 605–7.

Budyko, M. I. 1975. *Klimat i Życie.* Warsaw: Państwome Wydawnictwo Naukowe.

Buechner, H. K., and H. C. Dawkins. 1961. Vegetation change induced by elephants and fire in Murchison Falls National Park, Uganda. *Ecology* 42: 752–66.

Bullen, R. P., S. D. Webb, and B. I. Waller. 1970. A worked mammoth bone from Florida. *American Antiquity* 35(2): 203–5.

Bunn, H. T. 1986. Patterns of skeletal representation and hominid subsistence activities at Olduvai Gorge, Tanzania, and Koobi Fora, Kenya. *Journal of Human Evolution* 15: 673–90.

Bunn, H. T., L. E. Bartram, and E. M. Kroll. 1989. Variability in bone assemblage formation from Hadza hunting, scavenging, and carcass processing. *Journal of Anthropological Archaeology* 7: 412–57.

Buss, I. O. 1961. Some observations of food habits and behaviour of the African elephant. *Journal of Wildlife Management* 25: 131–48.

Butynski, T. M., and R. Mattingly. 1979. Burrow structure and fossorial ecology of the springhare *Pedetes capensis* in Botswana. *African Journal of Ecology* 17: 205–15.

Cahen, D., and J. Moeyersons. 1977. Subsurface movements of stone artifacts and their implications for the prehistory of central Africa. *Nature* 266: 812–15.

Calef, G. 1981. *Caribou and the Barren-Lands.* Toronto: Firefly Books.

Callow, P., and J. M. Cornford. 1986. *La Cotte de St. Brelade 1961–1978: Excavations by C. B. M. McBurney.* Norwich, U.K.: Geo Books.

Caloi, L., T. Kotsakis, M. R. Palombo, and C. Petronio. 1989. The dwarf elephants of Mediterranean islands. *Fifth International Theriological Congress, Abstract of Papers and Posters* 1: 141–2.

Campbell, A., and G. Child. 1971. The impact of man on the environment of Botswana. *Botswana Notes and Records* 3: 91–110.

Carbyn, L. N. 1974. *Wolf Predation and Behavioral Interactions with Elk and Other Ungulates in an Area of High Prey Diversity.* Parks Canada: Jasper National Park.

Carrington, R. 1957. *Mermaids and Mastodons: A Book of Natural and Unnatural History.* New York: Rinehart.

1958. *Elephants: A short Account of Their Natural History, Evolution, and Influence on Mankind.* London: Scientific Book Club.

Cassells, E. S. 1983. *The Archaeology of Colorado.* Boulder: Johnson Publishing.

Caughley, G. 1976. The elephant problem – an alternative hypothesis. *East African Wildlife Journal* 14: 265–83.

Chavaillon, J., J. L. Boisaubert, M. Faure, C. Guérin, J. L. Ma, B. Nickel, G. Poupeau, P. Rey, and S. A. Warsama. 1987. Le site de dépecage Pleistocène à *Elephas recki* de Barogali (République de Djibouti): Nouveaux résultats et datation. *C.R. Acad. Sci. (Paris)*, t. 305, Série II, pp. 1259–66.

Chavaillon, J., C. Guerin, J. L. Boisaubert, and Y. Coppens. 1986. Découverte d'un site de dépecage à *Elephas recki*, en République de Djibouti. *C.R. Acad. Sci. (Paris)*, t. 302, Série II, No. 5, pp. 243–246.

Cheatum, E. L. 1949. Bone marrow as an index of malnutrition in deer. *New York Conservation* 3(1): 19–22.

Child, G. 1965. *Report to the Government of Botswana on an Ecological Survey of Northeastern Botswana.* Rome: Food and Agriculture Organisation of the United Nations.

1968. *Behaviour of Large Mammals during the Formation of Lake Kariba.* National Museums of Rhodesia Kariba Studies. Salisbury (Harare, Zimbabwe): Trustees of the National Museums of Rhodesia.

1972. Observations on a wildebeeste die-off in Botswana. *Arnoldia Rhodesia* 5(31): 1–13.

Childes, S. L. 1984. The population dynamics of some woody specimens in the Kalahari sand vegetation of Hwange National Park. Unpublished M.Sc. thesis, University of Witswatersrand, South Africa.

Childes, S. L., and B. H. Walker, 1987. Ecology and dynamics of the woody vegetation on the Kalahari Sands in Hwange National Park, Zimbabwe. *Vegetatio* 72: 111–28.

Chmielewski, W., and H. Kubiak. 1962. The find of mammoth bones at Skaratki in the Lowicz district. *Folia Quaternaria* 9: 1–29.

Churcher, C. S. 1980. Did the North American mammoth migrate? *Canadian Journal of Anthropology* 1(1): 103–5.

Clark, J. D., and C. V. Haynes, Jr. 1970. An elephant butchery site at Mwanganda's Village, Karonga, Malawi, and its relevance for Palaeolithic archaeology. *World Archaeology* 1(3): 390–411.

Clutton-Brock, T. H. 1983. Selection in relation to sex. In D. S. Bendall, ed., *Evolution from Molecules to Men*, pp. 457–81. Cambridge University Press.

ed. 1988. *Reproduction Success.* University of Chicago Press.

Clutton-Brock, T. H., F. E. Guinness, and S. D. Albon. 1982. *Red Deer: Behavior and Ecology of Two Sexes.* University of Chicago Press.

Clutton-Brock, T. H., and P. H. Harvey. 1978. Mammals, resources and reproductive strategies. *Nature* 273: 191–5.

Coe, M. 1978. The decomposition of elephant carcasses in the Tsavo (East) National Park, Kenya. *Journal of Arid Environments* 1(1): 71–86.

COHMAP members. 1988. Climatic changes of the last 18,000 years: Observations and model simulations. *Science* 241: 1043–52.

Collins, M. O., ed. 1965. *Rhodesia: Its Natural Resources and Economic Development.* Salisbury, Rhodesia: M. O. Collins.

Colton, H. S. 1909. Peale's museum. *Popular Science Monthly* 75: 221–38.

Colyer, F., and A. E. W. Miles. 1957. Injury to and rate of growth of an elephant tusk. *Journal of Mammalogy* 38(2): 243–7.

Conybeare, A. 1972. The habitat preferences of kudu *Tragelaphus strepsiceros* Pallas in the Kalahari Sand Area of Wankie National Park, Rhodesia. Unpublished M.Sc. thesis, University of Rhodesia.

Conybeare, A., and G. Haynes. 1984. Observations on elephant mortality and bones in water holes. *Quaternary Research* 22: 189–200.

Cook, H. J. 1931. More evidence of the "Folsom Culture" race. *Scientific American* 144: 102–3.

Cooke, C. K. 1967. A preliminary report on the Stone Age of the Nata River, Botswana. *Arnoldia Rhodesia* 2(40): 1–10.

——— 1976. An open Middle Stone Age, Tshangula site near the Victoria Falls, Rhodesia. *Arnoldia Rhodesia* 8(2): 1–15.

——— 1979. The Stone Age in Botswana: A preliminary survey. *Arnoldia Rhodesia* 8(27): 1–32.

Cooke, H. B. S. 1957. Observations relating to Quaternary extinctions in east and south Africa. *Transactions of the Geological Society of South Africa* 60: 1–73.

Coombs, W. P. 1978. Theoretical aspects of cursorial adaptations in dinosaurs. *Quarterly Review of Biology* 53: 393–418.

Coope, G. R., and A. M. Lister. 1987. Late-glacial mammoth skeletons from Condover, Shropshire, England. *Nature* 330: 472–4.

Cooper, W. 1831. Notices of Big-Bone Lick. *Monthly American Journal of Geology and Natural Science* 1(4): 158–74.

Coppens, Y., V. J. Maglio, C. T. Madden, and M. Beden. 1978. Proboscidea. In V. J. Maglio and H. B. S. Cooke, eds., *Evolution of African Mammals*, pp. 336–67. Cambridge, Mass.: Harvard University Press.

Corfield, T. F. 1973. Elephant mortality in Tsavo National Park, Kenya. *East African Wildlife Journal* 11: 339–68.

Corner, R. G., and R. F. Diffendal, Jr. 1983. An Irvingtonian fauna from the oldest Quaternary alluvium in eastern Pumpkin Creek Valley, Morril and Banner counties, Nebraska. (University of Wyoming) *Contributions to Geology* 22(1): 39–43.

Cornwell, J. 1982. *Earth to Earth.* New York: Dell.

Coues, E., ed. 1897. *New Light on the Early History of the Greater Northwest. The Manuscript Journals of Alexander Henry, Fur Trader of the Northwest Territories; and of David Thompson... Vol. 1: The Red River of the North.* New York: Francis P. Harper.

Crader, D. C. 1983. Recent single-carcass bone scatters and the problem of "butchery" sites in the archaeological record. In J. Clutton-Brock and C. Grigson, eds., *Animals and Archeology. Vol. I: Hunters and Their Prey*, pp. 107–41. Oxford: BAR International Series No. 163.

——— 1984. Elephant butchery techniques of the Bisa of the Luangwa Valley, Zambia. Paper presented at the 49th annual meeting of the Society for American Archeology, 11–14 April, Portland, Oregon.

Cruxent, J. M. 1961. Huesos quemados en el yacimiento prehistórico Muaco, Edo. Falcón. *Instituto Venezolano de investigaciones Científicas, Departmento de Antropologia, Boletín Informitano* 2: 20–1.

——— 1970. Projectile points with Pleistocene mammals in Venezuela. *Antiquity* 44: 223–5.

Cumming, D. H. M. 1981. The management of elephant and other large mammals in Zimbabwe. In P. A. Jewell and S. Holt, eds., *Problems in Management of Locally Abundant Wild Mammals*, pp. 91–118. New York: Academic Press.

——— 1982. The influence of large herbivores on savanna structure in Africa. In B. J. Huntley and B. H. Walker, eds., *Ecological Studies. Vol. 42: Ecology of Tropical Savannas*, pp. 217–45. Berlin: Springer-Verlag.

Cumming, R. G. 1857. *Five Years' Adventures in the Far Interior of South Africa; with Notices of the Native Tribes and Savage Animals* (revised and condensed version). London: John Murray.

Czaplicka, M. A. 1914. *Aboriginal Siberia: A Study in Social Anthropology.* Oxford: Clarendon Press.

Dagg, A. I., and J. B. Foster. 1982. *The Giraffe: Its Biology, Behavior, and Ecology,* supplemental ed. Malabar, Fla.: R. E. Krieger.

Daly, M., and M. Wilson. 1988. *Homicide.* New York: Aldine De Gruyter.

Dansie, A. J., J. O. Davis, and T. W. Stafford, Jr. 1988. The Wizards Beach Recession: Farmdalian (25,500 yr B.P.) vertebrate fossils co-occur with early Holocene artifacts. In J. A. Willig, C. M. Aikens, and J. L. Fagan, eds., *Early Human Occupation in Western North America: The Clovis-Archaic Interface,* pp. 153–200. Nevada State Museum Anthropological Papers, No. 21.

Darwin, C. 1871. *The Descent of Man, and Selection in Relation to Sex.* New York: Appleton.

Dasmann, R. F., and A. S. Mossman. 1962. Abundance and population structure of wild ungulates in some areas of southern Rhodesia. *Journal of Wildlife Management* 26(3): 262–8.

Davis, L. C. 1987. Late Pleistocene/Holocene environmental changes in the Central Plains of the United States: The mammalian record. In R. W. Graham, H. A. Semken, Jr., and M. A. Graham, eds., *Late Quaternary Mammalian Biogeography and Environments of the Great Plains and Prairies. Illinois State Museum Scientific Papers* 22: 88–143.

Davis, O. K., J. I. Mead, P. S. Martin, and L. D. Agenbroad. 1985. Riparian plants were a major component of the diet of mammoths of southern Utah. *Current Research in the Pleistocene* 2: 81–2.

Davison, E. 1930–5. Monthly progress reports, Wankie Game Reserve (1930–5). Unpublished manuscript on file, Hwange National Park, Zimbabwe.

——— 1930. Report on Wankie Game Reserve to the end of the year ending 31 March 1930. Manuscript, letter to chief forest officer, Rhodesia Agriculture Department, Salisbury, on file, Hwange National Park main camp, Zimbabwe.

——— 1950. Wankie Game Reserve: Its Location and Development 1928–1950. Manuscript on file, Hwange National Park, Zimbabwe.

——— 1967. *Wankie: The Story of a Great Game Reserve.* Salisbury, Rhodesia: Regal Publishers.

Dean, M. C. 1987. Growth layers and incremental markings in hard tissues; a review of the literature and some preliminary observations about enamel structure in *Paranthropus boisei. Journal of Human Evolution* 16: 157–72.

de Baer (or von Baer), K. E. 1866. Fortsetzung der Berichte über die Expedition zur Aufsuchung des angekündigten Mammuths. *Bullétin de L' Académie Impériale des Sciences de St. Pétersburg* 10: 513–34.

de Blainville, H. D. 1845. *Ostéographie et Description Iconographique des Mammifères Récents et Fossiles.* Paris: Bertrand.

de Calesta, D. S., J. G. Nagy, and J. A. Bailey. 1977. Experiments on starvation and recovery in mule deer does. *Journal of Wildlife Management* 41: 81–6.

Denbow, J. R. 1984. Prehistoric herders and foragers of the Kalahari: The evidence for 1500 years of interaction. In C. Schrire, ed., *Past and Present in Hunter Gatherer Studies*, pp. 175–93. New York: Academic Press.

Deraniyagala, P. E. P. 1955. *Some Extinct Elephants, Their Relatives, and the Two Living Species*. Colombo: National Museum of Ceylon.

Derry, D. E. 1911. Damage done to skulls and bones by termites. *Nature* 86: 245–6.

Digby, B. 1926. *The Mammoth and Mammoth-Hunting in North-East Siberia*. London: Witherby.

Dillehay, T. D. 1986. The cultural relationships of Monte Verde: A late Pleistocene settlement site in the sub-antarctic forest of south-central Chile. In A. L. Bryan ed., *New Evidence for the Pleistocene Peopling of the Americas*, pp. 319–37. Orono, Maine: Center for the Study of Early Man.

1989. Researching early sites. In T. D. Dillehay, (ed.), *Monte Verde: A Late Pleistocene Settlement in Chile. Volume 1: Paleoenvironment and Site Context*, pp. 1–25. Washington, D.C.: Smithsonian Institution Press.

Dillehay, T. D., and M. Pino. 1986. Monte Verde: An Early Man site in south-central Chile. In R. Matos M., S. Turpin, and H. Eling, Jr. (eds.), *Andean Archaeology: Papers in Memory of Clifford Evans*, pp. 1–17 (University of California, Los Angeles, Institute of Archaeology Monograph 27). Los Angeles: University of California Institute of Archaeology.

Dillehay, T. D., M. Pino Q., E. M. Davis, S. Valastro, Jr., A. G. Varela, and R. Casimiquela. 1983. Monte Verde: Radiocarbon dates from an Early Man site in south-central Chile. *Journal of Field Archeology* 9: 547–9.

Douglas-Hamilton, I. 1972. On the ecology and behaviour of the African elephant. Unpublished Ph.D. thesis, Oxford University.

1973. On the ecology and behaviour of the Lake Manyara elephants. *East African Wildlife Journal* 11: 401–3.

1988. The Lake Manyara elephants. *Swara* 11(2): 12.

Douglas-Hamilton, I., and O. Douglas-Hamilton. 1975. *Among the Elephants*. New York: Penguin.

Dreimanis, A. 1967. Mastodons, their geologic age and extinction in Ontario, Canada. *Canadian Journal of Earth Sciences* 4: 663–75.

Drumm, J. 1963. *Mammoths and Mastodons: Ice-Age Elephants of New York*. University of New York, State Education Department, State Museum and Science Service Educational Leaflet No. 13.

Dubinin, V. B., and V. Ye. Garutt. 1954. O skelete mamonta iz del'ty reki Leny. *Zoologicheskiy Zhurnal* 33(2): 423–32.

Dubrovo, I. A. 1981. Die fossilen Elefanten Japans. *Quartärpaläontologie* 4: 49–84.

1982. Morfologiya skeleta Yuribeiskovo mamonta. In V. Ye. Sokolov, ed., *Yuribeiskii mamont*, pp. 53–99. Moscow: Nauka.

Duckworth, B. J. 1972. The distribution and movement of buffalo (*Syncerus caffer caffer*) herds in the Kalahari Sands area of Wankie National Park. Unpublished thesis for certificate in field ecology, University of Rhodesia.

Duffy, K. 1984. *Children of the Forest*. New York: Dodd, Mead.

du Toit, R. 1986. Comment (on "Elephants and Woodlands – What are the Issues?" by K. Lindsay). *Pachyderm, Newsletter of the African Elephant and Rhino Specialist Group* 7: 17–18.

Dutrow, B. L. 1980. Metric analysis of a Late Pleistocene mammoth assemblage,

Hot Springs, South Dakota. Unpublished M.Sc. thesis, Southern Methodist University.

Eberhardt, L. L. 1971. Population analysis. In R. H. Giles, Jr., ed., *Wildlife Management Techniques*, 3rd ed., pp. 457–95. Washington, D.C.: Wildlife Society.

Economos, A. C. 1981. The largest land mammal. *Journal of Theoretical Biology* 89: 211–15.

Edwards, W. E. 1967. The late-Pleistocene extinction and diminution in size of many mammalian species. In P. S. Martin and H. E. Wright, Jr., eds., *Pleistocene Extinctions: The Search for a Cause*, pp. 141–55. New Haven, Conn.: Yale University Press.

Eiseley, L. C. 1945. Myth and mammoth in archaeology. *American Antiquity* 11(2): 84–7.

1946. Men, mastodons, and myth. *Scientific Monthly* 62: 517–24.

Elder, W. H. 1970. Morphometry of elephant tusks. *Zoologica Africana* 5(1): 143–59.

Elias, S. A. 1986. Fossil insect evidence for late Pleistocene paleoenvironments of the Lamb Spring site, Colorado. *Geoarchaeology* 1: 381–7.

Elias, S. A., and L. J. Toolin. 1990. Accelerator dating of a mixed assemblage of late Pleistocene insect fossils from the Lamb Spring site, Colorado. *Quaternary Research* 33: 122–6.

Eltringham, S. K. 1982. *Elephants*. Poole, U.K.: Blandford Press.

Espinoza, E. O., M.-J. Mann, J. P. LeMay, and K. A. Oakes. 1990. A method for differentiating modern from ancient proboscidean ivory in worked objects. *Current Research in the Pleistocene* 7.

Ettensohn, F. R. 1976. New mastodon finds from southwestern Ohio, USA. *Ohio Journal of Science* 76(2): 71–2.

Evans, G. L. 1961. The Friesenhahn Cave. *Bulletin of the Texas Memorial Museum* 2: 5–22.

Ewer, R. F. 1968. *Ethology of Mammals*. London: Logos Press.

Falconer, H. C. 1863. On the American fossil elephant of the regions bordering the Gulf of Mexico (*Elephas columbi* Falconer), with general observations on the living and extinct species. *Natural History Review* 3: 43–114.

Falquet, R. A., and W. C. Haneberg. 1978. The Willard mastodon: Evidence of human predation. *Ohio Journal of Science, April Program and Abstracts Supplement* 78: 64.

FAO (Food and Agriculture Organization, United Nations). 1977. *Dietary Fats and Oils in Human Nutrition*. Food and Nutrition Paper No. 3. Rome: FAO.

Farrand, W. R. 1961. Frozen mammoths and modern geology. *Science* 133: 729–35.

Fatti, L. P., G. L. Smuts, A. M. Starfield, and A. A. Spurdle. 1980. Age determination in African elephants. *Journal of Mammalogy* 61: 547–51.

Felix, J. 1912. Das Mammuth von Borna. *Veröffentlichungen des Städtischen Museums für Völkerkunde zu Leipzig* 4: 1–55.

Field, C. R. 1971. Elephant ecology in the Queen Elizabeth National Park, Uganda. *East African Wildlife Journal* 9: 99–123.

Figgins, J. D. 1931. An additional discovery of the association of a "Folsom" artifact and fossil mammal remains. *Proceedings of the Colorado Museum of Natural History* 10(4): 23–4.

1933. A further contribution to the antiquity of man in America. *Colorado Museum of Natural History Proceedings* 12(2): 4–8.

Fiorillo, A. R. 1984. An introduction to the identification of trample marks. *Current Research* 1: 47–8.

Fisher, D. C. 1984a. Mastodon butchery by North American Paleo-Indians. *Nature* 308: 271–2.

1984b. Taphonomic analysis of late Pleistocene mastodon occurrences: Evidence of butchery by North American Paleo-Indians. *Paleobiology* 10: 338–57.

1987. Mastodont procurement by Paleoindians of the Great Lakes region: Hunting or scavenging? In M. H. Nitecki and D. V. Nitecki, eds., *The Evolution of Human Hunting*, pp. 309–421, New York: Plenum Press.

Fisher, D. C., and P. L. Koch, 1983. Seasonal mortality of late Pleistocene mastodons: Evidence for the impact of human hunting. *Geological Society of America Abstracts of Program* 15: 573.

Fisher, J. 1988. Present-day elephant butchery. Paper presented at the Society for American Archeology annual meeting, Phoenix, Arizona.

Flerov, K. K. 1931. Khobot mamonta (*Elephas primigenius* Blum.), naidennii v Kolymskom okruge. *Izvestia AN SSSR, seria* 7(2): 863–70.

Flint, R. F. 1971. *Glacial and Quaternary Geology.* New York: Wiley.

Flint, R. F., and G. W. Bond. 1968. Pleistocene sand ridges and pans in western Rhodesia. *Geological Society of America Bulletin* 79(3): 299–313.

Flower, S. S. 1943. Notes on age at sexual maturity, gestation period and growth of the Indian elephant, *Elephas maximus. Proceedings of the Zoological Society of London A* 113: 21–7.

1947. Further notes on the duration of life in mammals. V: The alleged and actual ages to which elephants live. *Proceedings of the Zoological Society of London* 117: 680–8.

Fothergill, R. J. 1953. *The Monuments of Southern Rhodesia.* Bulawayo: Commission for the Preservation of Natural and Historical Monuments and Relics.

Fowke, G. 1928. Archeological investigations. II. *U.S. Bureau of American Ethnology Annual Reports 1926–1927* 44: 399–540.

Fowler, C. W., and T. Smith. 1973. Characterizing stable populations: An application to the African elephant population. *Journal of Wildlife Management* 37: 513–23.

Frade, F. 1955. Ordre des proboscidiens (Proboscidea – Illiger 1811). In P.-P. Grassé, ed., *Traité de Zoologie* 17: 715–83.

Freeland, W. J., and D. H. Janzen. 1974. Strategies in herbivory by mammals: The role of plant secondary compounds. *American Naturalist* 108: 269–89.

Frison, G. C. 1970. The Glenrock Buffalo Jump, 48 CO 304. *Plains Anthropologist* 15(50): 1–66.

ed. 1974. *The Casper Site: A Hell Gap Kill on the High Plains.* New York: Academic Press.

1976. Cultural activity associated with prehistoric mammoth butchering and processing. *Science* 194: 728–30.

1978. *Prehistoric Hunters of the High Plains.* New York: Academic Press.

1986. Human artifacts, mammoth procurement, and Pleistocene extinctions as viewed from the Colby Site. In G. C. Frison and L. C. Todd, eds., *The Colby Mammoth Site: Taphonomy and Archaeology of a Clovis Kill in Northern Wyoming,* pp. 91–114. Albuquerque: University of New Mexico Press.

Frison, G. C., and D. Stanford, eds. 1982. *The Agate Basin Site.* New York: Academic Press.

Frison, G. C., and L. C. Todd. 1986. *The Colby Mammoth Site: Taphonomy and Archaeology of a Clovis Kill in Northern Wyoming.* Albuquerque: University of New Mexico Press.

Gambaryan, P. P. 1974. *How Mammals Run: Anatomical Adaptations* (translated by Hilary Hardin). New York: Wiley (Halstead Press).

Gamble, C. 1985. *The Palaeolithic Settlement of Europe.* Cambridge University Press.

Garlake, P. 1987. *The Painted Caves: An Introduction to the Prehistoric Art of Zimbabwe.* Harare, Zimbabwe: Modus Publications.

Garland, T., Jr. 1983. The relation between maximal running speed and body mass in terrestrial mammals. *Journal of Zoology (London)* 199: 157–70.

Garutt, V. Ye. 1954. Yuzhniy slon *Archidiskodon meridionalis* (Nesti) iz pliotsena severnovo poverezh'ya Azovskovo morya. *Trudi Komissiona po izucheniya chetvortichnovo perioda, Tom X, vipusk 2.*

1964. Das mammut *Mammuthus primigenius* (Blumenbach). In *Die Neue Brehm-Bücherei 331* (translated by Günther Grempe). Wittenberg: A. Ziemsen Verlag.

1965. Iskopayemiye slonov sibiri. In F. G. Markov et al., eds., *Antropogenoviy period v Arktike i Subarktike* [*k VII sessii Mezhdunarodnovo Kongressa Assotsiatsii po Izucheniyu Chetvertichnovo Perioda (INQUA)*], pp. 106–30. Trudi Nauchno-issledovatel'skovo Instituta Geologii Arktiki Gosudarstvennovo Geologicheskovo Komiteta SSSR Tom 143.

1981. Versuch der graphischen Rekonstruktion des Lebensbildes der Elefanten der Entwicklungslinie *Archidiskodon-Mammuthus. Quartärpaläontologie* 4: 19–25.

Garutt, V. Ye., and V. B. Dubinin. 1951. O skelete Taymirskovo mamonta. *Zoologicheskii Zhurnal* 30(1): 17–23.

Garutt, V. Ye., and I. V. Foronova, 1976. *Issledovaniye Zubov Vymershikh Slonov: Metodicheskiye Rekomendatsii.* Novosibirsk: Institut geologii i geophisiki SO AN SSSR.

Gauthier-Pilters, H., and A. I. Dagg. 1981. *The Camel: Its Evolution, Behavior, and Relationship to Man.* University of Chicago Press.

Geist, V. 1974. On the relationship of ecology and behavior in the evolution of ungulates: Theoretical considerations. In V. Geist and F. Walther, eds., *The Behavior of Ungulates and Its Relation to Management, Vol. I,* pp. 235–46. IUCN Publications, No. 24. Morges, Switzerland: IUCN.

1982. Behavior. In J. W. Thomas and D. E. Toweill, eds., *Elk of North America,* pp. 219–77. Harrisburg, Pa.: Stackpole Books.

Gekker, R. F. 1977. Istoriya obyavleniy Akademii Nauk o trupakh mamontov i nosorogov. *Palyeontologicheskiy Zhurnal* 1977(3): 140–4.

Ghiglieri, M. P. 1987. Sociobiology of the great apes and the hominid ancestor. *Journal of Human Evolution* 16: 319–57.

Gifford, D. P. 1981. Taphonomy and paleoecology: A critical review of archaeology's sister disciplines. *Advances in Archaeological Method and Theory* 4: 365–438.

Gifford-Gonzalez, D. P. 1989. Overview: Modern analogues: Developing an interpretive framework. In R. Bonnichsen and M. Sorg, eds., *Bone Modification*, pp. 43–52. Orono, Maine: Center for the Study of the First Americans.

Gillespie, J. M. 1970. Mammoth hair: Stability of *a*-keratin structure and constituent proteins. *Science* 170: 1100–2.

Glover, J. 1963. The elephant problem at Tsavo. *East African Wildlife Journal* 1: 30–9.

Goodman, M., J Shoshani, and M. Barnhart. 1979. Frozen mammoth muscle: Preliminary findings. *Paleopathology Newsletter* 25: 3–5.

Goodwin, A. J. H. 1929. The Middle Stone Age. *Annals of the South African Museum* 29: 95–145.

Gordon, J. 1956. Report to the chief game warden for the period 1 July 1956 to 31 August 1956. Manuscript on file, Hwange National Park, Zimbabwe.

Gordon, R. J. 1984. The !Kung in the Kalahari exchange: An ethnohistorical perspective. In C. Schrire, ed., *Past and Present in Hunter Gatherer Studies*, pp. 195–224. Orlando, Fla.: Academic Press.

Gorlova, R. N. 1982. Rastitel'nye markroostalki, obnaryzhennye v zheludochnokishechnom trakte yuribeiskovo mamonta. In V. Ye. Sokolov, ed., *Yuribeiskii mamont*, pp. 37–43. Moscow: Nauka.

Gove, P. B., ed. 1967. *Webster's Third International Dictionary of the English Language, Unabridged*. Springfield, Mass.: G. & C. Merriam Co.

Graham, R. W. 1976. Pleistocene and Holocene mammals, taphonomy and paleoecology of the Friesenhahn Cave local fauna, Bexar County, Texas. Unpublished Ph.D. thesis, University of Texas, Austin.

1979. Paleoclimates and Pleistocene faunal provinces in North America. In R. L. Humphrey and D. Stanford, eds., *Pre-Llano Cultures of the Americas: Possibilities and Paradoxes*, pp. 49–69. Washington, D.C.: Anthropological Society of Washington.

1981. Preliminary report on late Pleistocene vertebrates from the Selby and Dutton archeological/paleontological sites, Yuma County, Colorado. *Contributions to Geology* 20: 33–56.

1984. Taphonomy and demography of proboscidean remains from Friesenhahn Cave, Texas. Paper presented at the Society for American Archaeology annual meeting, 12–14 April, Portland, Oregon.

1986a. Taxonomy of North American mammoths. In G. C. Frison and L. C. Todd, eds., *The Colby Mammoth Site: Taphonomy and Archeology of a Clovis Kill in Northern Wyoming*, pp. 165–9. Albuquerque: University of New Mexico Press.

1986b. Descriptions of the dentitions and stylohyoids of *Mammuthus columbi* from the Colby site. In G. C. Frison and L. C. Todd, eds., *The Colby Mammoth Site: Taphonomy and Archaeology of Clovis Kill in Northern Wyoming*, pp. 171–90. Albuquerque: University of New Mexico Press.

1987. Late Quaternary mammalian faunas and paleoenvironments of the southwestern plains of the United States. In R. W. Grahm, H. A. Semken, Jr., and M. A. Graham, eds., *Late Quaternary Mammalian Biogeography and Environments of the Great Plains and Prairies,* pp. 24–87. Illinois State Museum Scientific Papers, Vol. 22.

Graham, R. W., C. V. Haynes, D. L. Johnson, and M. Kay. 1981. Kimmswick: A Clovis–mastodon association in eastern Missouri. *Science* 213: 1115–7.

Graham, R. W., and E. L. Lundelius. 1984. Coevolutionary disequilibrium and Pleistocene extinctions. In P. S. Martin and R. G. Klein, eds., *Quaternary Extinctions: A Prehistoric Revolution,* pp. 223–49. Tucson: University of Arizona Press.

Grassé, P.-P. 1955. *Traité de Zoologie: Anatomie, Systématique, Biologie.* Paris: Masson.

Grichuk, V. P. 1984. Late Pleistocene vegetation history. In A. A. Velichko, ed., *Late Quaternary Environments of the Soviet Union,* pp. 155–78. Minneapolis: University of Minnesota Press.

Grote, K., and F. Preul. 1978. Der mittelpaläolithische Lagerplatz in Salzgitter-Lebenstedt. Vorbericht über die Grabung und die geologische Untersuchung 1977. *Nachrichten aus Niedersachsens Urgeschichte* 47: 77–106.

Günther, E. W. 1975. Die Backenzähne der Elefanten von Ehringsdorf bei Weimar. *Abhandlungen des Zentralen Geologischen Instituts* 23: 399–452.

Gustafson, C. E., D. Gilbow, and R. Daugherty. 1979. The Manis mastodon site: Early Man on the Olympic Peninsula. *Canadian Journal of Archaeology* 3: 157–64.

Guthrie, R. D. 1968. Paleoecology of the large mammal community in interior Alaska during the late Pleistocene. *American Midland Naturalist* 79: 346–63.

1982. Mammals of the mammoth steppe as paleoenvironmental indicators. In D. M. Hopkins, J. V. Matthews, C. Schweger, and S. Young, eds., *Paleoecology of Beringia,* pp. 307–26. New York: Academic Press.

1984. Mosaics, allelochemics and nutrients. In P. S. Martin and R. G. Klein, eds., *Quaternary Extinctions: A Prehistoric Revolution,* pp. 259–98. Tucson: University of Arizona Press.

Guy, P. R. 1974. Feeding behaviour of the African elephant in the Sengwa Research Area of Rhodesia. Unpublished M.Sc. thesis, University of Rhodesia.

1975. The daily food intake of the African elephant, *Loxodonta africana* Blumenbach, in Rhodesia. *Arnoldia Rhodesia* 7(26): 1–8.

1976. The feeding behavior of elephant (*Loxodonta africana*) in the Sengwa Area, Rhodesia. *South African Journal of Wildlife Research* 6: 55–63.

1977. Coprophagy in the African elephant (*Loxadonta [sic] africana* Blumenbach). *East African Wildlife Journal* 15: 174.

Haglund, D. T., D. T. Reay, and D. Swindler. 1988. Carnivore-assisted disarticulation/dismemberment of human remains. *American Journal of Physical Anthropology* 75(2): 218–19.

Hallin, K. F. 1983. Hair of the American mastodon indicates an adaptation to a semiaquatic habitat. *American Zoologist* 23(4): 949.

1989. Wisconsin's Ice Age tuskers: Ice Age elephants and mastodonts. *Wisconsin Academy Review* 35(4): 6–10.

Hallin, K. F., and D. Gabriel. 1981. The first specimen of mastodon hair. *Geological Society of America 34th Annual Meeting of the Rocky Mountain Section, Abstracts with Program* 13(4): 199.

Hanks, J. 1979. *The Struggle for Survival.* New York: Mayflower.

Hanks, J., D. H. M. Cumming, J. L. Orpen, D. F. Parry, and H. B. Warren. 1976. Growth, condition, and reproduction in the impala ram (*Aepyceros melampus*). *Journal of Zoology (London)* 179: 421–35.

Hannus, L. A. 1983. The Lange-Ferguson site, an event of Clovis mammoth butchery with the associated bone tool technology: The mammoth and its track. Unpublished Ph.D. thesis, University of Utah.

1984. Flaked mammoth bone from the Lange/Ferguson site, White River Badlands area, South Dakota. Paper presented at the First International Conference on Bone Modification, 17–19 August, Carson City, Nevada.

1986. Butchering strategies at the Lange/Ferguson site: A Clovis period mammoth kill/butchering locality – White River Badlands, South Dakota. Paper presented at the Fifth International Conference of the International Council for Archaeozoology (ICAZ), 25–30 August, Bordeaux, France.

1990. The Lange-Ferguson site: A case for mammoth bone-butchering tools. In L. Agenbroad, J. Mead, and L. Nelson eds., *Megafauna and Man: Discovery of America's Heartland*, pp. 86–99 (*Mammoth Site of Hot Springs, South Dakota, Scientific Papers, Volume 1*). Hot Springs, South Dakota: Mammoth Site of Hot Springs.

Hansen, M. C., D. Davids, F. Erwin, G. R. Haver, J. E. Smith, and R. P. Wright. 1978. A radiocarbon-dated mammoth site, Marion County, Ohio. *Ohio Journal of Science* 78(2): 103–5.

Hansen, R. M. 1980. Late Pleistocene plant fragments in the dung of herbivores at Cowboy Cave. In J. D. Jennings, ed., *Cowboy Cave*, pp. 179–89. University of Utah Anthropological Papers No. 104.

Harington, C. R., H. W. Tipper, and R. J. Mott. 1974. Mammoth from Babine Lake, British Columbia. *Canadian Journal of Earth Sciences* 11: 285–303.

Harris, J. M., and G. T. Jefferson. 1985. *Rancho La Brea: Treasures of the Tar Pits.* Natural History Museum of Los Angeles County Science Series No. 31.

Haury, E. W. 1953. Artifacts with mammoth remains, Naco, Arizona. I: Discovery of the Naco mammoth and associated projectile points. *American Antiquity* 19(1): 1–14.

Haury, E. W., E. B. Sayles, and W. W. Wasley. 1959. The Lehner mammoth site, southeastern Arizona. *American Antiquity* 25(1): 2–30.

Haynes, C. V. 1964. Fluted projectile points: Their age and dispersion. *Science* 145(3639): 1408–13.

1966. Elephant hunting in North America. *Scientific American* 214: 104–12.

1967. Carbon-14 dates and Early Man in the New World. In P. S. Martin and H. E. Wright, eds., *Pleistocene Extinctions: The Search for a Cause*, pp. 267–86. New Haven, Conn.: Yale University Press.

1970. Geochronology of man–mammoth sites and their bearing upon the origin of the Llano Complex. In W. E. Dort and A. E. Johnson, eds., *Pleistocene and Recent Environments of the Central Plains*, pp. 77–92. Lincoln: University of Kansas Press.

1975. Pleistocene and Recent stratigraphy. In F. Wendorf and J. J. Hester, eds., *Late Pleistocene Environments of the Southern High Plains*, pp. 57–96. Fort Burgwin Research Center Publication No. 9, Fort Burgwin, New Mexico.

1976. Archeological investigations at the Murray Springs site, Arizona, 1968. *National Geographic Society Research Reports* (1968): 165–71.

1980. The Clovis culture. *Canadian Journal of Anthropology* 1(1): 115–21.

1981. Geochronology and paleoenvironments of the Murray Springs Clovis site, Arizona. *National Geographic Society Research Reports* 13: 243–51.

1982. Were Clovis progenitors in Beringia? In D. M. Hopkins, J. V. Matthews, C. E. Schweger, and S. B. Young, eds., *Paleoecology of Beringia*, pp. 383–98. New York: Academic Press.

1984. Stratigraphy and late Pleistocene extinction in the United States. In P. S. Martin and R. G. Klein, eds., *Quaternary Extinctions: A Prehistoric Revolution*, pp. 345–53. Tucson: University of Arizona Press.

1987. Clovis origin update. *The Kiva* 52(2): 83–93.

Haynes, C. V., and G. A. Agogino. 1966. Prehistoric springs and geochronology of the Clovis site, New Mexico. *American Antiquity* 31(6): 812–21.

Haynes, C. V., and E. T. Hemmings. 1968. Mammoth-bone shaft wrench from Murray Springs, Arizona. *Science* 159: 186–87.

Haynes, G. 1980a. Prey bones and predators: Potential ecologic information from analysis of bone sites. *Ossa* 7: 75–97.

1980b. Evidence of carnivore gnawing on Pleistocene and Recent mammalian bones. *Paleobiology* 6: 341–51.

1981. Bone modifications and skeletal disturbances by natural agencies: Studies in North America. Unpublished Ph.D. dissertation, Catholic University of America.

1982. Utilization and skeletal disturbances of North American prey carcasses. *Arctic* 35(2): 266–81.

1983a. A guide for differentiating mammalian carnivore taxa responsible for gnaw damage to herbivore limb bones. *Paleobiology* 9: 164–72.

1983b. Frequencies of spiral and green-bone fractures on ungulate limb bones in modern surface assemblages. *American Antiquity* 48(1): 102–14.

1984a. Tooth wear rate of northern *Bison*. *Journal of Mammalogy* 65(3): 487–91.

1984b. Taphonomic perspectives on late Pleistocene extinctions. *Current Research* 1(2): 49–50.

1985a. On watering holes, mineral licks, death, and predation. In D. Meltzer and J. I. Mead, eds., *Environments and Extinctions in Late Glacial North America*, pp. 53–71. Orono, Maine: Center for the Study of Early Man.

1985b. Age profiles in elephant and mammoth bone assemblages. *Quaternary Research* 24: 333–45.

1986. Spiral fractures and cutmark-mimics in noncultural elephant bone assemblages. *Current Research in the Pleistocene* 3: 45–6.

1987a. Where elephants die. *Natural History* 96(6): 28–33.

1987b. Elephant-butchering at modern mass-kill sites in Africa. *Current Research in the Pleistocene* 4: 75–7.

1987c. Proboscidean die-offs and die-outs: Age profiles in fossil collections. *Journal of Archaeological Science* 14(6): 659–68.

1988a. Longitudinal studies of African elephant death and bone deposits. *Journal of Archaeological Science* 15: 131–57.

1988b. Studies of elephant death and die-offs: Potential applications in understanding mammoth bone assemblages. In E. Webb, ed., *Recent Developments in Environmental Analysis in Old and New World Archaeology*, pp. 151–69. Oxford: BAR International Series No. 416.

1988c. Mass deaths and serial predation: Comparative taphonomic studies of modern large-mammal deathsites. *Journal of Archaeological Science* 15: 219–35.

in press. Late Pleistocene mammoth utilization in northeast Asia and North America. *Archaeozoologia.*

Haynes, G., and D. Stanford. 1984. On the possible utilization of *Camelops* by Early Man in North America. *Quaternary Research* 22: 216–30.

Heape, W. 1931. *Emigration, Migration, and Nomadism.* Cambridge University Press.

Heffner, R., and H. Heffner. 1980. Hearing in the elephant (*Elephas maximus*). *Science* 208: 518–20.

Heine, K. 1982. The main stages of late Quaternary evolution of the Kalahari region, southern Africa. *Paleoecology of Africa* 15: 53–76.

Heintz (Geyntz), A. Ye., and V. Ye. Garutt. 1964. Opredeleniye absolyutnovo vozrasta iskopayemikh ostatkov mamonta i sherstitstovo nosoroga iz vechnoy merzloty Sibiri pri pomoshchi radioaktivnovo ugleroda (C-14). *Dokladi Akademii Nauk SSSR* 154: 1367–70.

Helgren, D. M., and A. S. Brooks. 1983. Geoarchaeology at Gi, a Middle Stone Age and Later Stone Age site in the northwest Kalahari. *Journal of Archaeological Science* 10: 181–97.

Hemmings, E. T. 1970. Early Man in the San Pedro Valley, Arizona. Unpublished Ph.D. thesis, University of Arizona.

Hemmings, E. T., and C. V. Haynes. 1969. The Escapule mammoth and associated projectile points, San Pedro Valley, Arizona. *Journal of the Arizona Academy of Science* 5(3): 184–8.

Hempel, C. G. 1966. *Philosophy of Natural Science.* Englewood Cliffs, N.J.: Prentice-Hall.

Herz, O. F. 1902. Otchety nachal'nuka ekspeditsii imperatorskoy akademii nauk na Beryovzovku dlya raskopki trupa mamonta. *Izvestiya Imperatorskoy Akademii Nauk* 16(4): 137–74.

1904. Frozen mammoth in Siberia. In *Annual Report of the Board of the Smithsonian Institution...for the Year Ending June 30, 1903* (extracts from Herz 1902). Washington, D.C.: U.S. Government Printing Office.

Hester, J. J. 1972. *Blackwater Locality No. 1: A Stratified Early Man Site in Eastern New Mexico.* Fort Burgwin Research Center Publication No. 8, Fort Burgwin, New Mexico.

Hibbard, C. W. 1960. Pliocene and Pleistocene climates in North America. *Annual Report of the Michigan Academy of Sciences. Arts, and Letters* 62: 5–30.

Higgins, D. M. 1969. Aspects of the population dynamics, behaviour and ecology of the wildebeeste in the Wankie National Park with recommendations for management. Unpublished manuscript on file, Hwange National Park, Zimbabwe.

Hildebrand, M., and J. Hurley. 1985. Energy of the oscillating legs of a fast-moving cheetah, pronghorn, jackrabbit, and elephant. *Journal of Morphology* 184: 23–31.

Hill, A. P. 1979. Disarticulation and scattering of mammal skeletons. *Paleobiology* 5(3): 261–74.

Hill, W. C. O. 1953. The anatomy of the African elephant. In W. C. O. Hill, ed., *The Elephant in East Central Africa: A Monograph*, pp. 15–60. London: Rowland Ward.

Hodson, A. W. 1912. *Trekking the Great Thirst: Sport and Travel in the Kalahari Desert*. London: T. Fisher Unwin.

Hoffecker, J. F. 1987. Upper Pleistocene loess stratigraphy and Paleolithic site chronology on the Russian Plain. *Geoarchaeology* 2(4): 259–84.

Hoffman, R. R. 1985. Digestive physiology of the deer – their morphophysiological specialization and adaptation. In P. F. Fennessy and K. R. Drew, eds., *Biology of Deer Production*, pp. 393–407. Wellington: The Royal Society of New Zealand Bulletin No. 22.

Holen, S. R., and R. K. Blasing. 1990. Two late Pleistocene sites in the Central Plains: Preliminary statements. In *Society for American Archeology 55th Annual Meeting Program and Abstracts*, p. 104.

Holman, D. 1967. *The Elephant People*. London: John Murray.

Holman, J. A. 1986. The Dansville mastodon and associated wooden specimens. *National Geographic Research* 2(4): 416.

Holman, J. A., L. M. Abraczinskas, and D. B. Westjohn. 1988. Pleistocene proboscideans and Michigan salt deposits. *National Geographic Research* 4(1): 4–5.

Hooijer, D. A. 1978. The Indian elephant at Bronze Age Ras Shamra, Ugarit. *Ugaritica* 7: 187–8.

Hopkins, D. M. 1982. Aspects of the paleogeography of Beringia during the late Pleistocene. In D. M. Hopkins, J. V. Matthews, Jr., C. E. Schweger, and S. B. Young, eds., *Paleoecology of Beringia*, pp. 3–28. New York: Academic Press.

Hopkins, D. M., J. V. Matthews, Jr., C. E. Schweger, and S. B. Young, eds. 1982. *Paleoecology of Beringia*. New York: Academic Press.

Howell, A. B. 1944. *Speed in Animals*. University of Chicago Press.

Howell, F. C. 1966. Observations on the earlier phases of the European Lower Paleolithic. In J. D. Clark and F. C. Howell, eds., *Recent Studies in Paleoanthropology. American Anthropologist* 68(2): 88–201.

Howell, F. C., and L. G. Freeman. 1983. Ivory points from the earlier Acheulean of Spanish meseta. In *Homenaje al Professor Martin Almagro Basch*, pp. 41–61. Madrid: Ministerio de Cultura.

Howorth, H. H. 1887. *The Mammoth and the Flood: An Attempt to Confront the Theory of Uniformity with the Facts of Recent Geology*. London: Sampson Low, Marston, Searle, and Rivington.

Hsü, K. J. 1989. Catastrophic extinctions and the inevitability of the improbable. *Journal of the Geological Society, London* 146: 749–54.

Humphreys, H. F. 1926. Particulars relating to the broken tusk of a wild Indian elephant. *British Dental Journal* 47: 1400–7.

Ides, Y. E. 1706. *Three Years Travels From Moscow Over Land to China*. London.

Irving, L. 1966. Adaptation to cold. *Scientific American* 214(1): 94–101.

Irving, W. 1986. New dates from old bones. *Natural History* 87(2): 9–12.

Irving, W. N., and C. R. Harington. 1973. Upper Pleistocene radiocarbon-dated artifacts from the northern Yukon. *Science* 179: 335–40.

Irving, W. N., A. V. Jopling, and I. Kritsch-Armstrong. 1989. Studies of bone technology and taphonomy, Old Crow Basin, Yukon Territory. In R. Bonnichsen and M. Sorg (eds.), *Bone Modification*, pp. 347–79. Orono, Maine: Center for the Study of the First Americans.

Irwin, C., H. T. Irwin, and G. Agogino. 1962. Wyoming muck tells of battle: Ice Age man versus mammoth. *National Geographic* 121(6): 828–37.

Irwin, H. T. 1970. Archeological investigations at the Union Pacific mammoth kill site, Wyoming, 1961. *National Geographic Society Research Reports* (1961–2): 123–5.

Isaac, G. L. 1977. *Olorgesailie: Archaeological Studies of a Middle Pleistocene Lake Basin in Kenya*. University of Chicago Press.

Isaac, G. L., and D. C. Crader. 1981. To what extent were early hominids carnivorous? An archeological perspective. In R. S. O. Harding and G. Teleki, eds., *Omnivorous Primates: Gathering and Hunting in Human Evolution*, pp. 37–103. New York: Columbia University Press.

İşcan, M. Y., S. R. Loth, and R. K. Wright. 1987. Racial variation in the sternal extremity of the rib and its effect on age determination. *Journal of Forensic Sciences* 32(2): 452–66.

IUCN (International Union for Conservation of Nature and Natural Resources). 1985. *United Nations List of National Parks and Protected Areas*. Oland, Switzerland: IUCN.

Ivanov, A. N. 1977. V. N. Tatishchev – perviy issledovatel' mamonta v Rossii. In N. K. Vereshchagin, ed., *Mamontavaya Fauna i Sreda Yiyo Obitaniya v Antronogenye SSSR*, pp. 77–81. Trudi Zoologicheskovo Instituta No. 73.

Jachmann, H. 1985. Estimating age in African elephants. *African Journal of Ecology* 23(3): 199–202.

1988. Estimating age in African elephants: A revision of Laws' molar evaluation technique. *African Journal of Ecology* 26(1): 51–6.

Jacobs, S. 1989. Seasons of life in western New York. *Mammoth Trumpet* 5(4): 1, 4–5.

Jainundeen, M. R., G. M. McKay, and J. F. Eisenberg, 1972. Observations on musth in the domesticated Asiatic elephant (*Elephas maximus*). *Mammalia* 36: 247–61.

Janzen, D. H. 1975. *Ecology of Plants in the Tropics*. London: Arnold.

Jarman, P. J. 1972. Seasonal distribution of large mammal populations in the unflooded Middle Zambezi Valley. *Journal of Applied Ecology* 9(1): 283–99.

1974. The social organisation of antelope in relation to their ecology. *Behaviour* 48: 215–67.

Jennings, J. D. 1974. *Prehistory of North America*, 2nd ed. New York: McGraw-Hill.

Jochim, M. A. 1976. *Hunter-Gatherer Subsistence and Settlement: A Predictive Model*. New York: Academic Press.

1981. *Strategies for Survival*. New York: Academic Press.

Johnson, D. L., P. Kawano, and E. Ekker. 1980. Clovis strategies of hunting mammoth (*Mammuthus columbi*). *Canadian Journal of Anthropology* 1(1): 107–14.

Johnson, E. 1978. Paleo-Indian procurement and butchering patterns on the Llano Estacado. *Plains Anthropologist Memoirs* 14: 98–105.

1985. Current developments in bone technology. In M. B. Schiffer, ed., *Advances in Archaeological Method and Theory*, Vol. 8, pp. 157–235. Orlando, Fla.: Academic Press.

1989. Human-modified bones from early southern High Plains sites. In R. Bonnichsen and M. Sorg (eds.), *Bone Modification*, pp. 431–71. Orono, Maine: Center for the Study of the First Americans.

Johnson, E., and V. T. Holliday. 1985. A Clovis-age megafaunal processing station at the Lubbock Lake Landmark. *Current Research in the Pleistocene* 2: 17–19.

Johnson, E., and P. Shipman. 1986. Scanning electron microscope studies of bone modification. *Current Research in the Pleistocene* 3: 47–8.

Johnson, O. W., and I. O. Buss. 1965. Molariform teeth of male African elephants in relation to age, body dimensions, and growth. *Journal of Mammalogy* 40: 373–84.

Jones, R. L., and H. C. Hanson. 1985. *Mineral Licks, Geophagy, and Biogeochemistry of North American Ungulates*. Ames: Iowa State University Press.

Kapp, R. O., D. L. Cleary, G. G. Snyder, and D. C. Fisher. 1990. Vegetational and climatic history of the Crystal Lake area and the Eldridge mastodont site, Montcalm County, Michigan. *The American Midland Naturalist* 123: 47–63.

Kashchenko, N. 1896. *Ein von Menschen verzehrtes Mammuth*. Correspondenz-Blatt de deutschen Gesellschaft für Anthropologie, Ethnologie und Urgesch No. 45. Berlin.

1901. *Skelyet' mamonta so sledami upotrobleniya nekotorikh chastey tyela etovo zhivotnovo v pishchy sovremennym yemi chelovekom*. Zapiski Imperatorskoy Akademii Nauk VIII Series, Po Fiziko-Matematicheskomu Otdeleniyu, Tom XI, No. 7.

Kaufulu, Z. M. 1990. Sedimentary environments at the Mwanganda Site, Malawi. *Geoarchaeology* 5(1): 15–27.

Kaye, C. A. 1962. Early post-glacial beavers in southeastern New England. *Science* 138: 906–7.

Kearney, J. M. 1907. The ruins at Bumbusi. *Proceedings and Transactions of the Rhodesia Scientific Association* 7(1): 59–62.

Kelsall, J. P. 1968. *The Migratory Barren-Ground Caribou of Canada*. Ottawa: Queen's Printer.

Kiltie, R. A. 1984. Seasonality, gestation time, and large mammal extinctions. In P. S. Martin and R. G. Klein, eds., *Quaternary Extinctions: A Prehistoric Revolution*, pp. 299–314. Tucson: University of Arizona Press.

King, J. E., and J. J. Saunders. 1984. Environmental insularity and the extinction of the American mastodont. In P. S. Martin and R. G. Klein, eds., *Quaternary Extinctions: A Prehistoric Revolution*, pp. 315–39. Tucson: University of Arizona Press.

Kipling, R. 1921. *Rudyard Kipling's Verse: Inclusive Edition, 1885–1918*. Garden City, N.Y.: Doubleday, Page & Co.

Kitching, J. W. 1963. *Bone, Tooth, and Horn Tools of Palaeolithic Man: An Account of the Osteodontokeratic Discoveries in Pin Hole Cave, Derbyshire.* Manchester University Press.

Klein, R. G. 1973. *Ice-Age Hunters of the Ukraine.* University of Chicago Press.

1982. Age (mortality) profiles as a means of distinguishing hunted species from scavenged ones in stone age archeological sites. *Paleobiology* 8: 151–8.

1984. The large mammals of southern Africa: Late Pleistocene to Recent. In R. G. Klein, ed., *Southern African Prehistory and Paleoenvironments*, pp. 107–46. Rotterdam: A. A. Balkema.

1987. Problems and prospects in understanding how early people exploited animals. In M. H. Nitecki and D. V. Nitecki, eds., *Evolution of Human Hunting*, pp. 11–45. New York: Plenum Press.

Klein, R. G., and K. Cruz-Uribe. 1984. *The Analysis of Animal Bones from Archaeological Sites.* University of Chicago Press.

Klevezal, G. A., and S. E. Kleinenberg. 1967. *Age Determination of Mammals From Layered Structures in Teeth and Bone* (Israel Program for Scientists translation). Moscow: Nauka.

Klima, B. 1954. Palaeolithic huts at Dolní Vestonice, Czechoslovakia. *Antiquity* 109: 4–14.

1963. *Dolní Vestonice: Vyzkum Taboriste Lovcu Mamutu v Letech 1947–1952.* Prague: Nakladatelsví Cekeslovenské Akademie Ved.

Knight, M. H., A. K. Knight-Eloff, and J. J. Bornman. 1988. The importance of borehole water and lick sites to Kalahari ungulates. *Journal of Arid Environments* 15: 269–81.

Koch, A. C. 1839. Remains of the mastodon in Missouri. *The American Journal of Science and Arts* (first series) 37: 191–2.

Koch, P. L. 1988. The diet of Pleistocene proboscideans and its role in their extinction. In *The Geological Society of America 1988 Centennial Celebration Program*, p. A378.

Koch, P. L., D. C. Fisher, and D. Dettman. 1989. Oxygen isotope variation in the tusks of extinct proboscideans: A measure of season of death and seasonality. *Geology* 17: 515–19.

Koen, J. H. 1988. Trace elements and some other nutrients in the diet of the Knysna elephants. *South African Journal of Wildlife Research* 18(3): 109–10.

Koen, J. H., A. J. Hall-Martin, and T. Erasmus. 1988. Macro nutrients in plants available to Knysna, Addo, and Kruger National Park elephants. *South African Journal of Wildlife Research* 18(2): 69–71.

Kozlowski, J. K. 1986. The Gravettian in central and eastern Europe. In F. Wendorf and A. Close, eds., *Advances in World Archaeology, Vol. 5*, pp. 131–200. New York: Academic Press.

Kozlowski, J. K., H. Kubiak, E. Sachse-Kozlowska, B. van Vliet, and G. Zakrzewska. 1974. Upper Palaeolithic site with dwellings of mammoth bones, Kraków-Spadzista Street B. *Folia Quaternaria* 44: 1–110.

Kozlowski, J. K., A. Welc, and H. Kubiak. 1970. A Paleolithic site with mammoth remains at Nowa Huta. *Folia Quaternaria* 36:1.

Kubiak, H. 1980. The skulls of *Mammuthus primigenius* (Blumenbach) from Debica and Bzianka near Rzeszów, South Poland. *Folia Quaternaria* 51: 31–45.

1982. Morphological characters of the mammoth: An adaptation to the Arctic-steppe environment. In D. M. Hopkins, J. V. Mattews, Jr., C. E. Schweger, and S. B. Young, eds., *Paleoecology of Beringia*, pp. 281–9. New York: Academic Press.

Kubiak, H., and G. Zakrzewska. 1974. Fossil mammals. *Folia Quaternaria* 44: 77–110.

Kunz, G. F. 1916. *Ivory and the Elephant in Art, in Archaeology, and in Science.* Garden City, N.Y.: Doubleday, Page & Co.

Kurtén, B. 1964. Population structure in paleoecology. In J. Imbrie and N. Newell, eds., *Approaches to Paleoecology*, pp. 91–106. New York: Wiley.

1968. *Pleistocene Mammals of Europe.* Chicago: Aldine.

Kurtén, B., and E. Anderson. 1980. *Pleistocene Mammals of North America.* New York: Columbia University Press.

Kuznetsof (or Kuznezow), S. K. 1896. Fund eines Mammuthskeletes und menschlicher Spuren in der Nähe der Stadt Tomsk (Westsibirien). *Mittheilungen anthropologischen Gesellschaft in Wein* 26: 186–191.

Lahren, L., and R. Bonnichsen. 1974. Bone foreshafts from a Clovis burial in southwestern Montana. *Science* 186: 147–50.

Lamprey, H. F., P. E. Glover, M. I. M. Turner, and R. H. V. Bell. 1967. Invasion of the Serengeti National Park by elephants. *East African Wildlife Journal* 5: 151–66.

Lang, E. M. 1980. Observations on growth and molar change in the African elephant. *African Journal of Ecology* 18: 217–34.

Lang, H. 1925. Problems and facts about frozen Siberian mammoths (*Elephas primigenius*) and their ivory. *Zoologica* 4: 25–53.

Lark, R. M. 1984. A comparison between techniques for estimating the ages of African elephants (*Loxodonta africana*). *African Journal of Ecology* 22(1): 69–71.

Laws, R. M. 1966. Age criteria for the African elephant *Loxodonta a. africana. East African Wildlife Journal* 4: 1–37.

Laws, R. M., and I. S. C. Parker. 1969. Recent studies on elephant populations in East Africa. *Symposia of the Zoological Society of London* 20: 1–59.

Laws, R. M., I. S. C. Parker, and R. C. B. Johnstone. 1975. *Elephants and Their Habitats: The Ecology of Elephants in North Bunyoro, Uganda.* Oxford: Clarendon Press.

Leakey, M. 1971. *Olduvai Gorge. Vol. 3: Excavations in Beds I and II, 1960–1963.* Cambridge University Press.

Lee, R. B. 1979. *The!Kung San: Men, Women, and Work in a Foraging Society.* Cambridge University Press.

Leechman, D. 1950. An implement of elephant bone from Manitoba. *American Antiquity* 2: 157–8.

Leonhardy, F. C., ed. 1966. *Domebo: A Paleo-Indian Mammoth Kill in the Prairie-Plains.* Lawton, Oklahama: Museum of the Great Plains Contribution No. 1.

Leonhardy, F. C., and A. Anderson. 1966. The archeology of the Domebo site. In F. C. Leonhardy, ed., *Domebo: A Paleo-Indian Mammoth Kill in the Prairie-Plains*, pp. 14–26. Lawton, Oklahoma: Museum of the Great Plains Contribution No. 1.

Leuthold, W. 1977. *African Ungulates: A Comparative Review of Their Ethology and Behavioral Ecology*. Berlin: Springer-Verlag.

Leuthold, W., and J. B. Sale. 1973. Movements and patterns of habitat utilization of elephants in Tsavo National Park, Kenya. *East African Wildlife Journal* 11: 369–84.

Lewis, E. 1927. *Trader Horn*. Garden City, N.Y.: Garden City Publishing Co.

Lewis-Williams, J. D. 1984. Ideological continuities in prehistoric southern Africa: The evidence of rock art. In C. Schrire, ed., *Past and Present in Hunter Gatherer Studies*, pp. 225–52. Orlando, Fla.: Academic Press.

Lister, A. M. 1989. Proboscidean evolution. *Trends in Ecology and Evolution* 4[12(42)]: 362–3.

——— 1990. Late glacial mammoths in Britain. In N. Barton, D. Roe, and A. Roberts, eds., *The Late Glacial of Northwest Europe*. London: Council for British Archaeology.

——— in press. Pelvic sexing and skeletal maturation in the mammoth. In J. Shoshani and P. Tassy, eds., *Evolution and Paleoecology of the Proboscidea*. Oxford University Press.

Liu Tung-Sheng and Li Xing-guo. 1984. Mammoths in China. In P. S. Martin and R. G. Klein, eds., *Quaternary Extinctions: A Prehistoric Revolution*, pp. 517–27. Tucson: University of Arizona Press.

Lomax, A., and R. Abdul, eds. 1970. *Three Thousand Years of Black Poetry: An Anthology*. New York: Dodd, Mead.

Lomolino, M. V. 1985. Body size of mammals on islands: The island rule reexamined. *American Naturalist* 125: 310–16.

Lucas, F. A. 1902. *Animals Before Man in North America: Their Lives and Times*. New York: Appleton.

——— 1929. *Animals of the Past: An Account of Some of the Creatures of the Ancient World*, 7th ed. New York: American Museum of Natural History, Handbook Series No. 4.

Lydekker, R. 1908. *The Game Animals of Africa*. London: Rowland Ward.

MacCalman, H. R. 1967. The Zoo Park elephant site, Windhoek (1964–1965). *Palaeoecology of Africa* 2: 102–3.

McClellan, T. L. 1986. A Syrian fortress of the Bronze Age: el-Qitar. *National Geographic Research* 2: 418–40.

McCullagh, K. G. 1972. Arteriosclerosis in the African elephant. Part 1: Intimal atherosclerosis and its possible causes. *Atherosclerosis* 16: 307–35.

McDonald, J. N. 1984. The reordered North American selection regime and late Quaternary megafaunal extinctions. In P. S. Martin and R. G. Klein, eds., *Quaternary Extinctions: A Prehistoric Revolution*, pp. 404–39. Tucson: University of Arizona Press.

McHugh, T. (with the assistance of V. Hobson). 1972. *The Time of the Buffalo*. Lincoln: University of Nebraska Press.

MacInnes, C. M. 1930. *In the Shadow of the Rockies*. London: Rivington.

McKay, G. M. 1973. *Behavior and Ecology of the Asiatic Elephant in Southeastern Ceylon*. Smithsonian Contributions to Zoology No. 125.

McKnight, T. L. 1969. *The Camel in Australia*. Melbourne University Press.

McLean, D. M. 1986. Embryogenesis dysfunction in the Pleistocene/Holocene transition: Mammalian extinctions, dwarfing, and skeletal abnormality. In J. N. McDonald and S. O. Bird, eds., *The Quaternary of Virginia – A Symposium Volume*, pp. 105–20. Virginia Division of Mineral Resources Publication No. 75.

McMahon, T. A. 1973. Size and shape in biology. *Science* 179: 1201–4.

1975a. Allometry and biomechanics: Limb bones in adult ungulates. *The American Naturalist* 109(969): 547–63.

1975b. Using body size to understand the structural design of animals: Quadrupedal locomotion. *Journal of Applied Physiology* 39: 619–27.

McMillan, R. B. 1977. Man and mastodon: A review of Koch's 1840 Pomme de Terre expedition. In W. R. Wood and R. B. McMillan, eds., *Prehistoric Man and His Environments: A Case Study in the Ozark Highland*, pp. 81–96. New York: Academic Press.

Madden, C. M. 1981. Mammoths of North America. Unpublished Ph.D. dissertation, University of Colorado.

Maddock, L. 1979. The "migration" and grazing succession. In A. R. E. Sinclair and M. Norton-Griffiths, eds., *Serengeti: Dynamics of an Ecosystem*, pp. 104–29. University of Chicago Press.

Maglio, V. J. 1972. Evolution of mastication in the Elephantidae. *Evolution* 26(4): 638–58.

1973. Origin and evolution of the Elephantidae. *Transactions of the American Philosophical Society* (new series) 63(3): 1–149.

Mahboubi, M., R. Ameur, J. Y. Crochet, and J. J. Jaeger. 1984. Earliest known proboscidean from early Eocene of north-west Africa. *Nature* 308: 543–4.

Maloiy, G. M. O. 1972. Renal salt and water excretion in the camel *Camelus dromedarius*. *Symposia of the Zoological Society of London* 31: 243–59.

Malpas, R. C. 1978. The ecology of the African elephant in Rwenzori and Kabalega Falls National Parks. Unpublished Ph.D. thesis, Cambridge University.

Marean, C. W., and C. L. Ehrhardt. 1990. Paleoecology of extinct carnivores and implications for actualistic models. Paper presented at the Society of Africanist Archaeologists biennial conference, 22–25 March 1990, Gainesville, Fla.

Marks, S. A. 1976. *Large Mammals and a Brave People*. Seattle: University of Washington Press.

Marshall, L. G. 1984. Who killed Cock Robin? An investigation of the extinction controversy. In P. S. Martin and R. G. Klein, eds., *Quaternary Extinctions: A Prehistoric Revolution;* pp. 785–806. Tucson: University of Arizona Press.

Marshall, L. G., and R. S. Corruccini. 1978. Variability, evolutionary rates, and allometry in dwarfing lineages. *Paleobiology* 4: 101–19.

Martin, J. E. 1987. Paleoenvironment of the Lange-Ferguson Clovis kill site in the Badlands of South Dakota. In R. W. Graham, H. A. Semken, Jr., and M. A. Graham, eds., *Late Quaternary Mammalian Biogeography and Environments of the Great Plains and Prairies*, pp. 314–32. Springfield: Illinois State Museum Scientific Papers Vol. 22.

Martin, L. D., and A. M. Neuner. 1978. The end of the Pleistocene in North America. *Transactions of the Nebraska Academy of Sciences* 6: 117–26.

Martin, P. S. 1967. Prehistoric overkill. In P. S. Martin and H. E. Wright, Jr., eds., *Pleistocene Extinctions: The Search for a Cause*, pp. 75–120. New Haven, Conn.: Yale University Press.

1973. The discovery of America. *Science* 179: 969–74.

1984. Prehistoric overkill: The global model. In P. S. Martin and R. G. Klein, eds., *Quaternary Extinctions: A Prehistoric Revolution*, pp. 354–403. Tucson: University of Arizona Press.

Martin, P. S., and R. G. Klein, eds. 1984. *Quaternary Extinctions: A Prehistoric Revolution.* Tucson: University of Arizona Press.

Martin, P. S., and H. E. Wright, Jr., eds. 1967. *Pleistocene Extinctions: The Search for a Cause.* New Haven, Conn.: Yale University Press.

Martin, R. B. 1978. Aspects of elephant social organisation. *Rhodesia Science News* 12: 184–7.

1983. Zimbabwe completes tenth year of elephant radio-tracking. *(IUCN) African Elephant and Rhino Group Newsletter* 2: 5–7.

Matthiesson, P., and E. Porter 1974. *The Tree Where Man Was Born: The African Experience.* New York: Avon.

May, D. W. 1990. Geomorphology and late Quaternary stratigraphy of Paleoindian sites in the Medicine Creek Basin. In *Society for American Archeology 55th annual meeting program and abstracts*, p. 125.

Mead, J. I., and L. D. Agenbroad. 1990. Pleistocene dung and the extinct herbivores of the Colorado Plateau. *Cranium* (in press).

Mead, J. I., L. D. Agenbroad, O. K. Davis, and P. S. Martin. 1986. Dung of *Mammuthus* in the arid Southwest, North America. *Quaternary Research* 25: 121–7.

Mead, J. I., and D. I. Meltzer. 1984. North American late Quaternary extinctions and the radiocarbon record. In P. S. Martin and R. G. Klein, eds., *Quaternary Extinctions: A Prehistoric Revolution*, pp. 440–50. Tucson: University of Arizona Press.

Mech, L. D., and L. D. Frenzel, eds. 1971. *Ecological Studies of the Timber Wolf in Northeastern Minnesota.* USDA Forest Service Research Paper NC-52.

Mehl, M. G. 1966. The Domebo mammoth: Vertebrate paleomortology. In F. C. Leonhardy, ed., *Domebo: A Paleo-Indian Mammoth Kill in the Prairie-Plains*, pp. 27–30. Lawton, Oklahoma: Museum of the Great Plains Contribution No. 1.

1967. *The Grundel Mastodon.* Missouri Geological Survey and Water Resources Report of Investigations No. 35.

1975. Vertebrate paleomortology of the Cooperton site. In A. D. Anderson, ed., *The Cooperton Mammoth: An Early Man Bone Quarry. Great Plains Journal* 14(2): 130–73.

Mehringer, P. J., Jr. 1967. The environment of extinction of the late-Pleistocene megafauna in the arid southwestern United States. In P. S. Martin and H. E. Wright, Jr., eds., *Pleistocene Extinctions: The Search for a Cause*, pp. 247–66. New Haven, Conn.: Yale University Press.

Meier, A. H., and A. J. Fivizzani. 1980. Physiology of migration. In S. A. Gauthreaux, Jr., ed., *Animal Migration, Orientation, and Navigation*, pp. 225–82. New York: Academic Press.

Miller, S. J. 1989. Characteristics of mammoth bone reduction at Owl Cave, the Wasden site, Idaho. In R. Bonnichsen and M. Sorg (eds.), *Bone Modification*, pp. 381–93. Orono, Maine: Center for the Study of the First Americans.

Miller, S. J., and W. Dort, Jr. 1978. Early Man at Owl Cave: Current investigations at the Wasden site, eastern Snake River Plain, Idaho. In A. L. Bryan (ed.), *Early Man in America from a Circum-Pacific Perspective*, pp. 129–39. University of Alberta Department of Anthropology Occasional Paper No. 1. Edmonton: University of Alberta.

Miller, W. D. 1890a. Studies on the anatomy and pathology of the tusks of the elephant. *The Dental Cosmos* 32(5). 337–48.

  1890b. Studies on the anatomy and pathology of the tusks of the elephant (continued). *The Dental Cosmos* 32(6): 421–9.

Mingay, G. E. 1977. *Rural Life in Victorian England*. London: Heinemann.

Minshall, H. 1988. Comments on "Mammoth Rib Review." *Pacific Coast Archaeological Society Newsletter* 27(10): 3.

Mitchell, P. C. 1916. On the intestinal tract of mammals. *Proceedings of the Zoological Society of London* 86: 210–11.

Möbius, K. 1892. Die Behaarung des Mammuths und der lebenden Elephanten, vegleichend Untersucht. *Sitzungsherichte der Königlisch Preussischen Academie der Wissenschaften zu Berlin* 1: 527–38.

Mochanov, Yu. A. 1977. *Drevneyshiye etapi zaseleniya chelovekom Severo-Vostochnoy Azii*. Novosibirsk: Nauka.

Mochanov, Yu. A., S. Fedoseyeva, and A. Alekseyev. 1983. *Arkheologicheskiye pamyatniki yakutii: Basseyni Aldana i Olyokmi*. Novosibirsk: Nauka.

Moehlman, P. D. 1979. Behavior and ecology of feral asses (*Equus asinus*). *National Geographic Society Research Reports (1970)*, pp. 405–11.

Moeyersons, J. 1978. The behavior of stones and stone implements, buried in consolidating and creeping Kalahari sands. *Earth Surface Processes* 3: 115–28.

Mohr, E. 1971. *The Asiatic Wild Horse* (translated by D. M. Goodall). London: J. A. Allen & Co.

Morlan, R. E. 1980. *Taphonomy and Archaeology in the Upper Pleistocene of the Northern Yukon Territory*. Archaeological Survey of Canada paper No. 94.

  1986. Pleistocene archaeology in Old Crow Basin: A critical appraisal. In A. L. Bryan (ed.), *New Evidence for the Pleistocene Peopling of the Americas*, pp. 27–48. Orono, Maine: Center for the Study of Early Man.

Morlan, R. E., and J. Cinq-Mars. 1982. Ancient Beringians: Human occupation in the late Pleistocene of Alaska and the Yukon Territory. In D. M. Hopkins, J. V. Matthews, Jr., C. E. Schweger, and S. B. Young, eds., *Paleoecology of Beringia*, pp. 353–81. New York: Academic Press.

Morrison-Scott, T. C. S. 1947. A revision of our knowledge of African elephants' teeth with notes on forest and "pygmy" elephants. *Proceedings of the Zoological Society of London* 117: 505–27.

Moss, C. 1976. *Portraits in the Wild: Animal Behaviour in East Africa*. London: Hamish Hamilton.

  1977. The Amboseli elephants. *Wildlife News* 12: 9–12.

  1982. *Portraits in the Wild: Behavior Studies of East Africa Mammals*, 2nd ed. University of Chicago Press.

1983. Oestrous behavior and female choice, in the African elephant. *Behaviour* 86(3/4): 167–96.

1988. *Elephant Memories: Thirteen Years in the Life of an Elephant Family.* New York: Morrow.

Movius, H. L. 1950. A wooden spear of Third Interglacial age from lower Saxony. *Southwestern Journal of Anthropology* 6(2): 139–42.

Müller-Beck, H. 1982. Late Pleistocene man in northern Eurasia and the mammoth–steppe biome. In D. M. Hopkins, J. V. Matthews, Jr., C. E. Schweger, and S. B. Young, eds., *Paleoecology of Beringia*, pp. 329–52. New York: Academic Press.

Murie, A. 1934. *The Moose of Isle Royale.* University of Michigan Museum of Zoology Miscellaneous Publication No. 25.

Nelson, D., R. Morlan, J. Vogel, J. Southon, and C. Harington. 1986. New dates on northern Yukon artifacts: Holocene, not Upper Pleistocene. *Science* 232: 749–51.

Neville, A. C. 1967. Daily growth layers in animals and plants. *Biology Review* 42: 421–41.

Nilsson, T. 1983. *The Pleistocene: Geology and Life in the Quaternary Ice Age.* Dordrecht: D. Reidel.

Nordenskiold, A. E. 1882. *The Voyage of the Vega Round Asia and Europe, With a Historical Review of Previous Journeys Along the North Coast of the Old World.* New York: Macmillan.

Oberjohann, H. 1952. *Komoon! Capturing the Chad Elephant* (translated by R. de Terra). New York: Pantheon Books.

O'Brien, P. J. 1978. A mastodon from Stanton County, Kansas. *Kansas Academy of Science Transactions* 81(4): 371–3.

O'Connell, J. 1987. A year with the Hadza. Paper presented at the annual meeting of the Nevada Archeological Association, Fallon.

Okladnikov, A. 1953. *Sledi palyeolit v doline Reki Leni.* Materiali i Issledovaniya po Arkheologii SSSR, No. 39.

Olivier, R. C. D. 1978. On the ecology of the Asian elephant *Elephas maximus* Linn with particular reference to Malaya and Sri Lanka. Unpublished Ph.D. thesis, Cambridge University.

1982. Ecology and behavior of living elephants: Bases for assumptions concerning the extinct woolly mammoth. In D. M. Hopkins, J. V. Matthews, Jr., C. E. Schweger, and S. B. Young, eds., *Paleoecology of Beringia*, pp. 291–305. New York: Academic Press.

Olsen, S. J. 1972. Osteology for the archeologist. 3: The American mastodon and the woolly mammoth. *Howard University, Papers of the Peabody Museum of Archeology and Ethnology* 56(3): 1–47.

Olsen, S. L., and P. Shipman. 1988. Surface modification on bone: Trampling versus butchering. *Journal of Archaeological Science* 15: 535–53.

Oosenbrug, S., L. Carbyn, and D. West. 1980a. Wood Buffalo National Park wolf/bison studies program progress report No. 1. Edmonton: Canadian Wildlife Service.

1980b. Wood Buffalo National Park wolf/bison studies program progress report No. 2. Edmonton: Canadian Wildlife Service.

Orr, R. T. 1970. *Animals in Migration*. New York: Macmillan.

Osborn, H. F. 1925. *The Elephants and Mastodonts Arrive in America*. American Museum of Natural History Guide Leaflet Series No. 25(1).

1936. *Proboscidea: A Monograph of the Discovery, Evolution, Migration and Extinction of the Mastodonts and Elephants of the World. Vol. I: Moeritherioidea, Deinotherioidea, Mastodontoidea* (edited by M. R. Percy). New York: American Museum of Natural History.

1942. *Proboscidea: A Monograph of the Discovery, Evolution, Migration and Extinction of the Mastodonts and Elephants of the World. Vol. II: Stegodontoidea, Elephantoidea*. New York: American Museum of Natural History.

Owen-Smith, N. 1987. Pleistocene extinctions: The pivotal role of megaherbivores. *Paleobiology* 13(3): 351–62.

Pace, R. E. 1976. Haley mammoth site, Vigo County: A preliminary report. *Indiana Academy of Science* 85: 63.

Page, W. D. 1978. The geology of the El Bosque archaeological site, Nicaragua. In A. L. Bryan, ed., *Early Man in America from a Circum-Pacific Perspective*, pp. 231–60. Department of Anthropology, University of Alberta, Occasional Papers, No. 1.

Payne, K. B., W. R. Langbauer, Jr., and E. M. Thomas. 1986. Infrasonic calls of the Asian elephant (*Elephas maximus*). *Behavioral Ecology and Sociobiology* 18: 297–301.

Peale, R. 1803. *An Historical Disquisition on the Mammoth, or Great American Incognitum, an Extinct, Immense, Carnivorous Animal Whose Fossil Remains Have Been Found in North America*. London: E. Lawrence.

Pennycuick, C. J. 1979. Energy costs of locomotion and the concept of "foraging radius." In A. R. E. Sinclair and M. Norton-Griffiths, eds., *Serengeti: Dynamics of an Ecosystem*, pp. 164–84. University of Chicago Press.

Perry, J. S. 1954. Some observations on growth and tusk weight in male and female African elephants. *Proceedings of the Zoological Society of London* 124: 97–104.

Peters, R. H. 1983. *The Ecological Implications of Body Size*. Cambridge University Press.

Peterson, R. L. 1955. *North American Moose*. University of Toronto Press.

Peterson, R. O. 1977. *Wolf Ecology and Prey Relationships of Isle Royale*. National Park Service Scientific Monograph Series No. 11.

1986. *Ecological Studies of Wolves on Isle Royale. Annual Report 1985–86*. Houghton: Michigan Technological University.

Péwé, T. L. 1975. *Quaternary Geology of Alaska*. U.S. Geological Survey Professional Paper No. 835. Washington, D.C.: U.S. Government Printing Office.

Pfizenmayer, E. W. 1939. *Siberian Man and Mammoth*. London: Blackie & Sons.

Pianka, E. R. 1970. On r- and K-selection. *American Naturalist* 104: 592–7.

Pidoplichko, I. G. 1969. *Pozdnepalyeoliticheskiye zhilishcha iz kostey mamonta na Ukraine*. Kiev: Naukova Dumka.

Pienaar, U. de V. 1969. Predator–prey relationships amongst the larger mammals of the Kruger National Park. *Koedoe* 12: 108–76.

Pilgram, T., and D. Western. 1986. Inferring the sex and age of African elephants from tusk measurements. *Biological Conservation* 36: 39–52.

Pimlott, D. H., J. A. Shannon, and G. B. Kolenosky. 1969. *The Ecology of the Timber Wolf in Algonquin Provincial Park*. Ontario Department of Lands and Forests Research Report (Wildlife), No. 87.

Pohorecky, Z., and J. Wilson. 1968. Preliminary archeological report on the Saskatoon site. *Na'pao* 1(2): 35–70.

Poole, J. H. 1982. Musth and male–male competition in the African elephant. Unpublished Ph.D. thesis, Cambridge University.

1987. Elephants in musth, lust. *Natural History* 96(11): 47–55.

Poole, J. H., L. H. Kasman, E. C. Ramsay, and B. L. Lasley. 1984. Musth and urinary testosterone concentration in the African elephant (*Loxodonta africana*). *Journal of Reproduction and Fertility* 70: 255–60.

Poole, J. H., and C. Moss. 1981. Musth in the African elephant *Loxodonta africana*. *Nature* 292: 830–1.

Poplin, F. 1976. Les grands vertébrés de Gönnersdorf. Fouilles 1968. In G. Bosinski, ed., *Der Magdalénien-Fundplatz Gönnersdorf*, Vol. *1*, pp. 1–212. Weisbaden: Franz Steiner Verlag.

1978. Données de la grande faune sur le climat et l'environment. In G. Bosinski, ed., *Der Magdalénien-Fundplatz Gönnersdorf*, Vol. *2*, pp. 98–104. Weisbaden: Franz Steiner Verlag.

Porter, S. C., ed. 1983. *Late Quaternary Environments of the United States*. Vol. *1*: *The Late Pleistocene*. Minneapolis: University of Minnesota Press.

Potts, R., and P. Shipman. 1981. Cutmarks made by stone tools on bones from Olduvai Gorge, Tanzania. *Nature* 291: 577–80.

Praslov, N. D., and A. Rogachev, eds. 1982. *Paleolit Kostyonkovsko – Borshchevskovo rayona na Donu 1879–1979*. Leningrad: Nauka.

Puckette, W. L., and J. A. Scholtz. 1974. The Hazen mammoth *Mammuthus columbi*, Prairie County, Arkansas, USA. *Arkansas Academy of Sciences Proceedings* 28: 53–6.

Radmilli, A. M. 1982. Notizario: Castel di Guido (Roma). *Rivista di Scienze Preistoriche* 37: 307–8.

Rainey, F. G. 1940. Archeological investigations in central Alaska. *American Antiquity* 5(4): 299–308.

Rancier, J., G. Haynes, and D. Stanford. 1982. 1981 investigations at Lamb Spring. *Southwestern Lore* 8(2): 1–17.

Rasmussen, L. E., I. O. Buss, D. L. Hess, and M. J. Schmidt. 1984. Testosterone and dihydrotestosterone concentrations in elephant serum and temporal gland secretions. *Biology of Reproduction* 30: 352–62.

Rattray, J. M., R. M. M. Cormack, and R. R. Staples. 1953. The vlei areas of southern Rhodesia and their uses. *Rhodesia Agricultural Journal* 50: 465–83.

Redford, K. H., and J. G. Johnson. 1985. Hunting by indigenous peoples and conservation of game species. *Cultural Survival Quarterly* 9(1): 41–4.

Reese, D. S. 1982. Faunal remains as items of "trade" in Mediterranean archaeology (abstract). *American Journal of Archaeology* 86(2): 281–2.

1984. Strange and wonderful exotic fauna from sanctuary sites. *American School of Classical Studies at Athens Newsletter* (Fall): 13.

1985. Appendix VIII (D): Hippopotamus and elephant teeth from Kition. In V.

Karageorghis, ed., *Excavations at Kition V. Part II*, pp. 391–409. Nicosia: Department of Antiquities.

n.d. The dwarfed elephant of Cyprus. Unpublished manuscript in possession of the author.

Rhodes, R. S., III. 1984. *Paleoecology and Regional Paleoclimatic Implications of the Farmdalian Craigmile and Woodfordian Waubonsie Mammalian Local Faunas, Southwestern Iowa.* Illinois State Museum Reports of Investigations, No. 40.

Richardson, P. R. K. 1980. Carnivore damage to antelope bones and its archeological implications. *Palaeontologia Africana* 23:109–25.

Roberts, A. 1951. *The Mammals of South Africa.* Cape Town: Mammals of South Africa Trust.

Robinson, C. W., and P. D. Krynine. 1938. A new mastodon locality at Saltillo, Huntingdon County, Pennsylvania. *Pennsylvania Academy of Sciences Proceedings* 12: 93–6.

Robinson, K. R. 1966. The Iron Age site in Kapula vlei, near the Masuma dam, Wankie Game Reserve, Rhodesia. *Arnoldia Rhodesia* 2(39): 1–7.

Rogers, R. A., and L. D. Martin. 1989. Possible proboscidean bone artifact from the Kansas River. *Current Research in the Pleistocene* 6: 43–4.

Roth, V. L. 1982. Dwarf mammoths from the Santa Barbara, California, Channel Islands: Size, shape, development and evolution. Unpublished Ph.D. dissertation, Yale University.

1984. How elephants grow: Heterochrony and calibration of developmental stages in some living and fossil species. *Journal of Vertebrate Paleontology* 4(1): 126–45.

1989. Fabricational noise in elephant dentitions. *Paleobiology* 15(2): 165–79.

in press. Insular dwarf elephants – a case study in body mass estimation and ecological inferences. In J. Damuth and B. J. MacFadden, eds., *Body Size in Mammalian Paleobiology: A Conceptual and Empirical Tool.* Cambridge University Press.

Roth, V. L., and J. Shoshani. 1988. Dental identification and age determination in *Elephas maximus. Journal of Zoology (London)* 214: 567–88.

Rouse, I., and J. M. Cruxent. 1963. *Venezuelan Archaeology.* New Haven, Conn.: Yale University Press.

Ruddiman, W. F., and A. McIntyre. 1981. The mode and mechanism of the last deglaciation: Oceanic evidence. *Quaternary Research* 16: 125–34.

Ruddiman, W. F., and H. E. Wright, Jr. 1987. *The Geology of North America, Vol. K3: North America and Adjacent Oceans during the Last Deglaciation.* Denver: Geological Society of America.

Rugg, H., and L. Krueger. 1936. *Nature Peoples. Man and His Changing Society, Vol. II.* Boston: Ginn & Co.

Rushby, G. G. 1953. The elephant in Tanganyika. In W. C. O. Hill et al., eds., *The Elephant in East Central Africa: A Monograph*, pp. 126–42. London: Rowland Ward.

1965. *No More the Tusker.* London: W. H. Allen.

Rushworth, J. E. 1975. The floristic, physiognomic, and biomass structure of Kalahari sand scrub vegetation in relation to fire and frost in Wankie National Park, Rhodesia. Unpublished M.Sc. thesis, University of Rhodesia.

1978. Kalahari sand scrub – something of value. *Rhodesia Science News* 12(8): 193–5.

Ryder, M. L. 1974. Hair of the mammoth. *Nature* 249: 190–2.

Sampson, C. G. 1988. *Stylistic Boundaries Among Mobile Hunter-Foragers.* Washington, D.C.: Smithsonian Institution Press.

Santonja, M., and P. Villa. 1990. The lower paleolithic of Spain and Portugal. *Journal of World Prehistory* 4(1): 45–94.

Saunders, J. J. 1970. The distribution and taxonomy of *Mammuthus* in Arizona. Unpublished M.Sc. thesis, University of Arizona.

1977a. Lehner Ranch revisited. In E. Johnson, ed., *Paleo-Indian Lifeways. The Museum Journal* 17: 48–64

1977b. *Late Pleistocene Vertebrates of the Western Ozark Highland, Missouri.* Illinois State Museum Reports of Investigations, No. 33.

1979. A close look at ivory. *The Living Museum* 41(4): 56–9.

1980. A model for man–mammoth relationships in Late Pleistocene North America. *Canadian Journal of Anthropology* 1(1): 87–98.

1984. Late Pleistocene mastodons of North America. Paper presented at the Society for American Archeology annual meeting, 12–14 April, Portland, Oregon.

1985. A Clovis ivory artifact. *The Living Museum* 47(4): 50–2.

1987. Britain's newest mammoths. *Nature* 330: 419.

Saunders, J. J., C. V. Haynes, Jr., D. Stanford, and G. A. Agogino. 1990. A mammoth-ivory semifabricate from Blackwater Locality No. 1, New Mexico. *American Antiquity* 55(1):112–19.

Schäfer, W. 1972. *Ecology and Paleoecology of Marine Environments* (edited by G. Craig; translated by I. Oertel). University of Chicago Press.

Schebesta, P. 1932. *Bambuti: die Zwerge vom Kongo.* Leipzig: Brodhaus.

Schild, R. 1984. Terminal paleolithic of the north European Plain: A review of lost chances, potential and hopes. In F. Wendorf and A. Close, eds., *Advances in World Archeology, Vol. 3*, pp. 193–274. New York: Academic Press.

Schmidt-Nielson, K. 1960. *Animal Physiology.* Englewood Cliffs, N.J.: Prentice-Hall.

1984. *Scaling: Why Is Animal Size so Important?* Cambridge University Press.

Scholander, P. F., V. Walters, R. Hock, and L. Irving. 1950. Body insulation of some Arctic and tropical mammals and birds. *Biological Bulletin* 99(2): 225–36.

Schrire, C. 1980. An inquiry into the evolutionary status and apparent identity of San hunter-gatherers. *Human Ecology* 8: 9–32.

Scott, K. 1980. Two hunting episodes of middle Palaeolithic age at La Cotte de St. Brelade, Jersey. *World Archaeology* 12(2): 137–52.

1984. Modified mammoth bones from La Cotte de St. Brelade, Jersey, Channel Islands. In *Abstracts of the First International Conference on Bone Modification*, Carson City, Nevada, 17–19 August, pp. 30–1.

1986a. The large mammal fauna. In P. Callow and J. M. Cornford, eds., *La Cotte de St. Brelade 1961–1978: Excavations by C. B. M. McBurney*, pp. 109–37, Norwich, U.K.: Geo Books.

1986b. The bone assemblages of Layers 3 and 6. In P. Callow and J. M. Cornford, eds., *La Cotte de St. Brelade 1961–1978: Excavations by C. B. M. McBurney*, pp. 159–83. Norwich, U.K.: Geo Books.

Segre, A., and A. Ascenzi. 1984. Fontana Ranuccio: Italy's earliest Middle Pleistocene hominid site. *Current Anthropology* 25(2): 230–3.

Sellards, E. H. 1938. Artifacts associated with fossil elephant. *Geological Society of America Bulletin* 49:999–1010.

——— 1952. *Early Man in America*. Austin: University of Texas Press.

Selous, F. C. 1881. *A Hunter's Wanderings in Africa, Being a Narrative of Nine Years Spent Amongst the Game of the Far Interior of South Africa*. London: Richard Bentley & Son.

——— 1893. *Travel and Adventure in South-East Africa, Being the Narrative of the Last Eleven Years Spent by the Author on the Zambezi and Its Tributaries; With an Account of the Colonisation of Mashunaland and the Progress of the Gold Industry in That Country*. London: Rowland Ward.

Semenov, S. A. 1976. *Prehistoric Technology: An Experimental Study of the Oldest Tools and Artefacts From Traces of Manufacture and Wear* (translated by M. W. Thompson; 1st ed. 1957). New York: Barnes & Noble.

Semken, H. A., Jr., and C. R. Falk. 1987. Late Pleistocene/Holocene mammalian faunas and environmental change on the Northern Plains of the United States. In R. W. Graham, H. A. Semken, Jr., and M. A. Graham, eds., *Late Quaternary Mammalian Biogeography and Environments of the Great Plains and Prairies*, pp. 176–313. Illinois State Museum Scientific Papers, Vol. 22.

Sergin, V. Ya. 1987. *Struktura mezinskovo palyeoliticheskovo poseleriya*. Moscow: Nauka.

Shackleton, N. J., M. A. Hall, J. Line, and Cang Shuxi. 1983. Carbon isotope data in core V19-30 confirm reduced carbon dioxide concentration in the ice age atmosphere. *Nature* 306: 319–22.

Shackleton, N. J., and N. D. Opdyke. 1973. Oxygen isotope and paleomagnetic stratigraphy of equatorial Pacific core V28-238: Oxygen isotope temperatures and ice volumes on a 100,000 year and 1,000,000 year scale. *Quaternary Research* 3: 39–55.

Shilo, N. A., A. V. Loshkin, E. E. Titov, and Y. V. Shumilov. 1983. *Kirgilyachskii mamont*. Moscow: Nauka.

Shipman, P. 1975. Implications of drought for vertebrate fossil assemblages. *Nature* 257: 667–8.

——— 1986. Studies of hominid–faunal interactions at Olduvai Gorge. *Journal of Human Evolution* 15: 691–706.

Shipman, P., D. C. Fisher, and J. J. Rose. 1984. Mastodon butchery: Microscopic evidence of carcass processing and bone tool use. *Paleobiology* 10(3): 358–65.

Shipman, P., and J. Rose. 1983. Evidence of butchery and hominid activities at Torralba and Ambrona; an evaluation using microscopic techniques. *Journal of Archaeological Science* 10: 465–74.

Shoshani, J. (and 76 others). 1982. On the dissection of a female Asian elephant (*Elephas maximus maximus* Linneaus 1758) and data from other elephants. *Elephant* 2(1): 3–93.

Shoshani, J., D. C. Fisher, J. M. Zaniskie, S. J. Thurlow, S. L. Shoshani, W. S. Benninghoff, and F. H. Zock, 1989. The Shelton Mastodon Site: Multidisciplinary study of a late Pleistocene (Twocreekan) locality in southeastern Michigan. *University of Michigan Contributions from the Museum of Paleontology* 27(14): 393–436.

Shoshani, J., J. C. Hillman, and J. M. Walcek. 1987. "Ahmed," the logo of the Elephant Interest Group: Encounters in Marsabit and notes on his model and skeleton and model. *Elephant* 2(3):7–32.

Shoshani, S. L., J. Shoshani, and F. Dahlinger, Jr. 1986. Jumbo: Origin of the word and history of the elephant. *Elephant* 2(2):86–122.

Shovkoplyas, I. G. 1965. *Mezinskaya stoyanka: k istorii yerednyednyeprovskovo basseyna v pozdnyepalyeoticheskuyu epokhu.* Kiev: Akademiya Nauk Ukrainskovo SSR.

1976. Issledovaniya v Dobranicheka na Kievshchine – nekotoriye itogi issledovaniya. *Materialy i issledovaniya po Arkheologii SSSR 185: Palyeolit i Neolit SSSR 7:* 177–189.

Sikes, S. K. 1966. The African elephant, *Loxodonta africana:* A field method for the estimation of age. *Journal of Zoology (London)* 150:279–95.

1971. *The Natural History of the African Elephant.* London: Weidenfeld & Nicolson.

Silberbauer, G. B. 1981. *Hunter and Habitat in the Central Kalahari Desert.* Cambridge University Press.

Simpson, G. G. 1942. The beginnings of vertebrate paleontology in North America. *Proceedings of the American Philosophical Society* 86:130–88.

Simpson, G. G., and C. de Paula Couto. 1957. The mastodonts of Brazil. *American Museum of Natural History Bulletin* 112:125–90.

Sinclair, A. R. E. 1977. *The African Buffalo: A Study of Resource Limitation of Populations.* University of Chicago Press.

Sinclair, A. R. E., and J. J. R. Grimsdell. 1982. *Population Dynamics of Large Mammals.* African Wildlife Foundation Handbook on Techniques, No. 5. Nairobi: African Wildlife Foundation.

Sinclair, A. R. E., and M. Norton-Griffiths. 1979. *Serengeti Dynamics of an Ecosystem.* University of Chicago Press.

Sleigh, M. M. 1970. Department of antiquities report, Queen Victoria Museum, Salisbury. In *Report of the Trustees and Director of the National Museum of Rhodesia for the Year Ended 31st December, 1969.* Salisbury: Trustees of the National Museum.

Smith, F. 1980. The histology of the skin of the elephant. *Journal of Anatomical Physiology* (new series) 24(4): 493–503.

Smith, M. A. 1985. A morphological comparison of central Australian seedgrinding implements and Australian Pleistocene-age grindstones. *The Beagle, Occasional Papers of the Northern Territory Museum of Arts and Sciences* 2(1):23–38.

1986. The antiquity of seedgrinding in arid Australia. *Archaeology of Oceania* 21:29–39.

Smith, R. J. 1984. Allometric scaling in comparative biology: Problems of concept and method. *American Journal of Physiology* 246:R152–60.

Smith, S. J. 1986. *Rowland Ward's Records of Big Game: Africa and Asia,* 20th ed. San Antonio, Tex.: Rowland Ward.

Smithers, R. H. N., and V. J. Wilson. 1979. *Check List and Atlas of the Mammals of Zimbabwe Rhodesia*. Museum Memoir No. 9. Salisbury (Harare, Zimbabwe): Trustees of the National Museums and Monuments.

Soergel, W. 1912. Das Aussterben diluvialer Säugetiere und die Jagd des diluvialen Menschen. In *Festschrift zur XLIII Allgemeinen Versammlung der Deutschen Anthropologischen Gesellschaft, Heft 2*, pp. 1–81. Jena: Gustav Fischer.

Soffer, O. 1985. *The Upper Paleolithic of the Central Russian Plain*. New York: Academic Press.

1989 Storage, sedentism, and the Eurasian paleolithic. *Antiquity* 63: 719–32.

Sokolov, V. Ye. 1982a. *Mammal Skin*. Berkeley: University of California Press.

(ed.). 1982b. *Yuribeiskii mamont*. Moscow: Nauka.

Sokolov, B. Ye., and Ye. B. Sumina. 1981. Morfologiya volosyanovo pokrova mamontyonka. In N. K. Vereshchagin et al. (eds.), *Magadanskii Mamontyonok*, pp. 81–4. Leningrad: Nauka.

Sokolov, V. Ye., and Ye. B. Sumina. 1982. Morfologiya volosyanovo nokrova yuribeiskovo mamonta. In V. Ye. Sokolov, ed., *Yuribeiskii mamont*, pp. 99–103. Moscow: Nauka.

Sondaar, P. Y. 1977. Insularity and its effect on mammal evolution. In M. K. Hecht, P. C. Goody, and B. M. Hecht, eds., *Major Patterns in Vertebrate Evolution*, pp. 671–706. New York: Plenum Press.

Sondaar, P. Y., F. Aziz, G. v. d. Bergh, and S. T. Husain. 1989. The insular proboscideans of S. E. Asia. *Fifth International Theriological Congress, Abstract of Papers and Posters* 1: 161–2.

Spaulding, W. G., E. B. Leopold, and T. R. Van Devender. 1983. Late Wisconsin paleoecology of the American Southwest. In S. C. Porter, ed., *The Late Pleistocene* (Vol. 1 of *Late Quaternary Environments of the United States*, H. E. Wright, Jr., ed.), pp. 259–93. Minneapolis: University of Minnesota Press.

Speth, J. D. 1983. *Bison Kills and Bone Counts: Decision-Making by Ancient Hunters*. University of Chicago Press.

1987. Early hominid subsistence strategies in seasonal habitats. *Journal of Archaeological Science* 14:13–29.

Speth, J. D., and K. Spielmann. 1983. Energy source, protein metabolism, and hunter-gatherer subsistence strategies. *Journal of Anthropological Archaeology* 2: 1–31.

Staesche, U. 1983. Aspects of the life of middle Paleolithic hunters in the N. W. German lowlands, based on the site Salzgitter-Lebenstedt. In J. Clutton-Brock and C. Grigson, eds., *Animals and Archaeology. 1: Hunters and Their Prey*, pp. 173–81. Oxford: BAR International Series No. 163.

Stafford, T. W., Jr., A. J. T. Jull, K. Brendel, R. C. Duhamel, and D. Donahue. 1987. Study of bone radiocarbon dating accuracy at the University of Arizona NSF accelerator facility for radioisotope analysis. *Radiocarbon* 29(1): 24–44.

Stanford, D. 1979. The Selby and Dutton sites: Evidence for a possible pre-Clovis occupation in the High Plains. In R. Humphrey and D. Stanford, eds., *Pre-Llano Cultures of the Americas*, pp. 101–23. Washington, D.C.: Anthropological Society of Washington.

Stanford, D., R. Bonnichsen, and R. E. Morlan. 1981a. The Ginsberg experiment:

Modern and prehistoric evidence of a bone flaking technology. *Science* 212: 438–40.

Stanford, D., W. R. Wedel, and G. R. Scott. 1981b. Archeological investigations of the Lamb Spring site. *Southwestern Lore* 47(1): 14–27.

Steele, D. G., and D. L. Carlson. 1984. Mammoth remains recovered from the Brazos River, southcentral Texas. Paper presented at the Society for American Archeology 45th annual meeting, 12–14 April, Portland, Oregon.

1989. Excavation and taphonomy of mammoth remains from the Duewall-Newberry site, Brazos County, Texas. In R. Bonnichsen and M. Sorg (eds.), *Bone Modification*, pp. 413–30. Orono, Maine: Center for the Study of the First Americans.

Stewart, J. M. 1977. Frozen mammoths from Siberia bring the Ice Ages to vivid life. *Smithsonian* 8(9): 60–9.

1979. A baby mammoth, dead for 40,000 years, reveals a poignant story. *Smithsonian* 10(6): 125–6.

Stout, B. 1986. Discovery and C-14 dating of the Black Rock Desert mammoth. *Nevada Archeologist* 5(2): 21–3.

Strong, W. D. 1932. *Recent Discoveries of Human Artifacts Associated with Extinct Animals in Nebraska*. Washington, D.C.: Science Service Research Announcement No. 130.

Summers, R. 1971. *Ancient Ruins and Vanished Civilizations of Southern Africa*. Cape Town: T. A. Bulpin.

Sun Jianzhong, Wang Yuzhue, and Jiang Peng. 1981. A Paleolithic site at Zhou-jia-you-fang in Yushu County, Jilin Province. *Vertebrata PalAsiatica* 19(3): 281–91.

Sutcliffe, A. J. 1985. *On the Track of Ice Age Mammals*. London: British Museum (Natural History).

Sutherland, J. 1912. *The Adventures of an Elephant Hunter*. London: Macmillan.

Svoboda, J. 1989. Middle Pleistocene adaptations in central Europe. *Journal of World Prehistory* 3(1): 33–70.

Tabler, E. C. 1955. *The Far Interior: Chronicles of Pioneering in the Matabele and Mashona Countries, 1847–1879*. Cape Town: A. A. Balkema.

ed. 1971. [James Chapman's] *Travels in the Interior of South Africa 1849–1863: Hunting and Trading Journeys from Natal to Walvis Bay and Visits to Lake Ngami and Victoria Falls, Part 2*. Cape Town: A. A. Balkema.

Taggart, R. E., and A. T. Cross. 1983. Indications of temperate deciduous forest vegetation in association with mastodon remains from Athens County, Ohio (abstract). *Ohio Journal of Science* 83(2):26.

Talbot, L. M., W. S. A. Payne, H. P. Ledger, L. D. Verdcourt, and M. H. Talbot. 1965. *The Meat Production Potential of Wild Animals in Africa: A Review of Biological Knowledge*. Commonwealth Bureau of Animal Breeding and Genetics (Edinburgh) Technical Communication No. 16.

Tamers, M. A. 1966. Instituo Venezolano de Investigaciones Cientificos Natural Radiocarbon Measurements II. *Radiocarbon* 8: 204–12.

Tanaka, J. 1980. *The San, Hunter-Gatherers of the Kalahari: A Study in Ecological Anthropology* (translated by D. W. Hughes). University of Tokyo Press.

Tassay, P. 1982. Les principales dichotomies dans l'histoire des Proboscidea (Mammalia): Une approche phylogénétique. *Geobios Mémoire Spéciale* 6: 225–45.

Tassy, P., and M. Pickford. 1983. Un nouveau mastodonte Zygolophodonte (Proboscidea, Mammalia) dans le Miocene inférieur d'Afrique Orientale: Systématique et paléoenvironment. *Geobios* 16(1): 53–77.

Tatishchev, V. N. 1732. O mamontovikh kostyakh. *Primechaniya Na Vedomosti* 100–1: 401–8.

Taylor, R. D. 1987. Abundance and distribution of elephants in Matusadona National Park, Zimbabwe. *Transactions of the Zimbabwe Scientific Association* 63(6): 58–66.

Thomas, C. 1985. *Report of the Mound Explorations of the Bureau of Ethnology* (originally published 1894). Washington, D.C.: Smithsonian Institution Press.

Thomas, D. S. G. 1982. Evidence of Quaternary palaeoclimates in western Zimbabwe – a preliminary assessment. Paper presented to the Southern Africa Conference of the Commonwealth Geographical Bureau, 9–15 June 1982, Lusaka.

1983. Geomorphic evolution and river channel orientation in north west Zimbabwe. *Proceedings of the Geographical Association of Zimbabwe* 14: 12–22.

1984. Ancient ergs of the former arid zones of Zimbabwe, Zambia and Angola. *Transactions of the Institute of British Geography* (new series) 9: 75–88.

1985. Evidence of aeolian processes in the Zimbabwean landscape. *Transactions of the Zimbabwe Scientific Association* 62(8): 45–55.

1988. The nature and depositional setting of arid and semi arid Kalahari sediments, southern Africa. *Journal of Arid Environments* 14: 17–26.

Thomas, E. S. 1952. The Orleton Farms mastodon. *Ohio Journal of Science* 52(1): 1–5.

Thomas, P. J. 1975. The role of elephants, fire and other agents in the decline of *Brachystegia boehmii* woodland. *Journal of the South African Wildlife Management Association* 5: 11–18.

Thoreau, H. D. 1963. *A Week on the Concord and Merrimack Rivers* (originally published 1868). New York: Holt, Rinehart & Winston.

Thorington, R. W., and R. E. Vorek. 1976. Observations on the geographic variation and skeletal development of *Aotus*. *Laboratory Animal Science* 26: 1006–21.

Thorne, B. L., and R. B. Kimsey. 1983. Attraction of neotropical *Nasutitermes* termites to carrion. *Biotropica* 15(4): 295–6.

Tikhonov, A. N., and V. M. Khrabrii. 1989. Mamontyonok s Yamala. *Priroda* 6: 46–7.

Tilesius von Tilenau, W. G. 1812. De sceleto mammonteo Sibirico anno 1797 effosso. *Memoirs of the Imperial Academy of Sciences* (*St. Petersburg*) 5: 5.

Till, J. 1936–7. Monthly progress reports of assistant game warden, Wankie Game Reserve. Unpublished manuscript on file, Hwange National Park, Zimbabwe.

Tilson, R., F. von Blottnitz, and J. Henschel. 1980. Prey selection by spotted hyenas (*Crocuta crocuta*) in the Namib Desert. *Madoqua* 12(1): 41–9.

Todd, L. C., and G. C. Frison. 1986. Taphonomic study of the Colby site mammoth bones. In G. C. Frison and L. C. Todd, eds., *The Colby Mammoth Site: Taphonomy and Archaeology of a Clovis Kill in Northern Wyoming*, pp. 27–90. Albuquerque: University of New Mexico Press.

Tode, A. 1953. Einige archäologische Erkenntisse aus der paläolitischen Freiland-station von Salzgitter-Lebenstedt. *Eiszeitalter und Gegenwart* 3: 192–215.

Tode, A., F. Preul, K. Richter, W. Selle, K. Pfaffenberg, A. Kleinschmidt, and E. Guenther. 1953. Untersuchungen im Lössprofil von Rheindahlen/ Niederrheinische Bucht. *Quartär* 31–2: 41–67.

Tolmachoff, I. P. 1929. The carcasses of the mammoth and rhinoceros found in the frozen ground of Siberia. *Transactions of the American Philosophical Society* (new series) 23(1): 1–74.

Tooby, J., and I. DeVore. 1986. The reconstruction of hominid evolution through strategic modelling. In W. G. Kinzey, ed., *The Evolution of Human Behavior: Primate Models*, pp. 183–237. New York: SUNY Press.

Toots, H. 1965. Sequence of disarticulation in mammalian skeletons. *(University of Wyoming) Contributions to Geology* 4(1): 37–9.

Torrance, J. D. 1981. *Climate Handbook of Zimbabwe*. Salisbury: Zimbabwe Department of Meteorological Services.

Tseitlin, S. 1979. *Geologiya Paleolita Severnoy Azii*. Moscow: Nauka.

Turnbull, C. M. 1961. *The Forest People: A Study of the Pygmies of the Congo*. New York: Simon & Schuster.

Turnbull, W. D., and D. M. Martill. 1988. Taphonomy and preservation of a monospecific Titanothere assemblage from the Washakie Formation (Late Eocene), southern Wyoming. An ecological accident in the fossil record. *Palaeogeography, Palaeoclimatology, Palaeoecology* 63: 91–108.

van der Merwe, N., J. O. L. Thorp, and R. H. V. Bell. 1988. Carbon isotopes as indicators of elephant diets and African environments. *African Journal of Ecology* 26(2): 163–72.

van Hoven, W., and E. A. Boomker. 1985. Digestion. In R. J. Hudson and R. G. White, eds., *Bioenergetics of Wild Herbivores*, pp. 103–20. Boca Raton, Fla.: CRC Press.

van Hoven, W., R. A. Prins, and A. Lankhorst. 1981. Fermentative digestion in the African elephant. *South African Journal of Wildlife Research* 11: 78–86.

Vaufrey, R. 1955. Proboscidiens fossiles. In P.-P. Grassé, ed., *Traité de Zoologie Anatomie, Systématique, Biologie: Mammifères* 17: 784–847.

Velichko, A. A. 1984. Late Pleistocene spatial paleoclimatic reconstructions. In A. A. Velichko, ed., *Late Quaternary Environments of the Soviet Union*, pp. 261–85. Minneapolis: University of Minnesota Press.

Vereshchagin, N. K. 1967. Primitive hunters and Pleistocene extinction in the Soviet Union. In P. S. Martin and H. E. Wright Jr., eds., *Pleistocene Extinctions: The Search for a Cause*, pp. 365–98. New Haven, Conn.: Yale University Press.

1974. The mammoth "cemeteries" of north-east Siberia. *Polar Record* 17(106): 3–12.

1975. O mamontye s reki Shandrin. *Vestnik Zoologii (1972)*. 2: 81–4.

1977. Berelyokhskoye "kladbische" mamontov. *Trudi Zoologicheskovo Instituta* 72: 5–50.

ed. 1980. *Mlekopitayushchiye vostochnoy yevropi v antropogenye.* Trudi Zoologicheskovo Instituta, No. 93.

1981a. *Zapiski Paleontologa.* Leningrad: Nauka.

ed. 1981b. *Magadonskii mamontyonok Mammuthus primigenius (Blumenbach).* Leningrad: Nauka.

1982. Novaya gidanskaya (Yuribeiskaya) nakhodka mamonta. *Vestnik Zoologii* 3: 32–8.

Vereshchagin, N. K., et al. 1981. *Magadanskii Mamontyonok.* Leningrad: Nauka.

Vereshchagin, N. K., and G. F. Baryshnikov. 1982. Paleoecology of the mammoth fauna in the Eurasian Arctic. In D. M. Hopkins, J. V. Matthews, Jr., C. E. Schweger, and S. B. Young, eds., *Paleoecology of Beringia*, pp. 267–79. New York: Academic Press.

1984. Quaternary mammalian extinctions in northern Eurasia. In P. S. Martin and R. G. Klein, eds., *Quaternary Extinctions: A Prehistoric Revolution*, pp. 483–516. Tucson: University of Arizona Press.

Vereshchagin, N. K., and I. A. Dubrovo. 1979. Paleontologicheskoye opisaniye nakhodki. *Priroda* 1978(1):21.

Vereshchagin, N. K., and A. I. Nikolayev. 1982. Raskopki Khatangskovo mamonta. In N. K. Vereshchagin and I. Ye. Kuz'mina, eds., *Mamontovaya Fauna Aziatskoy Chasti SSSR*, pp. 3–17. Trudi Zoologicheskovo Instituta, No. 111.

Vereshchagin, N. K., and A. N. Tikhonov. 1986. Issledovaniya bivnyey mamontov. In N. K. Vereshchagin and I. Ye. Kuz'mina, eds., *Mlekopitayushchiye Chetvertichnoy Fauny SSSR*, pp.3–14. Trudi Zoologicheskovo Instituta, No. 149.

Verney, J. W. 1938–9. Monthly reports of assistant game warden J. W. Verney, September 1938–October 1939. Unpublished manuscript on file, Hwange National Park, Zimbabwe.

Voice, M. 1983. Drought. *Australian Natural History* 21(1): 2–9.

Voigt, E. A. 1982. Ivory in the early Iron Age of South Africa. *Transvaal Museum Bulletin* 18: 17–20.

Vollosovich, K. A. 1909. Raskopki Sanga-Yurakhskovo [*sic*] mamonta v 1908 g. *Izvestiya Imperatorskoy Akademii Nauk, 6th Series* 3: 437–58.

Volman, T. P. 1984. Early prehistory of southern Africa. In R. G. Klein, ed., *Southern African Prehistory and Paleoenvironments*, pp. 169–220. Rotterdam: A. A. Balkema.

von Schrenck, L. 1880. Der erste Fund einer Leiche von *Rhinoceros merckii* Jaeg. *Mémoirs de L'Académie Impériales des Sciences de St.-Pétersburg, 7th Series* 27(7).

Voorhies, M. 1969. *Taphonomy and Population Dynamics of an Early Pliocene Vertebrate Fauna, Knox County, Nebraska.* Contributions to Geology, Special Paper No. 1, University of Wyoming.

Vrba, E. S. 1980. The significance of bovid remains as indicators of environment and predation patterns. In A. K. Behrensmeyer and A. P. Hill, eds., *Fossils in the Making*, pp. 247–71. University of Chicago Press.

Vyhnanek, L., and M. Stloukal. 1988. Harris's lines in adults – an open problem. Paper presented at 12th International Congress of Anthropological and Ethnological Sciences, Zagreb, Yugoslavia.

Wankie National Park Management Plans. 1971. Unpublished manuscript on file, Hwange National Park, Zimbabwe.

Warnica, J. M. 1966. New discoveries at the Clovis site. *American Antiquity* 31(3): 345–57.

Warren, J. C. 1855. *The Mastodon Giganteus of North America*. Boston: John Wilson.

Waters, M. R. 1989. Late Quaternary lacustrine history and paleoclimatic significance of Pluvial Lake Cochise, southeastern Arizona. *Quaternary Research* 32(1): 1–11.

Watson, J. A. L., and H. M. Abbey. 1986. The effects of termites (*Isoptera*) on bone: Some archaeological implications. *Sociobiology* 11(3): 245–54.

Watts, W. A. 1983. Vegetational history of the eastern United States 25,000 to 10,000 years ago. In S. C. Porter, ed., *The Late Pleistocene* (Vol. I of *Late Quaternary Environments of the United States*, H. E. Wright, Jr., ed.), pp. 294–310. Minneapolis: University of Minnesota Press.

Webb, S. D., G. S. Morgan, R. C. Hulbert, Jr., D. S. Jones, B. J. MacFadden, and P. A. Mueller. 1989. Geochronology of a rich early Pleistocene vertebrate fauna, Leisey Shell Pit, Tampa Bay, Florida. *Quaternary Research* 32(1): 96–110.

Weber, T. 1988. Ein eemwarmzeitlicher Waldelefanten-Schlachtplatz von Gröbern, Kr. Gräfenhainichen. *Ausgrabungen und Funde* 33: 181–8.

Weigelt, J. 1927. *Rezente Wirbeltierleichen und ihre paläobiologische Bedeutung*. Leipzig: Verlag von Max Weg.

Weir, J. S. 1972. Spatial distribution of elephants in an African national park in relation to environmental sodium. *Oikos* 23: 1–13.

Weir, J. S., and E. Davison. 1965. Daily occurrences of African game animals at water holes during dry weather. *Zoologica Africana* 1(2): 353–68.

Weissner, P. 1983. Style and social information in Kalahari San projectile points. *American Antiquity* 48: 253–76.

Wellington, J. H. 1955. *Southern Africa: A Geographical Study. Vol. 1: Physical Geography*. Cambridge University Press.

Western, D. 1979. Size, life history, and ecology in mammals. *African Journal of Ecology* 17: 185–204.

1980. Linking the ecology of past and present mammal communities. In A. K. Behrensmeyer and A. P. Hill, eds., *Fossils in the Making: Vertebrate Taphonomy and Paleoecology*, pp. 41–54. University of Chicago Press.

Western, D., C. Moss, and N. Georgiadis. 1983. Age estimation and population age structure of elephants from footprint dimensions. *Journal of Wildlife Management* 47(4): 1192–7.

Weyerhaeuser, R. 1980. Lake Manyara elephant research. *Elephant* 1(4): 164–8.

Wheelock, N. D. 1980. Environmental sodium as a factor in the behavior and distribution of African elephants. *Elephant* 1(4): 169–77.

Whitaker, R. H. 1975. *Communities and Ecosystems*, 2nd. ed. New York: Macmillan.

Whitlow, R. 1984. Some morphological characteristics of Dambo features in Zimbabwe. *Transactions of the Zimbabwe Scientific Association* 62(1): 1–15.

Whitmore, F. C., Jr., K. O. Emery, H. B. S. Cooke, and D. J. P. Swift. 1967. Elephant teeth from the Atlantic Continental Shelf. *Science* 156: 1477–80.

Willey, G. R. 1966. *An Introduction to American Archaeology. Vol. I: North and Middle America.* Englewood Cliffs, N.J. Prentice-Hall.

Williamson, B. R. 1975a. Seasonal distribution of elephant in Wankie National Park. *Arnoldia Rhodesia* 7(11): 1–16.

    1975b. The condition and nutrition of elephant in Wankie National Park. *Arnoldia Rhodesia* 7(12): 1–20.

    1976. Reproduction in female African elephant in the Wankie National Park, Rhodesia. *South African Journal of Wildlife Research* 6: 89–93.

    1979a. A history of wildebeeste and their movements in Wankie National Park. Unpublished manuscript on file, Hwange National Park, Zimbabwe.

    1979b. Aspects of the ecology of wildebeeste (*Connochaetes taurinus taurinus* Burchell) in Wankie National Park, Rhodesia, Unpublished manuscript on file, Hwange National Park, Zimbabwe.

Williamson, D., J. Williamson, and K. J. Ngwamotsoko. 1988. Wildebeeste migration in the Kalahari. *African Journal of Ecology* 26(4): 269–80.

Wilmsen, E. N. 1989. *Land Filled With Flies: A Political Economy of the Kalahari.* University of Chicago Press.

Wilson, D., and P. Ayerst. 1976. *White Gold.* New York: Taplinger.

Wilson, M. 1974. The Casper local fauna and its fossil bison. In G. C. Frison, ed., *The Casper Site: A Hell Gap Bison Kill on the High Plains*, pp. 125–71. New York: Academic Press.

Winterhalder, B., and E. A. Smith. 1981. *Hunter-Gatherer Foraging Strategies: Ethnographic and Archaeological Analyses.* University of Chicago Press.

Wittry, W. L. 1965. The institute digs a mastodon. *Cranbrook Institute of Science News Letter* 35(2): 14–19.

Wobst, H. M. 1977. Stylistic behavior and information exchange. In C. E. Cleland, ed., *Papers for the Director: Research Essays in Honor of James B. Griffin*, pp. 317–42. University of Michigan Museum of Anthropology Anthropological Papers, No. 61.

Wood, A. E. 1952. Tooth-marks on bones of the Orleton Farms mastodon. *Ohio Journal of Science* 52(1): 27–8.

Wood, J. G. 1868. *The Natural History of Man, Being an Account of the Manners and Customs of the Uncivilized Races of Men. Vol. 1: Africa.* London: Routledge, Warne, & Routledge.

Wormington, H. M. 1957. *Ancient Man in North America*, 4th rev. ed. Denver: Colorado Museum of Natural History, Popular Series No. 4.

Wylie, A. 1985. The reaction against analogy. In M. B. Schiffer, ed., *Advances in Archaeological Method and Theory, Vol. 8*, pp. 63–111. Orlando, Fla.: Academic Press.

Yefremov (Efremov), I. A. 1940. Taphonomy: New branch of paleontology. *Pan-American Geologist* 74(2): 81–93.

Yellen, J. E. 1977. *Archaeological Approaches to the Present: Models for Reconstructing the Past.* New York: Academic Press.

Yellen, J. E., and H. C. Harpending. 1972. Hunter-gatherer populations and archaeological inference. *World Archaeology* 4: 244–53.

Yevseyev, V. P., I. A. Dubrovo, N. V. Rengarten, and A. Ya. Stremyakov. 1982. Mestonakhozhdeniye yuribeyskovo mamonta: Geologiya, tafonomiya, paleontologiya. In V. Ye. Sokolov, ed., *Yuribeiskii mamont*, pp. 5–19. Moscow: Nauka.

Young, A. 1976. *Tropical Soils and Soil Survey*. Cambridge University Press.

Yudichev, Yu. F., and A. N. Averikhin. 1975. Diagnoz cherez 40 tisyach lyet. O prichinakh gibeli Shandrinskovo mamonta i usloviyakh, sposobstvovavshikh sokraneniyu yivo vnutrenikh organov. *Zemlya Sibirskaya, Dal'nyevostochnae* 8: 57–8.

Zalyenskii, V. 1903. Ostyeologicheskiya i odontographicheskiya izsleyedovaniya nad mamontov (*Elephas primigenius* Blum) i slonami (*El. indicus* L i *El. africanus* Blum). *Nauchniye Rezultati Ekspeditsii, Snaryozhennoy Imperatorskoy Akademiyei Nauk dlya Raskopki Mamonta, Naidennovo na Reke Berizovke v 1901 Godu* 1: 1–124.

——— 1909. Mikroskopicheskiye issledovaniya nekotorikh organov mamonta, naidennovo na reke Beryozovke. *Nauchniye Rezultati Ekspeditsii, Snaryozhennoy Imperatorskoy Akademiyei Nauk dlya Raskopki Mamonta, Naidennovo na Reke Berizovke v 1901 Godu* 2: 21–36.

Zavala D. M. 1987. Leyendas y realidades de Marcahuasi. *Boletin de Lima* 53(9): 14–18.

Index